MORE RAVES FOR
OIL & HONOR . . .

"A must for the oil industry and stock-market crowd and highly instructive entertainment for everyone else."—*Publishers Weekly*

"Probing . . . ironic . . . authoritative . . . *Oil & Honor* is a strong book."—*Newsday*

"It has the sweep and characterization of a James Michener novel as it builds toward the verdict."
—*Across the Board*

"A splendid account of the legal battle that erupted over Pennzoil's claim that Texaco improperly filed its proposed merger with Getty Oil . . . Petzinger's book puts into focus how it all came about and does so in an entertaining way."—*Fort Worth Morning Star-Telegram*

"Reads more like fiction than like a narrative of actual events leading to the headline-making litigation . . . an inside look at the rarified world of takeover artists, slick investment bankers, and crafty lawyers."—*Library Journal*

OIL & HONOR!

"A stunningly detailed reconstruction of the highest-stakes competition in corporate history—the bitter $11.1 billion battle for Getty Oil . . . financial journalism at its very best."—*Kirkus*

"*Oil & Honor* is a super read. The cast of characters itself is attention grabbing. There is the entire Getty family (at least a mini-series), Laurence Tisch, controversial Chairman of the Board of CBS, Boone Pickens, a five-star takeover player, Vice President George Bush, the damndest bunch of Texas lawyers and Wall Street financial gunfighters ever to tramp across the pages of non-fiction, and a bevy of high-rollers . . . But wait, the author adds the final ingredients—excellent writing, dramatic flair and a climax worthy of a freshly tapped geyser."
—*Rave Reviews*

"*Wall Street Journal* reporter Petzinger has done an admirable job of giving play to both sides in this gargantuan grudge match . . . masterfully sets forth the facts as seen from both sides, and brings the legal issues into sharp focus."—*Business Week*

OIL & HONOR

THE TEXACO-PENNZOIL WARS

THOMAS PETZINGER, Jr.

BERKLEY BOOKS, NEW YORK

OIL & HONOR

A Berkley Book / published by arrangement with
G. P. Putnam's Sons

PRINTING HISTORY
G. P. Putnam's Sons edition / May 1987
Berkley edition / August 1988

ISBN: 0-425-11172-5

A BERKLEY BOOK® TM 757,375
Berkley Books are published by The Berkley Publishing Group,
200 Madison Avenue, New York, New York 10016.
The name "BERKLEY" and the "B" logo
are trademarks belonging to Berkley Publishing Corporation.

PRINTED IN THE UNITED STATES OF AMERICA

10 9 8 7 6 5 4 3 2 1

To Cynthia and Beatrice
And to my parents

· AUTHOR'S NOTE ·

This book is based on more than fifty lengthy interviews with the principal characters in the story and some two hundred interviews conducted as part of my coverage for *The Wall Street Journal,* as well as about fifty thousand pages of sworn trial testimony, depositions and affidavits, and published reports. None of the dialogue or description has been invented. To preserve the complete independence of the project, no one involved in the story was given the opportunity to review the manuscript, although several of the parties cooperated in an extensive fact-checking exercise. I urge readers to consult the notes in the back of the book for a fuller description of my sources.

· PREFACE ·

To a unique degree, *The Wall Street Journal* fosters teamwork in journalism. Never have I appreciated that more than now.

Although only one name appears on the cover of this book, it is in many ways the product of dozens of *Journal* reporters and editors, beginning with Norman Pearlstine, the managing editor. Without his encouragement, I would never have begun it and could never have completed it.

My colleagues in the Houston bureau—George Getschow, Steve Frazier, Matt Moffett, Dianna Solis, Paulette Thomas, James Tanner and Lorretta Gasper—were unflagging in their support. Without the criticism, writing and editing assistance I received from Frazier, Getschow and Moffett, I would still be laboring over the manuscript.

Exceptional reporting by my colleagues in Houston and elsewhere provided me not only with an abundance of facts and insights, but also with the inspiration to undertake this project in the first place. The work of Allanna Sullivan and James B. Stewart in New York in covering the story, as well as the trailblazing coverage of the problems at Getty Oil by Stephen J. Sansweet in Los Angeles, has been especially important. But the work of literally dozens of other *Journal* staffers has assisted me in ways that they could never imagine.

I could never identify all the outstanding journalists from out-side *The Wall Street Journal* whose work I have also relied upon. Only a small number appear in the notes in the back of the book; to them, and to the many others I have neglected to identify, I express my thanks for fine work.

T. J. Simoneaux, Howard Shatz and Laurie Abraham performed invaluable research for me. Cynthia Petzinger was my best re-searcher and critic, besides being a tower of strength through it all.

My agent, Alice Fried Martell, maintained my spirit at mo-ments when it was perilously near exhaustion. Neil Nyren, my editor at G. P. Putnam's Sons, afforded me greater patience than any author deserves. Fred Sawyer, my copy editor, saved me from myself on occasion.

For their special interest and assistance I wish to thank Sarah McGill of Baker & Botts, Greg Good and Larry Bingamin of Texaco, Bob Harper of Pennzoil and Hunter Martin of POGO Producing. To them and the many lawyers, executives and others who are part of the story, I extend my thanks not just for your time and trust, but for your understanding that I could not write exactly the book that any of you might have liked. I hope that each of you recognizes the extent of your contribution to my effort to make the book as balanced and fair as possible.

Finally, to Thomas and Jean Petzinger, Charles and Rebecca Petzinger, Elizabeth Ann Petzinger, Robert and Joyce Bloom and Kristian and Sally Bloom—and above all to Cynthia and Bea-trice—I acknowledge my greatest debt, for your many in-dulgences, tireless assistance and love.

THOMAS PETZINGER JR.
Houston, Texas
December 17, 1986

· PART 1 ·

THE MEN

In the Oklahoma oilfields, I'd learned that the bigger the bully, the better the brawl was likely to be, that no matter how tough the opponent who backed you into a corner, there was always a good chance of outslugging, outboxing or outlasting him in the infighting.

—J. Paul Getty

· 1 ·

For as long as ten hours they waited—three hundred people packed elbow to elbow, hip to hip, in a seedy, turn-of-the-century courtroom built to accommodate roughly one-tenth their number. All decorum was laid aside. Women fanned their legs with the hems of their dresses and men pulled their ties lower by the hour. Anyone going to the bathroom took the risk of never regaining his place.

The mob had begun forming long before daybreak. Julie Iannotti, a corporate secretary, had rolled out of bed at 4 A.M. to catch a bus in the dark, taking no chance that she would fail to reserve a seat for her boss. So many newspaper reporters occupied the jury box that some crouched underneath and in between the swivel chairs. A pregnant legal clerk spent all day standing on a folding chair. Like rock-and-roll groupies or movie fans turned out for the spectacle of opening night, no one wanted to miss a single gesture or inflection by the star of the attraction.

In addition to the curiosity seekers came the fortune seekers, dozens of them from as far as 1700 miles away, each plotting to become the first to buy or sell stock on the outcome of the hearing. In the hallway, near the gumball machine and the row of mismatched filing cabinets, the earliest arrivals spent the day defending their control of the telephone booths. James Rhodes, a stockbroker from San Antonio, shrewdly slipped a ten-dollar bill to a courthouse clerk to lock up access to a private office phone—

only to have his link to the outside world outdone by a junkyard dealer, a New York securities analyst and more than a dozen other stock-market speculators equipped with cellular telephones and walkie-talkies.

After a career spent calling the courtroom to order for the likes of whiplash and leaky-roof lawsuits, Bailiff Carl Shaw basked in heroism; the night before, he had nabbed two intruders sifting through the papers in the court bench, hoping to glean a clue about the ruling from any of the judge's notes, doodles or scrawls. That morning, in a special precaution, Bailiff Shaw had presided over a bomb sweep through the courtroom.

But as the moment of the judge's decision drew nearer, any semblance of security fell apart. A shoving match erupted as still more gawkers and traders tried to shoulder their way into the room. A phalanx of television cameramen defied the house rules by invading the gridlocked courtroom and panning the circuslike scene for evening-news footage. As the crowd continued to grow, clogging the entrance and jamming the hallways, Sheriff's Deputy D. B. Diaz threw up his hands and gave up trying to search for weapons.

It was December 10, 1985, in Houston—doomsday in a case called *Pennzoil v. Texaco*.

Three weeks earlier, a jury of eight women and four men had ordered Texaco Incorporated to pay $10.53 billion to Pennzoil Company, a sum greater than the gross national product of 116 of the world's countries. In a stunning display of courtroom ingenuity that escaped virtually all public notice while it was happening, Pennzoil convinced the jury that Texaco had fraudulently induced Getty Oil Company to break a binding merger contract with Pennzoil, stealing a deal that would have given Pennzoil a billion barrels of choice oil reserves. The jury's verdict was remarkable even apart from the money involved; by its own admission, Pennzoil had failed to secure a signed, formal contract for the deal. Yet after hearing four and one half months of testimony spanning twenty-four thousand pages of court transcript, the jury had taken barely two days not only to decide the facts, but to impose a fine ninety times larger than the greatest judgment ever to survive an appeal.

To the throng assembled in State Court #151, only one thing could be more incredible than the jury verdict itself: that any judge would have the nerve to uphold it.

"The most devastating specter of disaster in all of legal history!" a Texaco lawyer had called it in court. "The award given by this jury is beyond the comprehension of most human beings!'' the company declared in a legal brief. Legal scholars and editorial writers had leaped to Texaco's defense, deriding the verdict as a "miscarriage of justice," "hang-'em high" jurisprudence, a "fluke," a "travesty," an "absurdity" and every other adjective they could conjure. Surely, the world thought, the judge would exercise his discretion to reverse the jury award, or at least to shrink it by several orders of magnitude.

At 5:15 P.M., Judge Solomon Casseb Jr., a seventy-year-old former divorce lawyer, resolutely climbed the bench, clutching a four-page document. A hush fell over the courtroom as he began reading—and stumbling over the words. Pennzoil, he said, shall receive from Texaco "the sum of seven thousand—"

He caught himself.

"—seven *billion*, five hundred thirty million, as its actual damages." *Plus* $3 billion in punitive damages, *plus* interest, for a grand total of $11,120,976,110.83.

Judge Casseb signed the judgment order with a 39-cent ballpoint pen and in barely a moment was gone, leaving the courtroom, and the world, in a state of disbelief.

With Texaco standing on the threshold of bankruptcy, the greatest financial crisis in corporate history had begun. How could it have happened?

The moment at which Texaco committed its $11 billion "crime"—a scene that would be retold to the jury from nearly a dozen perspectives—had occurred on a chilly evening twenty-three months earlier on the Upper East Side of New York. Texaco Chairman John K. McKinley was taking a five-second elevator ride in the opulent Hotel Pierre, and it was all the time he had to make a multibillion-dollar decision.

Even for the chairman of the nation's fifth-largest corporation, it was an excruciating moment. At age sixty-three, the former organic chemist had climbed to the top of Texaco from the company's grease laboratory by moving with painstaking caution, by emphasizing deliberation over decisiveness. Yet when the elevator doors flung open shortly before midnight, McKinley had told no one—nor, indeed, was he certain himself—just what price he would offer for the biggest corporate takeover in history.

Awaiting him was Gordon Getty, the curly-headed, fifty-year-old poet, composer and opera singer who had been recently, though inaccurately, described as the richest man in America. Gordon had spent his adulthood brooding over the likes of Emily

Dickinson and Antonin Dvorak, pursuits that had hardly qualified him to manage the industrial empire built by his late father, J. Paul Getty. In fact, Gordon's career with Getty Oil had been every bit as brief as it was embarrassing. But a bizarre chain of family tragedy and strife—a suicide, a brain tumor, a case of drug addiction and bitterness over a messy divorce—had eliminated Gordon's four brothers from positions of leadership, not just in the company but within the family as well. Gordon alone had been thrust into control over the family's 40% ownership of Getty Oil.

Texaco's McKinley walked through the serpentine, gray-colored hallway, entered the suite and turned to the expectant-looking heir.

"I am prepared to offer—"

"I accept!" Gordon gushed, and as he caught the startled reaction of everyone else he quickly looked back to McKinley. "Oh! You're supposed to give the price first!"

Laughter swept the room and the tension melted away, but an odd discomfort soon returned. For everyone knew that only four days earlier, in the same hotel room, Gordon had agreed to a merger with J. Hugh Liedtke, a barrel-chested oilman whom no one was anxious to offend. Liedtke, the chairman of Pennzoil, was one of the most ferocious battlers of the oil patch, a corporate chieftain whose assertiveness had earned him the nickname Chairman Mao. He was the only oilman ever claimed as a mentor by T. Boone Pickens, the archnemesis of Big Oil. He had once dared to call the bluff of J. Paul Getty himself in a deal that gave birth to Pennzoil. He had helped to start Vice President George Bush out in business and remained on close terms with him.

"One tough hombre," Gordon would call Liedtke.

Liedtke's deal to go into partnership with Gordon Getty had fulfilled his lifelong dream of elevating Pennzoil into the ranks of Big Oil. But the morning after John McKinley's midnight meeting with Gordon, Liedtke would awaken in the Waldorf-Astoria Hotel, only a few blocks from the Pierre, to the nightmare that Texaco had taken it all away.

"They made off with a billion barrels of our oil!" he would boom.

One truism has guided the oil industry for nearly a century: Oil is best when it is still in the ground. When you owned all the oil-in-the-ground you could possibly obtain, everything else usually took care of itself. Want to borrow money? How much oil have you got in the ground? the banker asked. Want to sell some shares of stock? Same question from the investor. Oil occupied a special place in the world of international commodities. It was not like gold, which met its highest purpose dangling from someone's ears, or like wheat, which had to be consumed quickly or not at all. If you were an oilman, oil-in-the-ground was your portfolio, your principal. Oil flowing from a well was only the dividend, preferably to be spent finding more oil in the ground.

Texaco never fully grasped this rule, choosing to make its fortune from refineries and filling stations rather than from finding oil fields. When a gusher came in, Texaco was often the first to lay a pipeline for buying everyone else's oil but often the last to put down a well of its own. Even when it tried, Texaco had little flair for finding oil. As any *real* oilman knew, the only place Texaco could find oil was in a filling station.

Nobody, however, could refine and sell oil products with the mastery of Texaco. Texaco had Milton Berle and the Texaco Star Theater, Bob Hope in his Texaco Fire Chief helmet, and, unlike any other oil company, gasoline stations in all fifty states. Everything was fine as long as oil was plentiful enough to keep the refineries humming. But by the time John McKinley had become chairman of Texaco, nobody had much oil to spare, and Texaco's supplies were shorter than anyone else's. It was theoretically possible that within a decade, Texaco's oil fields would be played out, dried up, bereft of crude. The Big Red Texaco Star was in danger of disappearing from the heavens—unless, of course, Texaco could take over all the oil fields of a company like Getty.

If Texaco was behind its time coming to grips with the oil-in-the-ground rule, Hugh Liedtke of Pennzoil was ahead of his. From his earliest days in the oil patch, Liedtke had a knack for finding oil fields that flowed slowly but for a long time, where the oil had to be coaxed from the earth. This was just fine for Pennzoil, which, after all, sold oil by the *quart,* in little yellow cans. The only problem was that this patient strategy prevented Pennzoil from becoming a major oil company, from joining the ranks of Exxon, Mobil—or Getty.

Texaco and Pennzoil alike realized at about the same time that Getty Oil would make their cups runneth over. Neither had the time to try drilling for that much oil, but more importantly, neither had the money. Ten billion dollars may seem like a lot to spend buying oil-in-the-ground, and indeed it is more money than every wage earner in Detroit—put together—takes home in a year. But by one method of reckoning, Texaco *saved* $60 billion over many years by purchasing Getty Oil instead of exploring for the same amount of oil-in-the-ground—and that is nearly as much as every-one in New York City earns in a year.

Amid the differences in corporate strategies, a collision of ethical values would erupt in the Houston courtroom. That was because every important event in the fracas—from the midnight dealmak-ing on the Upper East Side of New York to the trial itself less than two years later in Houston—involved a battle between two groups: the Old Good-Ol'-Boys of the oil fields and the New Good-Ol'-Boys of Wall Street.

Under the Old Good-Ol'-Boy rules, you always dealt honorably with your friends; as for everyone else, well, they had better watch their step. No industry honored personal friendship more highly than the 20th-century oil industry, and no industry permit-ted such ruthless treatment of outsiders. You dealt with comrades on a handshake—"my word is my bond," went the motto of the All-American Wildcatters Association, of which Pennzoil's Liedtke was a prominent member—but with *outsiders* on a con-tract. "If I give my word, no one can break it," an oilman once told J. Paul Getty, "but if I sign this contract, my lawyers can break it."

The New Good-Ol'-Boys of the 1980s lived by a much more modern code. Until the dotted line was signed, you remained free to stick it to anyone, even your friends if you had to. In fact, your friends expected it, and sometimes they even respected you for it. Wall Street permitted such anything-goes behavior for a perfectly good reason: The person who outmaneuvered everyone else—whether to sell at the highest price or buy at the lowest—got the best deal for whichever pension funds, insurance companies, widows, orphans, small-time speculators and big-money investors he happened to represent. In this culture, nothing was final until the last *hereas* and *wherefore* and *thereupon* had been recorded. There were "deals," and then there were "done deals."

Investment banking and takeover lawyering had become the ultimate yuppie attainments by the time of *Pennzoil v. Texaco*, and they were occupations that could flourish only in the every-man-for-himself culture that permeated the Wall Street of the eighties. An entire industry devoted to the fostering of corporate combinations had been born, offering cunning, daring and pure wit at levels that most large corporations could never muster in the heat of a takeover battle. Oil companies, for instance, spent too much time worrying about pipelines, tankers, geology and OPEC to be bothered with "poison pills," "bear hugs," "two-tier bootstrap mergers'' and the other maneuvers that the New York take-over community had come to package and sell as products.

Investment bankers and merger lawyers functioned like all-purpose quarterbacks: calling the play, taking the snap, handing off the ball to the chairman of the board and then blocking for him the whole way down the field—and the next week, they were quarterbacking for another team. Their ambitions were simple: completing the deal and then collecting their fee from the done deal. At the top of their profession—which they often reached by their mid-thirties—takeover professionals commonly pulled down salaries approaching or exceeding $1 million a year. They could buy summer homes in the Hamptons that they never had time to visit and drape Galanos gowns on the wives they rarely saw. But it was worth it for the thrill of doing the deal.

Along with the institutions of Big Oil and big takeovers, *Pennzoil v. Texaco* would involve another distinctly American phenomenon: the use of juries to judge complex civil disputes.

Congressmen, insurance lobbyists and legal scholars had already begun to decry the jury system as the one true defect of democracy, blaming it for everything from the high incidence of caesarean births to the presence of engine brakes on lawnmowers. Jury verdicts had reached such proportions that they began pushing giant corporations into the sanctuary of bankruptcy court: Johns Manville Corporation, for instance, as a result of its asbestos-related claims, and A. H. Robins because of the health problems linked to the Dalkon Shield contraceptive.

When the Old Good-Ol'-Boys of Texas hauled the New Good-Ol'-Boys of Wall Street before a hometown jury, the outcome was almost inevitable.

Hugh Liedtke of Pennzoil entrusted the case to one of his best drinking buddies and fishing companions—a contingency-fee lawyer named Joe Jamail. If anybody could arouse a jury's passion over the loss of a billion barrels of oil, it was Jamail. In a career spent representing the victims of intersection collisions and oil-rig accidents, Jamail had extracted more million-dollar judgments and settlements than anyone in the history of the bar. He was pretentiously unpretentious—a Brillo-haired lawyer in boots who spent long afternoons drinking with shrimpers at Galveston Bay and who could turn the air blue with expletives. He was vintage Texas, a close friend of country-music singer Willie Nelson, who had Jamail's wife, Lee, in mind when he wrote the song ''Good Hearted Woman.''

> You're a good-hearted woman
> In love with a good-timin' man

A less likely lawyer was hardly imaginable for the case. Jamail
was a self-admitted dunce with numbers—except, as he liked to
joke, when it came to dividing by thirds to calculate his fees. Far
from adopting the modern science of jury research in a big case,
Jamail occasionally consulted the zodiac to resolve a toss-up between two jury candidates. He had never handled anything remotely close to a takeover-related case, much less one involving
the largest corporate acquisition in history. But all that counted
was Jamail's way with a jury. He would try *Pennzoil v. Texaco*
not as a matter of securities law, fiduciary duty or any other
highfalutin' issue of the takeover game, but as a matter of honor.

"This is a case of promises," he would tell the jury, "and what
those promises meant to Pennzoil, and what they ultimately meant
to Texaco. . . . The question ultimately that you are going to have
to decide is what a promise is worth, what your word is worth,
what a handshake is worth, what a contract is worth. Because
that's what a contract is: a promise."

There was no mistaking it, Jamail would assure the jury. "It is
the most important case ever brought in the history of America."
For the first time, he said, a jury could "send a message" that the
public demands a higher standard of behavior from Wall Street.

The lawyers and investment bankers in the Getty deal were
among the profession's best—a few were considered *the* best in
their fields—but in the courtroom Jamail would cast them as
blackguards of American business. The crime perpetrated upon
Pennzoil, Jamail would tell the jury, was nothing less than "a
conspiracy between Texaco and a group of New York investment
bankers and New York lawyers"—with emphasis on the words
New York. Before the first New Yorker ever took the witness
stand, Jamail had conditioned the jury to expect the characters in a
police lineup. "You are going to get to see them eye to eye," he
would say. "*You* size them up."

Among those watching from the jury box would be Jim Shannon, a thirty-two-year-old dyed-in-the-wool survivor of the sixties
with a receding hairline and forelocks that fell into his eyes.
Shannon had dropped out of high school to join what he called the
"rock-and-roll wing" of Democratic Party politics—helping to
publish an underground newspaper, managing music groups,
counseling draft dodgers. In later years, while working as a

"communications specialist" for the city of Houston, Shannon's political projects ranged from collaborating with the distinguished Congresswoman Barbara Jordan on a public-service advertisement to distributing hate literature against a conservative political candidate. His writing activities ranged from sophisticated model-cities projects to screenplays luridly entitled *Death Drives a Pontiac* and *Blood on the Wing*. When summoned for jury duty in *Pennzoil v. Texaco*, Shannon was deeply into writing a novel about a blues guitarist.

"A one-man jury," Texaco's lawyers would eventually call Shannon. That was a bit of hyperbole, but it was true that of all the factors weighing against Texaco in the case, Shannon loomed large.

"We weren't going to put our imprint of approval on this trash," he would explain.

The jurors would be vilified as vigilantes, attacked from all corners for exercising passion and prejudice. Texaco would drag two of them back into court, seeking to probe details of their private lives in an attempt to prove that they had been swayed by family bias. But outrage and prejudice, and even Jim Shannon, could not begin to account for the $11 billion judgment.

The jurors could reach their verdict and impose their damages based solely on what the lawyers played out on the stage before them, and Pennzoil's production was far more artful. Moreover, it was the judges in the case who had final say over what the jury was permitted to see and hear, and the judges who instructed the jury on how to apply the law.

As it turned out, Pennzoil's Joe Jamail had a way with judges as well as juries. The first judge in the case became gravely ill; when he died, Joe Jamail would serve as a pallbearer. When a new judge took the bench, he happened to be a boyhood friend of Jamail.

"When you get out of law school, you think the right side is going to win a case," Richard Miller, Texaco's chief trial lawyer, would say when it was all over. "Then you learn the side with the best lawyer is going to win. Then you learn that if the judge has an interest in the case, you're fucked."

To that end, Texaco too tried to foster a special relationship in the case, but would fail at duplicating the rapport it believed that Pennzoil had. Today, some Texaco lawyers believe that Pennzoil stole the trial, every bit as strongly as Pennzoil believed Texaco had stolen its deal.

* * *

Pennzoil certainly had reached an accommodation of some sort with Getty Oil—no one disputed that. But did it have a *contract*? Gordon Getty, for one, still doesn't know for sure, one way or the other. "I think it was a close call," he would say in an interview a year after the $11 billion verdict.

It was a cornerstone of Texaco's defense to argue that Pennzoil had crassly pandered to the insecurity and personal ambition of Gordon Getty, offering him the executive position in Getty Oil that he had long been denied by his father and others. "They found out Gordon Getty's intense desire to exceed his father," Miller of Texaco would tell the jury, "a desire that came from his earlier failures—and his earlier inability to satisfy his now-dead Daddy."

As it turned out, Gordon *was* the central figure in the case, even if he never attended a moment of the trial. Before Pennzoil and Texaco had come along, Gordon was steeped in one of the most extraordinary battles for control ever to erupt within a major corporation—against professional managers who considered themselves over him to be the spiritual inheritors of J. Paul Getty. Gordon would launch a bitter power play against his father's successors, who would respond by exploiting the divisions that existed within the Getty family itself—and in the process assure the self-destruction of the company they sought to defend.

Yet for the significance of his role in the case, Gordon's character was also the least understood. Within Getty Oil and the Getty family, in the war rooms of Pennzoil and Texaco, in the Houston courtroom and elsewhere, Gordon would become the subject of more amateur psychoanalysis than a soap opera heroine. Depending upon who was talking, Gordon would be defamed as either foolish or devious, indecisive or impulsive, a son who wished to prove himself to his dead father or one who didn't give a damn about his dead father's wishes.

This much was certain: Nearly everything that threw Getty Oil into the takeover arena—and ultimately what landed Texaco in the soup—somehow traced its way back to the founding father, an industrialist who was said to be the world's richest, and who surely was among the meanest. The jury would receive only a selective recitation of the facts concerning J. Paul Getty, but it was his invisible hand—or, more aptly, his fist—that shaped the most crucial facts in the case.

Long after the jurors had walked out of the courtroom for the last time, some would remain haunted by the knowledge that the most important figure in the drama had been a man dead for almost ten years.

· 2 ·

The Harvard Military Academy gave schoolboys plenty of attention, and it chafed at the teenage J. Paul Getty. The 5 A.M. bugle call, the boot-and-button polishing, the unforgiving inspections—even in his parents' overstrict household he was unaccustomed to such regimentation. When the confinement grew too oppressive, the young cadet found relief only by going AWOL, and one such outing proved particularly rewarding. He arranged a rendezvous with a "willing damsel" he had met at a heavily chaperoned dance a week earlier. The engagement proved to be such a "memorable milestone" that nearly seventy years later he would write, "This initiatory experience was everything it was cracked up to be—and then some."

In his later teens, J. Paul Getty kept his nighttime appointments by sneaking aboard his father's Chadwick and silently coasting down the driveway to a sufficiently safe distance before starting the engine. Returning without detection required the greater challenge of racing down the street, killing the engine and coasting back *up* the driveway. The system fulfilled J. Paul's objectives until the night he found the Chadwick chained to the garage floor.

His appetite for women was undiminished by adulthood. In 1916, after making his first million in the oil fields of Oklahoma, he traded in his third-hand Model T for a Cadillac Roadster and moved to Southern California for what he would later describe as "a total immersion course in the pleasure principle." He was tall,

muscular and immaculate at age twenty-three—strikingly debo-
nair, with deep blue eyes and an arrogant smirk, his hands thrust
deep in his jacket and his thumbs hanging from the pockets. He
called this period his retirement. His father, however, called it his
"postadolescent hibernation," and indeed, within two years, the
tomcatter was again a wildcatter, slowly building his first million
into his first billion. But the thrill of the chase that he enjoyed in
the oil patch also continued to rule his romantic life. On occasion
it was difficult to distinguish between his business affairs and his
love affairs. He christened one of his earliest ventures Lorena Oil
Company, borrowing the name from "a very charming young
lady of my acquaintance."

It was J. Paul Getty's unwavering interest in boardrooms and
bedrooms—to the exclusion of nearly everything else—that ultimately caused strife to grip both of his far-flung legacies: his
family and his company. Because he could remain faithful only to
his business enterprises, Getty went through five marriages that
produced four sons on two continents—not the kind of situation
prone to promoting sibling harmony. (A fifth son, Timmy, died of
a brain tumor at age twelve.) And although he greatly wished his
sons to follow his footsteps in the oil business, J. Paul strove so
hard to avoid pampering them that he wound up inflicting special
treatment of the wrong kind.

"*I* was not born with a silver spoon in my mouth," J. Paul once
declared, but in many respects he was. His father, George F.
Getty, was a well-to-do attorney in Minneapolis, who became one
of the foremost oil pioneers of this century after he traveled into
the heart of Indian territory to collect a $2500 insurance payment
in 1903. Oil had just been discovered in the red-dirt prairies of the
Osage Indian Nation, and it was immediately apparent to George
Getty, as it was not to everyone, that the recent invention of the
horseless carriage guaranteed a profitable future for oil. George
also began studying the formative science of geology, which held
that it was more productive to drill around hills, valleys and
inclines than under doodlebugs and divining rods. George Getty
plunked down $5000 for a land lease near the town of Bartlesville,
Oklahoma, and went on to drill fifty wells. He found oil in forty-
seven of them, and the Getty family fortune was born.

When J. Paul struck out on his own, he happily accepted a stake
from his father that included the then-kingly sum of $100 a month
for living expenses and, more importantly, 70% of whatever
money it took for him to buy a lease and drill it. It was, in short,
only by having a father of great means that J. Paul Getty was able

to make his first millions. In his early success, as a mere millionaire, J. Paul graciously acknowledged this debt, giving his father top billing in the title of his first book, *The Oil Business of George F. and J. Paul Getty,* published in 1941. But later in life, after being proclaimed the world's richest man, J. Paul was writing books for Playboy Press (*How to be Rich, How to Be a Successful Executive*) in which he bent over backward to distance himself from his father's contributions. It was fine to share the credit for a few million, but not for a billion.

J. Paul Getty's unwillingness to acknowledge that fathers may help their sons would become a heavy burden indeed to his own sons.

The first, named George II for his grandfather, was conceived on the honeymoon, when J. Paul was thirty-one and his bride, Jeanette Demont, eighteen. The task of building the oil company he operated with his father proved so time-consuming that the marriage was off within two months of George's birth, moving J. Paul to acknowledge that no wife "enjoys feeling that she is being neglected for an oil rig." His second marriage—a summer fling in Mexico—produced no children. On his third go-around J. Paul married another eighteen-year-old, Fini Helmle of Berlin. While Fini was pregnant and living with her parents in Germany, J. Paul was in New York witnessing the stock-market crash firsthand. He could not bear to leave before analyzing all the bargain-priced oil stocks created by the crash, but he did manage to reach Berlin just before the birth of their child, his second son, Ronald. When that marriage, too, was quickly scuttled, his wife extracted such a lucrative divorce settlement that J. Paul would forever exclude Ronald from the family fortune—a fact that would have a significant bearing on the eventual battle between Texaco and Pennzoil for control of Getty Oil.

J. Paul's fourth marriage produced Eugene Paul and Gordon, the third and fourth sons. Their mother was Ann Rork, a sweet-faced brunette whose father, Sam Rork, was a talent manager who represented the "It Girl," Clara Bow, and who did a little acting himself. Ann's parents initially forbade her to date J. Paul, reasoning that a girl of fourteen ought not to confine her interests to a man twice her years. But Ann's father was wiped out by the Depression by the time she was in her early twenties and suddenly the thrice-divorced, forty-year-old millionaire seemed a fitting groom. Initially, J. Paul and Ann declared their vows in the privacy of a New York hotel room, but when the U.S. State Depart-

ment challenged the wedding date sworn on their passport as
something less than a legal ceremony, J. Paul and Ann rushed off
to Cuernavaca for a ceremony conducted in Spanish, which she
did not understand.

When their second child, Gordon Peter Getty, was born on
December 20, 1933, his parents' marriage was already crumbling
beyond repair. While pregnant with Eugene Paul, Ann became so
despairing of her husband's attention to business and to other
women that she attempted suicide by swilling a bottle of iodine. J.
Paul's months of absence during both of her pregnancies led her to
conclude in her divorce petition that "the sight of a pregnant
woman was repulsive to him." When J. Paul finally made it to the
hospital, hours late, on the day of Gordon's birth, he told Ann that
he was too rushed to remain longer than a minute. When she
inquired whether he might like to view his new son, he replied, "I
would, if the nurse would hurry up and bring him in." As Ann's
divorce papers recounted it, J. Paul's entire reaction to the sight of
the newborn Gordon consisted of, "Huh, he looks like you."
Then he was off to San Diego for a yacht trip.

In the depths of the Depression, Gordon and his brother spent their
boyhood year wanting for very little. Their mother's sensational,
headline-grabbing divorce petition had won her a relatively rich,
Depression-era settlement and with the boys' own income the
household got by on about $40,000 a year—a sum that permitted
Ann to hire a maid, a cook and a gardener. (The divorce settlement might have been much higher, except that her wedding in
Cuernavaca had involved the signing of some papers, also in
Spanish, in which she had unwittingly disclaimed joint ownership
of her husband's estate, then valued at more than $20 million.)
Ann had been a starlet of minor repute in the "talkies," and
eventually set up her household in a Los Angeles residence that
had once belonged to Harold Lloyd, the slapstick silent-movie
star. At age four, Gordon undertook what became his greatest
passion in adulthood—music composition—while taking lessons
on a piano that once belonged to Chico Marx.

What Eugene Paul and Gordon did not have was a father in any
remotely conventional sense. Their mother married three times
further, a fact that prevented them from forming any deep attach-
ments to a single stepfather, although one of them conveyed to
Gordon an appreciation of Shakespeare. Their biological father
remained a distant and seldom-seen figure. Like a Little Lord
Fauntleroy, Gordon was once disappointed to learn that the elec-

tric motor kit sent by his father at Christmas had actually been selected by a secretary, and to make matters worse, Gordon never did succeed in getting the thing to work.

Some people who knew all four of the Getty boys considered Eugene Paul and Gordon to be the brightest of the lot, but thereafter the similarites between them ended. Eugene Paul grew up glib and exceptionally good-looking, a hit with the girls. His kid brother grew up as "Gordo"—six-foot-three, dark, bespectacled and gawkish, a painfully serious and aloof boy with an air that struck some people as superior, even arrogant. In another generation he might even have been called a nerd, for in addition to his serious music studies, Gordon had acquired a childhood passion for poetry. While the other boys were playing with tin soldiers, Gordon was assiduously memorizing all seventy stanzas of Thomas Babington Macaulay's 19th-century poem "Horatius at the Bridge." When the sons visited their father in Tulsa in 1943, the Old Man's diary entry for the day included the notation that Gordon, then age ten, "read a poem he had written about the good qualities of Negroes." Gordon's more conventional adolescent interests included baseball. "Never made the team," he says. "But always among the last cut."

It wasn't until his teen years that Gordon ever saw what ordinary family life was like, and even then it was not in his own house. His mother had moved with her boys to San Francisco, where Gordon was virtually adopted by the family of his high school friend Bill Newsom. As Newsom would later recall, Gordon delighted at the mirth and bedlam of the Irish-Catholic household teeming with children, whose domestic pursuits included wrestling, breaking vases, playing Monopoly and sitting around the dinner table listening to Mr. Newsom talk about business and politics; Gordon loved him, and called him "Boss."

In the world of business, however, Gordon would establish decidedly less favorable relationships with his real bosses.

Eugene Paul and Gordon took their first assignments in their father's vast corporate empire in the late 1950s, about the time that *Fortune* shocked the world—as well as the Getty wives and children—by certifying J. Paul Getty as America's wealthiest man. At the time, Eugene Paul was already twice a young father, Gordon a newly discharged second lieutenant from the quartermaster corps in Ft. Lee, Virginia. (Poor eyesight had saved him from the infantry.) Together, the sons of the world's richest man began their corporate careers as nozzle jockeys at the Flying A filling station on U.S. 101, just north of the Golden Gate Bridge.

James McDonald, an executive in the Old Man's San Francisco office, used to stop at the station on his way to work from San Rafael and check up on the boys. "Eugene Paul was a real go-getter," McDonald would recall learning from the station manager. "He met customers well, got his hands dirty and really got into the business of helping run a service station." But as for Gordon: "Music was his bag and he wasn't about to get his hands dirty." The station manager, according to McDonald, "felt that he would not go anywhere in this world. Surely not the corporate world."

But the brothers did get thrust into the world when their father gave them their first taste of management experience in 1958. Eugene Paul was transferred into a life of excitement in Rome, where he was installed in a senior position at Getty Oil Italiana S.p.A. Gordon, meanwhile, was transferred to a place called the Neutral Zone, a forbidden stretch of desolate and undeveloped sand jointly claimed by Saudi Arabia and Kuwait. They called it the Zone for short.

For Gordon, the pressure was on. Ten years earlier, J. Paul had leaped ahead of Big Oil to obtain the rights to drill for the Saudis' share of the mineral in the Zone. Obtaining the drilling concession had been one of the Old Man's proudest achievements, and one of his most costly; the terms extracted by King Ibn Saud included not only a guaranteed payment of $1 million a year, regardless of how much oil was produced, but also a new house for the Saudi emir who supervised the territory. Gordon's big brother, George, had already completed a highly successful assignment in the Zone, bringing in a record-setting oil well in an area which had produced nothing but dusters. Following his achievements in the Zone, George had embarked on a fast-track career to the top of his father's U.S. operations.

For his part, Gordon became Ping-Pong champion of the Zone and played excellent bridge, but he brought in no gushers. It was, in fact, almost an impossible assignment for a twenty-five-year-old of Gordon's temperament and experience. His duties included the management of a multiracial work force in one of the world's most politically volatile areas, as well as an assignment to collect payments for all the household supplies that Getty Oil had delivered free of charge to the personal residence of the emir. Allegations were rife that company managers were collecting kickbacks on their purchases of water and lettuce. In this atmosphere, the incumbent executives became convinced that Gordon had arrived in the Zone as a personal spy for his father, and Gordon's aloofness made him seem all the more suspicious. What the managers mistook as Gordon's arrogance, however, was probably the result of his inimitable seriousness—and of his having formative musical scores perpetually chiming through his head. On an upright piano he had shipped in from Kuwait, Gordon had begun composing *All Along the Valley,* music for an *a capella* male chorus, with lyrics from the Tennyson work of the same name. ("All along the valley, by rock and cave and tree/The voice of long ago was a living voice to me.")

Then a terrible thing happened—terrible certainly in Saudi territory. A company employee who had crashed his pickup truck into a pipeline "disappeared in a puff of smoke," as Gordon recalls it, rather than incur the risk of facing the Hammurabi-style justice that prevailed on the Saudi side of the Zone. While the long arm of Saudi law was reaching out to find the supervisor of the fugitive, Gordon sanctioned *his* departure across the border into Kuwait. When the authorities came around with a warrant for the arrest of the errant supervisor, Saudi justice saw fit that Gordon should be sent to the slammer in Dhahran. "It was a visitation of the sins of whoever you can't catch, on those you can," Gordon would recall. "When they couldn't find him, they found me." True to form, Gordon rose above the occasion and took advantage of his week-long confinement by catching up on his reading, completing *The Complete Works of Keats* as well as *Measure for Measure* and other dark comedies by Shakespeare.

Rumors swept the company's management ranks that during his arrest Gordon had flung insults against the Saudi Royal family. It was said that the Old Man could bargain for his freedom only by relinquishing a drilling concession in the offshore waters of the Zone, one that predictably became a huge producer for another company. These stories were probably the exaggerations of executives who doubtless felt some rivalry toward the son of the company's patriarch. But regardless of whatever handsprings the Old Man had to turn in order to free his son, J. Paul would never forgive him for committing such an indiscretion. He would later write his son:

> Your failure to be duly respectful to the king's representatives was immature, undiplomatic and bad business for Getty Oil Co. Face and prestige are very important in the Orient and to have a Getty arrested and confined by Saudi authorities was harmful to

the morale of the company's employees. . . . I was anxious for
your safety and about the company's relations with the Saudi
government. It was all very distressing. Thanks largely to my
intervention with the governor of the Neutral Zone your release
from confinement was obtained upon the promise by me that you
would immediately leave the Zone and Saudi Arabia and not
return. This episode naturally affected my opinion as to your
business judgment when in an executive position.

When J. Paul transferred Gordon to Tulsa after the debacle in the
Zone, it was like the circus coming to town. Gordon's big brother
George also had served there a decade earlier, but in relative
obscurity; Gordon's arrival, on the other hand, occurred while the
"world's richest man" moniker was still freshly attached to his
father's name. Gordon was assigned to a middle-level management
position at his father's Spartan Aircraft Company, a company being
converted to postwar production of mobile homes. Tulsa's curiosity
about Gordon was heightened by the announcement that his arrival
was intended to build Spartan Aircraft into "the General Motors of
the house-trailer industry." "Gordon Getty, with his feet on the
ground floor, hopes to see the Spartan mission accomplished," the
Tulsa World proclaimed.

After Gordon had spent a few days on the job, the paper dispatched a reporter and photographer to the plant. "I'm an English
major, but I've always found it easy to understand business," the
twenty-seven-year-old proclaimed. Gordon disclosed that besides
having lived in a house trailer in the Zone, his qualifications
included the assembly of "his own stereophonic set—with five
speakers!" "People might be interested to know that I built the
entire rig for only six hundred dollars," he said.

Gordon proved adept at conceiving "off-the-wall" business
strategies, as he would later describe them, and many of his ideas
succeeded. But in the Age of the Organization Man Gordon
flunked the business basics. He was intimidated by the multibut-
ton telephone on his desk. He kept strange hours and took long,
unannounced absences. When Gordon *was* at work, he was said to
be meddlesome instead of subordinate, and invariably, conversations *with* him turned into lectures *from* him, most often on music.
After soliciting a comment on Gordon's performance from Captain Max Balfour, the former flying ace who managed the factory,
the *Tulsa World* reported that "Capt. Balfour credits Gordon as
being talented as a pianist and a baritone." But Captain Balfour

began telling his friends around Tulsa that "either he goes or I go," and soon Gordon was gone again.

"Everyone loved Captain Balfour," Gordon would remark some twenty-five years later, "but he was a very emotional man who brooded over things. To know Max Balfour was to be on and off his shit list."

And so it was with the Old Man, except that by this time Gordon was on the list a great deal more often than he was off. As his father would write:

> Your work at Tulsa and elsewhere has shown that you have good intellectual equipment, but I have heard complaints that you were overly independent, would not obey orders of your superiors or do much planning. . . .
>
> In my opinion you are a fine young man in most respects. I love you and am proud of you. I do feel, however, that for some reason or other you are very immature in relation to business. You seem to have kept a rebellious attitude toward authority.
>
> I realize that none of us is perfect. I myself am not perfect. However, I don't think a lack of maturity is one of my shortcomings. . . .
>
> Love,
> Father

Gordon, it seemed, had a way of getting lost in his own solipsistic mind, where he could dream up melodies or dream up the solution to a business problem unperturbed by the constraints of the real world. His father wasn't the only family member to observe this odd trait. "At 28," Gordon's mother Ann would write her ex-husband, J. Paul, "Gordo has retained the perspective of an eight year old: that the nursery is the world and the sun beams for the benefit of the occupants of his orbit."

After Gordon's mishap in the Middle East and his crash landing at Spartan Aircraft, the Old Man had glumly concluded that Gordon would never measure up as a businessman to his older brothers. The number-one son, George II, had climbed to the top of the U.S. oil operations. The second, Ronald, although disinherited from the family trust as a result of his mother's expensive divorce settlement, had obtained an important job heading up his father's marketing operations in Germany. The third son, Eugene Paul, had worked his way into the upper ranks of Getty Oil's Italian operation. But as for Gordon, "He seems to take for granted that

he is in a privileged position,'' the Old Man said in a letter to his
firstborn, George II.

> He feels that he is entitled to a job with one of the Getty group of
> companies whenever he wants a job. . . . Gordon is inclined to
> work as he chooses and when he chooses. He is a poor admin-
> istrator, he is very immature in his viewpoints. He is very intel-
> ligent, but I doubt that he has the qualities that make a good
> businessman.

In a deposition taken several years later, the Old Man would
express the opinion that, besides being arrogant, impulsive and
detached, Gordon had ''good intellectual equipment'' and a
strong head for numbers—and that all these characteristics made
him a thought-provoking critic. Thus, in his effort to find
something to do with Gordon, the Old Man engaged him as a
personal consultant. ''You know how Gordon consults,'' Eugene
Paul had told his father. ''He just states his opinion and he is not
interested in yours unless you agree with him''—the same quality,
in short, prized by any self-respecting consultant.

The arrangement satisfied everyone for as long as Gordon studied safe subjects, such as the ventilation system of the Hotel Pierre
in New York, which his father had purchased in the Depression.
But when Gordon began snooping around the domain of the professional oilmen in his father's organization, he deepened the
already broad mistrust of his abilities and intentions. There was,
for instance, the morning that Gordon outfitted himself in a proto-
preppie sportcoat and saddle shoes and showed up unannounced to
undertake a study of the company's huge, money-losing refinery
in Delaware. Gordon happened to pick the day that a VIP tour and
luncheon had been scheduled for Sir Phillip Southwell, chairman
of the Anglo-owned Kuwait Petroleum Company. The refinery
managers watched in horror as Gordon unremittingly held forth on
Kuwait, a country that few Westerners knew better than Sir Phillip.

While politely listening to Gordon, the dignitary threw a discreet wink at one of the company men, who would tell the story
for the next twenty years. As the accounts of Gordon's foibles
passed from executive to executive they were embellished with
each retelling, unfairly depicting Gordon as everything but the
idiot brother.

Nobody in the Getty organization seemed to take Gordon less
seriously than his half brother George, whose corporate rank was

second only to the Old Man. When Gordon filed a consulting report, George would distribute it to his assistants with orders to find the fault in it. Once, Gordon solicited a three-month consulting engagement by explaining to his brother that he only wished to prove he was capable of solving a complex business problem in a short period of time; George replied that this would be "almost like Hitler or Napoleon trying to conquer the world in ninety days."

Certainly, Gordon brought much of the criticism on himself. According to the memos he sent to his brother, he seemed intent on obtaining the most dignified title and the highest salary possible, regardless of the value of the work he was performing. He insisted on scheduling his assignments to suit his schedule rather than the company's. When he objected to the terms of his engagement, he reacted with outrage rather than conciliation. "I must be expected to decline playing a ridiculous role and thereby wasting (the company's) time and mine," he wrote to his brother George, "however valueless mine may be."

And there was something exceedingly odd about the reports and recommendations that Gordon would produce on his consulting assignments. His analysis was always creative and occasionally brilliant, yet his recommendations were often contradictory or utterly impractical. He conducted an unassailable analysis showing that his father was operating too many oil storage and loading facilities, but the solution—to consolidate them at the most economical location—failed to take account of the zoning restrictions of that area, according to the recollections of company managers. He showed how the company could improve profits by selling its gasoline stations to distributors, but his recommendation came on the heels of another report in which he seemed to urge the opposite strategy. "In view of your basic conclusions on selling service stations," his brother George asked in a memo, "why do you want to *buy* service stations?"

An unmistakable character trait had emerged in Gordon's business career. As George described it in a 1965 letter to the Old Man:

I think it is a fine thing that Gordon has the gift of creative thinking and enjoys doing it. Nevertheless, I cannot imagine how he can ever become a well-rounded and seasoned business executive if he never engages in the "pick and shovel work" that so much of the time is required to translate plans and programs into established results.

George, however, did not send such letters to his father solely out of love and concern for his half brother. George, in fact, appeared to go out of his way to keep the Old Man abreast of all of Gordon's foul-ups. When Gordon angrily refused a $750-a-month assignment from his brother, supposedly by saying that he could do better "bumming off friends in San Francisco," George wrote a letter tattling on his brother that very day. George also saw to it that his father received a full accounting of the matter of Gordon and his cars.

Gordon had trouble with cars. By his own admission in later years, he did not particularly care about cars. He found it easy to forget about them and to him they mostly looked the same. In 1966, George mailed his father a four-page, single-space dossier recounting three incidents involving his brother and his autos. He informed the Old Man of the time that Gordon had his car impounded after abandoning it at an airport on the East Coast. Then George recalled another story about when Gordon was arranging to depart from Spartan Aircraft in Tulsa. A company official, leaving nothing to chance, arranged for Gordon to trade in his Buick at a local dealership and fly back to San Francisco, where he would pick up a new car. Gordon, however, never showed up at the Tulsa dealership. When the company official tried to fetch Gordon from his apartment, he found books and clothes scattered everywhere, causing him to panic that the son of the nation's richest man had been kidnapped after putting up a valiant struggle. After the Oklahoma State Police had spent a week on the lookout, Gordon showed up in San Francisco, having neglected to tell anyone that he planned to make the trip by car after all. Finally, George mentioned the car that Gordon had left at Los Angeles International Airport—for six months. When a company official was dispatched to collect the car, he found that the registration had expired more than a year earlier.

"Gordon certainly does not seem to have been in touch with reality," George wrote. "I think you will agree with me, Father, that these three episodes involving Gordon and his automobile give us a great deal to wonder about."

Clearly, there was much more than cars, job titles and salaries at stake in the relationship of the Getty boys with their father. There was control of the family trust, which was where the real wealth resided—all $4 billion of it, by the time that Pennzoil and Texaco would come along—as well as the real power over the Getty Oil Company.

· 3 ·

Shortly after Gordon's flame-out at Spartan Aircraft in Tulsa, the image of another young Oklahoman was shining brightly in the eyes of J. Paul Getty. He was J. Hugh Liedtke, a budding oilman who in the 1950s had parlayed his Tulsa connections into a small petroleum empire that had already made him a millionaire. Now, in 1962, thanks to some close ties reaching all the way to J. Paul Getty himself, Liedtke was about to create a new company called Pennzoil.

John Hugh Liedtke was born in Tulsa on February 10, 1922, when it was still the oil capital of the world. Few families were more closely identified with Tulsa's oil roots than the Liedtkes. Hugh's father, William, the son of Prussian immigrants who settled in one of the many German *émigré* communities in Texas, had graduated from the University of Texas Law School and moved to Oklahoma when it was still Indian territory, helping to develop some of the earliest laws governing the ownership and extraction of petroleum. At the turn of the century, about the time that J. Paul Getty's father made his first oil strike in the Osage Nation, J. Hugh Liedtke's father was living in the Creek Nation, where he was elected the youngest delegate to the constitutional convention that made Oklahoma a state. William Liedtke, then age twenty-four, was a populist who became known as the convention's ''boy orator,'' and later spent several years as a judge.

Although Judge Liedtke railed against the power of giant corporations, he later became one of the highest-ranking lawyers in Gulf Oil Corporation. As a result, Hugh Liedtke and his younger brother, William Jr., grew up as children of privilege in Tulsa's most exclusive neighborhood.

By all accounts Judge Liedtke was one of Tulsa's most dignified and respected citizens—a man noted in the community for an almost obsessive commitment to principle. He regularly instructed a household servant to smash the family's empty beverage bottles before they were collected by the trash company; the reason was that federal law at the time prohibited reusing bottles, and Judge Liedtke wouldn't think of having *his* empties become instruments of such wrongdoing. Many of Oklahoma's wealthy made a practice of draining their bank accounts at year-end to avoid state personal-property taxes; Judge Liedtke proudly refused to do so. He raised his boys in an almost Elizabethan environment, enrolling them in a prep school run by Augustinian priests even though the family was Protestant. For years they were forced to study piano, practicing in shifts beginning at 7 A.M. until it was time to leave for school at 9 A.M. The boys' mother, Mabel Claire, also imbued the household with a certain dignity and refinement, regularly entertaining lady friends for readings of Browning and Shakespeare and staging dinner parties for the jurists, lawyers and marshals who regularly called upon Judge Liedtke from all over the country.

Even in so genteel a household, of course, boys will be boys. Bill Liedtke managed to get expelled from Spanish class. Hugh Liedtke liked to tease his little cousin with magic tricks. Once, in his later teens, while his parents were out celebrating New Year's Eve, Hugh held an "air raid party," periodically reenacting the bombing of London by playing siren music and turning out the lights, giving the young couples sitting on the couches the opportunity to get better acquainted—and his prankster kid brother the chance to sneak up and snap photos by using a flash attachment.

Hugh Liedtke got his first exposure to the oil patch as a fourteen-year-old, and he thrilled at the manly rough and rollick seldom seen by well-off city boys from Tulsa. He spent the summer with his father in the newly discovered oil fields around Flora, Illinois, a town gripped by such a boom that Hugh had to spend his first several nights sleeping in a bathtub at the Star Hotel. While his father feverishly sorted through land titles and mineral leases at the local courthouse and law offices, Hugh spent the day in the oil fields watching the roughnecks clanging chains and drill collars against sections of drill pipe. Judge Liedtke had put his son in the charge of a crusty old crew boss, who enjoyed ordering Hugh to wade through cold, muddy farm ponds setting trotlines so that the boss could eat fresh fish for breakfast. In the evening, Hugh hung out in the local watering hole, watching the workers shoot snooker and drink like thirsty camels.

At the tender age of sixteen, Hugh desperately wanted to prove himself in a real oil field job, and got his chance spending the summer digging a pipeline ditch across the Kiamichi Mountains in southeast Oklahoma. "He was a boy who wanted to make it among men," says Hugh's cousin, J. Hugh Roff Jr., whose parents put Hugh up for the summer while he was working the pipeline gang. Roff recalls Hugh coming in from a day of work proudly showing off the blisters on his hands and the grease splattered on his khakis. "He was 'the rich kid from Tulsa,'" Roff says of his cousin, "and he wanted to measure up." It was, however, the last time Hugh Liedtke would work with his hands. "I learned something that summer," he would recall. "I didn't want to spend my life on the work end of a shovel or pick." But a pattern had been set: Overreaching—trying to prove something— would become a hallmark of Hugh Liedtke's career.

In all the romantic lore of the Liedtke brothers, one story—a war story—would receive the most frequent telling.

Their father had sent them to Amherst College in Massachusetts, a school that neither had heard of, where Hugh majored in philosophy and won the department prize with an honors thesis called "Religion and the Limits of Knowledge in the Philosophies of Hume, Santayana and Dewey." Hugh then spent a year taking a compressed master's program in business administration at Harvard University. But by the time the brothers had obtained commissions in the U.S. Navy, at the height of the war for the South Pacific, they had still not decided in which direction they would follow their father's footsteps—whether into the oil business or the legal profession. Hugh had become a lieutenant supervising bomb-loadings on an aircraft carrier, and Bill was the gunnery officer on a tank-landing transport. But by chance, the brothers twice crossed paths, in Saipan and Okinawa.

"We'd have a few drinks and decide what we'd do if we ever got out of that place," Hugh would recall, "and then we did it." Like their father, they would deal in oil *and* law.

They enrolled together at the University of Texas Law School, where they rented the servants' quarters behind the home of a politician named Lyndon Johnson, who was living in Washington as a freshman U.S. senator. In the main residence of the LBJ spread lived an up-and-coming politician named John Connally and his wife, Nellie. It was miraculous that the Connallys would remain lifelong friends of the Liedtkes, for during those high-pressure law school years the boys made a practice of hosting rather raucous soirees, forcing the Connallys to ring the buzzer in the servants' quarters to call for some quiet. The Liedtke brothers and their dates were able to bring one such party to conclusion only by gathering up the remaining stock of gin and hiding it in the bushes. The next morning, while the heavy-headed Hugh was retrieving the bottles from the yard, he looked up to see John Connally staring through a window in disbelief. "He must have thought we were crazy," Liedtke would recall.

When the Liedtkes set up shop in the West Texas tumbleweed town of Midland in 1949, the city was in the throes of meta-morphosis that few communities outside of Texas had ever experienced. Hugh walked into the Scharbauer Hotel and found the lobby teeming with calves penned up for the local fat show. At the same time, the throng of fortune seekers flocking to Midland following the recent discovery of oil there caused such a shortage of office space that the new law firm of Liedtke & Liedtke had to settle for a room in the basement of the nearby Crawford Hotel. Hugh and Bill launched their practice by buying royalty interests on behalf of a family friend in Tulsa, who paid them a 5% commission on every deal they handled.

It was a heady time and place for aspiring oil barons. Unlike the oil fields discovered twenty years earlier in East Texas, where big landowners and Big Oil predominated, the oil basin around Midland was covered by small ranches, enabling independent lease brokers and small-time oil operators to swap drilling rights in smaller denominations, and dream of making it to the big time themselves.

Before long the Liedtkes struck up a friendship with a Yalie and newly discharged Navy flier named George Bush. The son of U.S. Senator Prescott Bush, George had left the service at Corpus Christi, Texas, so anxious to establish a reputation independent of his father that he refused to return East. In an interview thirty years ago, he explained that he had no interest in "cut-and-dried jobs, with everybody just like everybody else, getting a job with

Dad's help and through Dad's friends.'' George Bush was happy, however, to rely on Dad's friends out West. Prescott Bush arranged a meeting for his son with the president of Dresser Industries, who gave George a job as an oil-field supply salesman in nearby Odessa. Before long Bush went into business for himself dealing in oil leases, and eventually took an office in the same building as the Liedtke brothers, with whom he became a close friend.

These war veterans and would-be oil barons formed a clique, taking turns hosting backyard barbeques and Saturday-night dances in their living rooms. They led devil-may-care lives, skipping from town to town in a rickety old plane with a pilot everyone knew as Smilin' Jack, who navigated by flying low enough to read the names painted on the water towers. They formed a touchfootball team called the Midland Misfits, which played every Sunday while the wives and kids watched and while everyone drank beer. Visiting teams quickly learned that the Misfits gave themselves an advantage by permitting forward laterals and by playing on a field with narrower boundaries than any other team was accustomed to. The home-field rules proved particularly useful when the Lubbock Leftovers arrived one Sunday with Bobby Layne of the Detroit Lions playing quarterback and Glenn Davis, the star halfback from Army, who had trouble completing his pass patterns before running out of bounds.

The Midlanders played by their own rules in business, as well, operating under a creed which held that your word was everything. In the morning you could do a deal on a handshake over a cup of coffee at the Scharbauer Hotel restaurant, move to the next table and sell part of the deal with another handshake. During lunch at the local beer hall they would help themselves to a bowl of ''red'' (chili), drink a few beers, shoot a little pool and wander over to the county section maps hanging on the wall to discuss a deal. And why not? These were friends doing business together, and in the oil fields of Midland and elsewhere, where the deal was everything, there often wasn't time to wait for the paperwork. ''Things worked by trading off one thing against another, by making deals together,'' Fred Chambers, a Midland friend of Bush and the Liedtkes, would later say. ''It was usually a question of your word; things moved fast, and we skipped the paperwork.''

How soon y'all need that rig? the contractor asked. *Hell, man, I need it yesterday! Just move it on down here and send me the paperwork later*, the oilman replied. *We been gettin' two grand a day for that rig*, the contractor would say. *Listen here, fella*, the oilman answered. *That sucker's gonna be sittin' on a hole for ten weeks—eighteen hundred and not a penny more!*

Miscommunication, and sometimes outright bad faith, caused problems and lawsuits, of course—but no more often, it seemed, than when the paperwork had been completed ahead of time. A man's word was a greater bond than a contract. *If I give my word no one can break it. But if I sign this contract my lawyers can break it.*

In this environment, where the advantage belonged to the imaginative and the quick, nobody outperformed Hugh Liedtke in either category. He began trading oil-producing properties in a way that permitted the eventual owner to defer his tax liabilities until the field was depleted, much like taxes in IRA accounts are deferred to someone's retirement. These deals proved particularly popular with the Liedtkes' growing stable of well-heeled investors back in Tulsa. Hugh Liedtke would spend the day prowling for a land deal, all night completing the thirteen-hour trip back to Tulsa to collect the money and the next day returning to Midland in hopes that no one else had caught wind of the deal. After months of such commuting he decided that if he could raise a big chunk of money all at once—$1 million, say—they could "inventory" these deals rather than drive to Tulsa every time a new one came along.

George Bush went to New York and raised $500,000 through his family connections; the Liedtkes went to Tulsa and raised the same amount through theirs. They wanted a catchy name for their new oil company, one that would appear either first or last in the telephone listing. They looked no further than the nearest movie marquee, where *Viva Zapata!* was playing with Marlon Brando in the role of the Mexican revolutionary. After Zapata Petroleum Company was formed in 1953, Hugh Liedtke would get a kick out of telling people that depending upon your point of view, Zapata was either a patriot or a bandit.

For a while, the investors must have wondered the same thing about Bush and the Liedtkes, who had taken the $1 million and decided to gamble it all on a single oil field. It was a barren stretch of West Texas sagebrush country called the Jameson Field, where six widely dispersed wells had been producing for several years. Bush and the Liedtkes were convinced the wells were "connected." *Those wells can't possibly connect at that distance*, the company's most experienced Tulsa investor told Hugh Liedtke.

"Goddamn, they do!" he cried. Bush's backers from the East, who were unacquainted with the dicey nature of the business, were especially wary. "The New York guys were just about to pee in their pants," Hugh Liedtke would recall. The wells, fortunately, *were* connected, and the newly founded Zapata Petroleum put down 127 producers without ever encountering a dry hole.

But to everyone's dismay, the field began dwindling in about three years. There was still plenty of oil-in-the-ground—decades' worth, in fact. But like a can of whipped cream that loses its oomph before all the cream is gone, the field had exhausted the natural pressure that made the oil flow. All was not lost, however. The Liedtkes began pumping water through the underground formations, a costly practice that *almost* gave the field the appearance of having an inexhaustible supply of oil-in-the-ground. Liedtke became preoccupied, almost obsessed, with the success of the "enhanced recovery" venture, viewing it as everything but a perpetual-motion machine—something that could keep the heart of an old oil field beating long past its prime.

Despite their success, Bush and the Liedtkes soon became convinced that the oil fields of the U.S. had become so thoroughly picked that the "elephant" fields remaining to be discovered lay offshore, where the federal government had recently begun auctioning mineral rights. In 1954 they launched a for-hire drilling company, Zapata Off-Shore, in the midst of a stock-market boom that made it easier than ever to raise money from the public. "The stock market lent itself to speculation," Bush would explain, "and you could get equity capital for new ventures."

By the late 1950s the young oilmen were millionaires, but conflicts had erupted between the Western and Eastern investors backing the two companies. The "New York guys," as Hugh Liedtke called them, wanted to move headlong into the offshore drilling-services business while the Westerners wished to concentrate on exploration and production. So in 1959, Bush's backers bought out the Liedtkes' backers in the drilling company and the Liedtkes' backers bought out the interest of Bush's backers in the producing company, and the two groups parted ways. The divorce was amicable as far as Bush and the Liedtkes were concerned, as subsequent events would prove. But the split-up made it plain that where Eastern capitalists were concerned, Hugh Liedtke refused to be intimidated.

*   *   *

Among the connections that Hugh Liedtke had developed over the years, none were as strong as those leading to the Getty organization. Hugh's parents maintained a close friendship with Jack Roth, one of Old Man Getty's most trusted executives. The Liedtkes' nextdoor neighbor had been Bill Skelly, the son of an oil field teamster who built up one of the greatest independent oil companies in the country, Skelly Oil, along with Spartan Aircraft—companies over which J. Paul Getty had obtained control. Liedtke would later recall that even his "wife's oldest brother's wife's father" had been a high-ranking executive in the Getty empire.

Indeed, not long after Gordon Getty had visited his father in Tulsa and recited his poem about "the good qualities of Negroes," the Old Man was making his first acquaintance with Hugh Liedtke. The occasion was a cocktail party thrown by the Skelly family to honor Hugh's engagement to a neighbor girl named Betty Lyn Dirickson. And there he stood in the flesh among the finery of Tulsa who had turned out to honor the future bride and groom: J. Paul Getty.

Liedtke had also solidified his Getty connections during the Midland years. He socialized with George Getty II, who, unlike his kid brother Gordon, put in some time in West Texas while pursuing the fast-track upward in the Old Man's empire. One of Hugh's close friends from Tulsa, Bill Bovaird, had gone to work as an oilfield-tool distributor in Midland after marrying the daughter of Jack Roth, the Old Man's trusted Tulsa aide.

Now, in 1960, these connections would prove the most valuable that Liedtke had ever developed.

Forty-nine years earlier, the government had busted John Rockefeller's Standard Oil Trust into twenty-seven different companies, each intended to be independent of the next. Over the years, through two world wars and a dozen cycles of economic contraction and expansion, seventeen of Standard's progeny would go out of existence through mergers or failures, leaving just ten survivors, including the forerunners of Exxon, Mobil, Amoco, Arco, Chevron and Sohio. Among these giant survivors was a little outfit called South Penn Oil Company, located in the heart of the world's first oil field, in Oil City, Pennsylvania.

Liedtke came upon the South Penn name in 1960, while flipping through *Moody's Industrial Manual* looking for a place to spend the dough he collected selling his share of Zapata Off-Shore to the Bush interests. It was almost incidental that South Penn controlled a regional brand of motor oil called Pennzoil, for South Penn had what Hugh Liedtke really wanted: oil-in-the-ground. It owned a piece of the famous Bradford field—the first billion-barrel oil field ever discovered—where it was coaxing oil from the ground with water-flooding techniques similar to those that Liedtke was using in Midland. But South Penn was languishing, failing to keep pace with the latest innovations in secondary recovery and ignoring opportunities to expand into new oil fields. "It looked like kind of a dead situation," Liedtke thought.

He decided that *he* ought to be running South Penn, and he quickly identified a path into the company: According to the entry in *Moody's,* Old Man Getty controlled nearly 10% of South Penn.

Liedtke made his pitch to his friends Jack Roth and George Getty, who brought back word from the Old Man: When you own as much of South Penn as I do, maybe then I'll see about putting you in charge. "I don't have that kind of money," Liedtke protested. Further word from the Old Man: "You just buy as much as you can, and when you run out of money, I'll let you know." Liedtke put together an investment partnership and before long had bought 10% of South Penn—the same amount held by J. Paul. Liedtke had held up his end of the bargain.

On the strength of Liedtke's heavyweight backing, the board of South Penn eventually agreed to oust the company's chairman and chief executive officer—a professorial lawyer of about sixty years named John Selden—and install the thirty-nine-year-old Liedtke as boss.

There was only one problem: Nobody apparently informed John Selden of all the arrangements. Selden publicly announced that Liedtke had assured him that no takeover was planned. In fact, Selden was certain that Liedtke had assured him that he was buying South Penn stock "for investment purposes only"—an expression that was often a prelude to a hostile takeover in the 1980s, but that generally meant nothing more than it implied in the 1960s. When Selden finally awoke to the fact that Liedtke had actually been brought in to take over the company, he implored the board to have Liedtke fired. But the directors wouldn't think of throwing out the imaginative and aggressive young oilman—who also happened to be carrying a mandate from J. Paul Getty. Before long, Selden was out altogether, and management control belonged to Liedtke alone.

"It was pretty rough for the incumbents," says Ed Boyle, one of the South Penn directors who sided with Liedtke. Selden, who today is in his eighties, bristles at the very mention of Hugh Liedtke's name and refuses to discuss him in any way. "He doesn't like me worth a damn," Liedtke acknowledges today.

But it was a small price to pay, for J. Hugh Liedtke was now, in a sense, working for the larger-than-life J. Paul Getty. And best of all, Hugh Liedtke controlled South Penn's oil-in-the-ground, as well as the yellow can of Pennzoil, whose name he adopted for the corporation.

Twenty-six years later, after Hugh Liedtke had shocked the world in a legal battle called *Pennzoil v. Texaco*, a bunch of wildcatters would gather together to roast their old friend Hugh Liedtke and retell the story of how he had come into his own as an oilman. The master of ceremonies told the true story of the boyhood fascination held by Hugh Liedtke for a rather odd furnishing in the home of his nextdoor neighbors, the Skellys. Mrs. Skelly had installed a yellow chair frame over the toilet, creating a golden throne effect—a "yellow can."

This yellow can was to become to Hugh Baby what Rosebud was to Citizen Kane. It would symbolize power, something to be obtained. . . .

He wanted to hit the long ball. Then it was right there in front of him: South Penn Oil, his friend J. Paul Getty, the yellow can. Hugh Baby was on it like a duck on a June bug: negotiation, manipulation, high-powered trading. Pennzoil became his; Hugh Baby was its top man. From Midland to Pittsburgh to Houston, Pennzoil moved with Hugh Baby at the helm. Hugh Baby brought in Arnold Palmer, and he told the world about Hugh Baby's bright yellow can.

The other big plum lay out there ripe for the plucking: Getty Oil—a cool billion barrels that could move Pennzoil up with the *really* big boys. . . .

There was one other principal character on a collision course with Gordon Getty and Hugh Liedtke.

In 1947, when Liedtke was making his first acquaintance with J. Paul Getty, a quiet and somewhat aloof young chemist named John McKinley was busily taking apart molecules of fats and soaps at the Texas Company grease lab in the dreary refinery town of Port Arthur, Texas. He was a quiet, demurring man with a gentle Southern drawl, whose long face, large ears and slightly squinted eyes bore him an uncanny resemblance to Lyndon

Johnson. Still a decade from adopting its retail-trade name—
Texaco—as its corporate name, the Texas Company of postwar
years was the perfect place for a scientist such as McKinley to
launch a career.

John Key McKinley was born in Tuscaloosa, Alabama, on
March 24, 1920, the grandson of a Confederate soldier and the
son of a onetime deputy sheriff who maintained the law with a
sawed-off shotgun from the back of a mule. But by the time of
John's birth, his father, forty-year-old Virgil McKinley, had long
since turned in his badge for a career in academia, graduating
from Columbia and returning to Tuscaloosa to become a professor
of industrial management at the University of Alabama. Fittingly,
the professor married his wife in the chemistry lab at Columbia.

Young John became very much the product of his academic
surroundings. "Our living room was filled with my father's
friends—chemists, psychologists, literary critics—talking and
debating the issues of the day," McKinley would recall. Upstairs,
the spare bedrooms were full of student boarders, and the sound of
classical music was often in the air. John himself was a retiring
and intently serious student whose activities at Tuscaloosa High
School included the Stamp Club, Latin Club and Science Club.

John McKinley remained in his cozy campus environment
through his own collegiate years, majoring in chemical engineering, attending classes year-round, remaining at home to complete
a master's in organic chemistry. Everything about McKinley was
cerebral, analytical. "I used to refer to it as a 'steel-trap mind,'"
says Harlan Meredith, a lifelong friend. "He had a tremendous
capacity for remembering what was told to him, and he could
repeat it back to you almost verbatim at a later date." McKinley
would spend the rest of his life regretting the single "C" that
blotched his academic record.

There was something painfully proper about every activity that
engaged the young McKinley. In addition to his membership in a
laundry list of academic honoraries, McKinley was active in
Quadrangle, a religious organization devoted to "the development
of four-square men"; Excelsior, a club whose purposes included
"enlarging reasoning facilities"; and the Senior Cabinet of the
YMCA. Fraternity life, he would later remark, gave him the
opportunity to learn "human relations and social amenities"
rather than the pure pleasure of raising hell. In a 1941 yearbook
photo of a chemistry organization, McKinley appears sitting
ramrod straight in a bunch of slouchers.

None of his college associations would leave a greater imprint on McKinley than the military ones. As an ROTC cadet he was inducted into the highly exclusive Scabbard & Blade, a military society, and became the cadet colonel and brigade commander on campus. And when the U.S. entered the war, McKinley's military aptitude elevated him to the rank of major. He operated 40mm antiaircraft guns and earned a Bronze Star for bravery for his role in the crossing of the Roer River in Germany.

Military-style protocol and unwavering intellectual precision—John McKinley's strongest personality traits—were the characteristics that a young man needed most to go far in the Texas Company, an organization that in less than fifty years had grown practically into a nation unto itself from its founding atop a hill only a few miles from the grease lab where McKinley began his career.

The Texas Company was born on the theory that it is better to sell someone else's oil than to go out and find some for yourself. Things might have been different if Joseph "Buckskin Joe" Cullinan, a rough-hewn Irishman from the oil fields of western Pennsylvania, had accepted an invitation in 1900 to invest in a wildcat well being drilled on a little hill called Spindletop in Beaumont, Texas. The discovery of oil the following year in the geological "salt dome" beneath the hill touched off the most hell-bent oil boom ever seen before or since, transforming the sleepy town along the Gulf of Mexico into a mecca of such frenzied speculation that it quickly became known as "Swindletop."

While grousing over his failure to get in on the ground floor, Buckskin Joe surveyed the carnage of the boom and quickly sensed opportunity of a different kind. A forest of some four hundred and forty wooden derricks had sprung up, each so close to the next that you could walk across the field without ever touching the ground. And it was a good thing, for giant puddles of crude had accumulated beneath the platforms for the lack of any market. Between the glut of oil and the resulting contamination of the local water supply, oil sold for a mere 3 cents a barrel, potable water for 5 cents a cup. Cullinan decided the great riches were available to the man who could buy the oil at Spindletop for 3 cents a barrel, run it through a "teakettle" refinery and sell it elsewhere for many times that much.

Cullinan obtained backing for his venture from the Lapham Brothers of New York, who controlled the nation's leather industry; from a high-rolling Chicago railroader and steel baron named John "Bet-a-Million" Gates and a variety of local interests, including former Texas Governor James Hogg. Cullinan's Texas Company—so named to preserve the hometown character of an operation financed with so much Yankee money—built dozens of storage tanks, miles of pipeline, loading docks and a refinery. It began moving crude oil and refined products by train and pipeline to the crude-short refineries of the North and to the sugar grinders along the Mississippi. And while Standard Oil continued to dominate the market for illuminating and fuel oil, Cullinan and his backers quickly identified an all-new market in the newly introduced horseless carriage. Seventeen years after opening its first filling station in 1911, the Texas Company would attain the singular achievement of operating outlets in all forty-eight states, and eventually all fifty.

But the Texas Company would do most of its growing without Buckskin Joe, who quit the company within a decade of founding it. Seventy-five years later, when Texaco was cast as an arrogant out-of-towner in a Houston courtroom, the circumstances of Buckskin Joe's bitter resignation would stand as one of the most ironic events in the company's history.

He had moved the company's headquarters from Beaumont to the swampy and underdeveloped city of Houston, about one hundred miles away. "Houston," he wrote one of the company's Eastern backers, "seems to me to be the coming center of the oil business of the Southwest." At the same time, Buckskin Joe perceived a need to begin amassing huge new tracts of oil land to step up the company's sputtering search for oil-in-the-ground. But the company's backers in New York balked at the expensive exploration plans and set up an executive committee that began tightening the reins on Buckskin Joe's activities. Worse still, the moneymen had established a de facto management headquarters in New York. Buckskin Joe, by this time a happily adopted Texan, began soliciting proxies to have the New York brigade thrown out of power, reminding the company's stockholders that "its original management, its corporate attitude and activities were branded with the name *Texas* and Texas ideals." The uncertain outcome of the power struggle caused such anxiety and suspense that even the rebellion of Pancho Villa along the nearby Mexican border was suddenly relegated to second billing in the Houston newspapers. Alas, Buckskin Joe lost the proxy fight by a wide margin, causing the *Houston Chronicle* to demand a state investigation of the

company's flight into "the maelstrom of Manhattan." The newspaper warned: "New York water and Texas oil are not going to
mix very well."

For his part, Buckskin Joe spent his later years as an eccentric
defender of Irish and other ethnic causes, flying a pirate's skull-
and-crossbones over the Petroleum Building in Houston "as a
warning to privilege and oppression."

The relocation of Texas Company headquarters forever changed
its corporate character. Most major U.S. oil companies outside of
the Standard Oil group maintained their roots in the oil patch—
Phillips Petroleum in Oklahoma, Gulf Oil in western Pennsylvania and Getty Oil, Union Oil and Occidental Petroleum in
California—but virtually alone did the Texas Company make its
home amid the financial titans of Wall Street. The company's
presence in New York certainly furthered its access to the capital
markets, facilitated the management of its growing overseas interests and eased its pioneering involvement in national broadcast
advertising. But back in the oil patch, Texaco was rapidly becoming known as an outsider. Every other major oil company happily
subleased land to local independents, but the high-and-mighty
Texas Company refused to share the wealth. As a company ruled
by strait-laced men of finance rather than by the good old boys of
the oil patch, it became notorious for squeezing the hardest on any
purchase of supplies. When the Petroleum Club of Houston
moved into new quarters on the top floor of the Humble Oil (now
Exxon) Building in 1962, every major oil company was providing
financial assistance—"Everybody but Texaco," a club official
complained, "and I guess they never will." It was the only major
oil company without representation in the All-American Wildcatters Association, a group of crusty old oilmen—including the
Liedtke brothers of Pennzoil—who went out of their way to make
Texaco the butt of their jokes.

Part of Texaco's arrogance had nothing to do with its New York
address, however. The fact was that Texaco didn't *need* to do
deals with anyone else in the oil patch. The reason was that
Texaco owned Louisiana.

In a deal that would loom large in the events to come, the Texas
Company of the 1930s cast its lot with the immense political
machine of former Louisiana Governor Huey Long, in one fell
swoop fulfilling its need for a huge supply of crude oil to keep all
those refining and filling stations primed. Shortly before his as-

sassination in 1935, the beloved populist Long, then a U.S. senator, had set up a secret venture called Win or Lose Corporation, which obtained the mineral rights to nearly all state-owned land in south Louisiana. Because south Louisiana was almost half marshland—and because the state owned all land that sat under water—Win or Lose couldn't lose: Long's political dynasty wound up owning some one *million* acres of oil leases. Then along came the Texas Company, which subleased almost the entire lot. Long had just been assassinated, but his cronies and relatives, including U.S. Senator Russell Long, would go on to enjoy tens of millions of dollars in royalty payments from Texaco's Win or Lose leases.

Beneath those mosquito-infested swamps and grasslands lay dozens of gigantic, underground salt domes, each one a geological sibling of the Spindletop field. South Louisiana became the Saudi Arabia of the West, with Texaco controlling practically all of it. By the 1950s Texaco had become the nation's number-one oil producer, deriving its largest share of income from the isolated bayous of the Pelican State. No oil company in the nation owned more oil-in-the-ground.

Ever the marketer, Texaco also discovered a way to make money from the natural gas that poured out with the oil from its Louisiana wells. For years the oil industry had wastefully flared gas that came from the ground along with Louisiana oil, illuminating south Louisiana with torches that looked like giant birthday candles. But in the 1950s and 1960s, Texaco drew dozens of sprawling chemical and other processing plants to Louisiana by offering to sell the gas under long-term contracts for as little as 25 cents per thousand cubic feet. Before long, one of the world's great petrochemical and power-plant corridors sprang up along the Mississippi between Baton Rouge and New Orleans. Dow Chemical, Union Carbide, Allied, Goodyear, Borden, Monsanto—practically all the chemical giants settled along the riverbanks. The chemical companies sucked up Texaco's natural gas like soda through a straw, transforming it into fertilizer for Iowa, rubber feedstocks for Akron and enough bulk plastic to keep an entire generation of baby-boomers in sandwich bags and polyester shirts. Year by year Texaco deepened its addiction to Louisiana oil and gas, lacing the swamps with pipelines of ever-increasing diameters and committing huge supplies—not only to its industrial customers, but to the sprawling retail network that was selling gasoline under a constellation of Big Red Texaco stars.

Nearly all of this growth occurred under one man.

Throughout most of the postwar era, Texaco was in the indomitable grip of Gus Long, an Annapolis graduate who had quit the Navy when he realized he would never make admiral and who rose to the top of Texaco through his strict adherence to the company's push-the-product philosophy. Gus Long (no relation to Huey) saw to it that Texaco took the lead in lining the nation's new interstate highways with filling stations as soon as the roadwork began in 1956. He elbowed his way into the commercial and residential heating business by acquiring fuel-oil distributors, and no delivery was impossible to complete. During one of the worst snowstorms in New York history, Long was sitting in his office high atop the Chrysler Building when he was notified that his dinner meeting with stock analysts had been canceled because another oil company was unable to replenish the building's heating oil. Long personally arranged to deliver an alternate supply. Of course, the meeting went on as scheduled.

The corporate culture established by Long fostered a single-minded commitment to sales and productivity, but it discouraged the freewheeling, risk-taking spirit necessary to succeed in the oil-prospecting business. Long was a notorious penny-pincher who was said to review every expenditure over $10,000—a ridiculously low amount in an industry where wells could cost hundreds of thousands of dollars, even in those days. Long filled the organization with former military officers, particularly Navy men, who did everything but salute and "aye aye" at his commands. Other oilmen called him "the businessman's Patton" and Texaco "the meanest company in the world." One oilman told *Dun's Business Monthly:* "The way they work, it's something like taking 40 lashings every morning—and enjoying it."

John McKinley flourished in Gus Long's Texaco. He ascended from grease research to the "cracking" lab, discovering some fourteen new ways to reconstitute the hydrocarbon molecules of crude oil into the very gasoline additives that were helping to make Texaco the number-one streetcorner retailer in America. McKinley was a consummate company man, and his appearances at company research meetings drew the attention of top management. By 1956 he had traded in his lab coat for a suit coat, moving into a top job in the company's worldwide research headquarters in Beacon, New York. John McKinley was on the fast track.

For McKinley and Texaco, the oil business was beautiful—so long as there remained enough inexpensive oil-in-the-ground to keep all those refineries and filling stations operating at full tilt. It was an article of faith within Texaco that whatever oil the company needed it could find in Louisiana, if nowhere else. ''Finding new reserves is absolutely essential,'' Texaco told its employees in a 1965 newsletter, ''and Texaco keeps at it just as assiduously in oil-rich Louisiana as it does everywhere.'' Other oil companies were beginning to recognize the gravity of the disaster that would occur if ever the nation's oil began running dry. But when a trade group banded together in the 1960s in an advertising campaign to warn of just such a possibility, Texaco, conspicuously, refused to support it.

Texaco was setting itself up for a hard, hard fall.

· 4 ·

When Gordon Getty called on his father for consulting assignments, the Old Man was living outside of London in a seventy-two-room Tudor mansion called Sutton Place, the very castle where Henry VIII had had his first fateful meeting with Anne Boleyn. Most sons calling on an estranged billionaire father who lived with a dozen Alsatian guard dogs might fairly find the experience intimidating. Indeed, George, the oldest Getty son, would once remark that "coming to see him is like visiting Mount Olympus." Gordon's comment, however, was that his father's fortress showed "a remarkable amount of architectural integrity."

It was during one such visit that Gordon first expressed curiosity about the family trust, a subject that his father understandably treated as a deep, dark secret. For the truth was that J. Paul Getty actually was no billionaire. The majority of the fortune attributed to him by the press existed in a trust established to benefit not only him, but also Gordon, his brothers and all their born and unborn children. It wasn't even J. Paul who had set up the trust. It was his *mother*. It was only in 1960, at age twenty-seven, that Gordon finally asked about his inheritance, and his father gave him a five-word response: "Your grandmother created a trust." When Gordon inquired further, his father told him that the trust documents were a matter of public record, on file with the Los Angeles County Recorder's office.

For three years, Gordon let the matter rest. He was, after all, picking up a few thousand dollars here and there as a consultant for the company. He had a modest annuity that his mother had purchased for him. In addition, at age thirty, he received 4225 shares (worth about $125,000) in his father's U.S. oil company from a comparatively small trust that his grandmother had also set up for her grandchildren. He had invested some money in a land-development project in Reno, but his partner had declared bankruptcy.

Thus, while Gordon was hardly living a life of poverty in the early 1960s, neither was he living like the son of the nation's only purported billionaire. And while his three brothers continued their ascents, Gordon had altogether abandoned the idea of an executive career in his father's oil empire—an enterprise toward which he had failed to develop the slightest emotional attachment. Gordon was so unsentimental about the family business that he unloaded those few thousand shares almost immediately after inheriting them—only to watch the stock quickly quadruple in value. Getty Oil would have to limp by without Gordon Getty.

"That started me thinking what alternatives I had in life," he would later say. "And the thoughts finally circulated around to the trust."

Although it was created to satisfy business objectives rather than bloodlines, the Trust of Sarah C. Getty Dated December 31, 1934, as it was legally constituted, would live for fifty years as the Getty family's primary instrument of revenge. J. Paul Getty's father was a devout Christian Scientist who recoiled at the romantic imbalance of his *bon vivant* son. Thus, when George Getty I died in 1930, the thirty-eight-year-old J. Paul was shocked to find that he had been left about 3% of his father's estate—a paltry $500,000, with the remaining $15.5 million, an astonishing sum for the day, going to J. Paul's mother, Sarah. It wasn't the money that mattered, and it wasn't even the principle. The fortune was entirely in *stock,* meaning that control of the oil business owned by J. Paul and his father had now passed to his mother. J. Paul now had to answer to his mother's stuffed-shirt *lawyers.* This was a major problem, for the stock-market crash had just hit, and J. Paul had been counting on using the enormous repository of credit available to his father's oil company to finance a stock-buying binge on Wall Street.

Fifty years later, during another downturn in oil stocks, the likes of T. Boone Pickens of Mesa Petroleum, J. Hugh Liedtke of Pennzoil and John McKinley of Texaco would adopt the strategy

that taking over other oil companies was preferable to drilling for
oil when the value of the target company's oil-in-the-ground was
not nearly reflected in its stock price. Oil shares in the 1980s
would be like pawn tickets—collect enough of them and you own
every guitar in town, cheap. People would hail this stroke of
genius as "drilling for oil on Wall Street," as if the concept had
just been minted. It was J. Paul Getty, however, who invented it.

"I considered that it was foolish to buy oil properties with 100-
cent dollars," he would explain, "when you could buy them
indirectly"—on Wall Street, that is—"with 50-cent dollars." But
as for Sarah, "she did not see that point." In 1930 people weren't
putting money in the bank, much less in the stock market. People
were putting money in mattresses. Sarah called her son's plan
"reckless . . . a risky and hazardous proceeding." She was convinced that he would gamble away his father's fortune even before
her grandsons, let alone their unborn children, would ever see it.
Your father was no plunger! she cried. J. Paul considered his
mother's squeamishness to be "womanlike," but Sarah's cadre of
advisors urged her to stand firm. The idea of seizing control of
another company—against that company's wishes!—was heretical, *unethical,* they told her.

After months of battle between mother and son, J. Paul persuaded Sarah to "give up the worries of the business," as he
would inventively recall. Sarah agreed to sell her stock back to the
family oil company, which put her mattress-stuffing lawyers out
of the picture and left J. Paul free to call the shots. Quickly, he
stepped up his efforts to "drill for oil on Wall Street," choosing
as his primary target Tide Water Associated Oil, the concern
whose refineries and oil-in-the-ground eventually would become
the cornerstone of Getty Oil Company.

Soon, however, J. Paul encountered another obstacle: The company still owed money to Sarah for the stock it had bought from her.
Meanwhile, J. Paul was hocking the company to buy Tide Water
stock on margin. Unless he pursuaded his mother to cancel the
company's debt to her, the stock purchases would grind to a halt.
Time was of the essence, for J. Paul had just learned of a major
shareholder who was willing to sell a huge block of Tide Water
shares, practically enough to give him control of the company.

On a distinctly unfestive Christmas Day in 1934, with Sarah
bargaining from a position of strength, she proposed a solution.
She would take back those notes and put them in a trust to be
managed by her son. In effect, he would now be his own creditor,
freeing his company to begin borrowing more money for the Tide

Water takeover. Eventually, the trust's notes would be paid off altogether with stock in the family oil company, making the trust the largest shareholder.

But J. Paul had to abide by several severe conditions:

1. He had to put a great deal of the company stock he personally held into his mother's trust, assuring that much of *his* fortune would also be preserved for Sarah's great-grandchildren.

2. To perpetuate the family fortune as long as possible, the entire principal amount—the corpus—would remain in the trust until not only J. Paul was dead, but each of his children. Only then could the assets of the trust be divided among Sarah Getty's great-grandchildren, and by that time the 21st century would have rolled around.

3. Of whatever money the trust earned on a yearly basis—from dividends on company stock, for instance—J. Paul could receive only an amount in proportion to his mother's contribution to the trust. Any money earned on J. Paul's contribution would in turn go to his children—George II, Eugene Paul and Gordon. However, Ronald, the second son, would never receive more than $3000 in a year. J. Paul had assured his mother that Ronald would be more than amply provided for from the rich divorce settlement his mother had obtained. (He would not be, as it turned out.)

J. Paul waffled at his mother's proposal, which she presented on a take-it-or-leave-it basis. It meant that no matter how big Getty Oil became—no matter how valuable the trust became, even $1 billion, say—J. Paul would be the trustee of the fortune, but its sole owner in name only. Yet the urgency of the Tide Water takeover left little choice; only by capitulating to his mother's proposal would he have any hope of buying the big block of Tide Water stock that was rumored to be for sale. J. Paul gave in.

On December 31, 1934, after finally signing the eighteen-page document creating the Trust of Sarah C. Getty, J. Paul climbed into his tuxedo and went to a New Year's Eve party at William Randolph Hearst's castle at San Simeon. The following morning, as he was sleeping off the splendor, word came from his lawyer: That big block of Tide Water stock had been purchased. Getty Oil had just multiplied in size. One of history's first hostile takeovers would set the stage for history's biggest takeover of any kind, and much more.

The idea that he might be entitled to something from his grandmother's pot of gold struck Gordon as an epiphany, in a syllogism: "It suddenly occurred to me that maybe a beneficiary is entitled to benefits,'' he would explain. As odd as it may seem, Gordon's mind really seemed to work by such cold reasoning. But once he seized upon the logic, his logic gave way to passion.

Gordon prevailed upon a friendly family lawyer to locate a copy of the trust instrument and quickly realized he was entitled to one-third of all the trust's income that his father didn't receive. Eugene Paul was to get another third; George the last third. Yet Gordon and his brothers had received virtually no income from the trust, and with good reason: The trust itself had no income. By the 1960s, the trust only owned shares of the oil company, and the oil company paid no cash dividends. No cash dividends, no trust income. Instead of sharing profits with the trust and the public investors who now held a minority interest in the company, the Old Man had hoarded virtually every penny and plowed it right back into the earth, in search of more oil-in-the-ground. It was a unique policy among major oil companies, one that made Getty Oil grow at an astonishing rate. This policy did not upset Gordon's brothers, who had promising careers with the company. But it made Gordon mad.

"I believe Getty Oil is extremely well able to pay a dividend," Gordon wrote his father in 1963. "This income to me should not under any circumstances be less than $50,000 per year." *Don't listen to him,* Gordon's brother George implored his father. J. Paul Getty, however, was petrified to consider the publicity that would flourish if Gordon dared to file a lawsuit over the issue— and there was no predicting what rash action Gordon might take. "So far as I know, the Kennedy family has never had a suit," his father wrote him. "Neither has the Rockefeller family." To avoid an open breach, the Old Man relented by having his company declare a dividend of 10 cents a share. Although it was a puny amount in relation to the company's total earnings, the Old Man resentfully told Gordon that the cash distribution would probably cause the company to lose ground in its effort to keep pace with the rest of Big Oil. He also wrote to him: "I would be very pleased if you would follow the example of your brothers and work for a living."

Now that the trust finally had some income on its assets, J. Paul fulfilled his mother's instruction that he keep for himself only that portion of the income equal to her original contribution. According to the disbursement record, he paid himself 79.2923389% of the trust's dividend income. The boys' shares of the remaining 20.7076611% worked out to $46,217 each.

J. Paul Getty took one other action as trustee in 1962. On

Gordon's twenty-ninth birthday, in retaliation against his demand for a dividend, he blotted Gordon's name from the list of his sons eligible to succeed him as trustee over the family estate. "You are, so to speak, the lone wolf," he had written Gordon. Half brother George, who applauded the action, told Gordon that the Old Man never wanted to lay eyes on him again. "So it looks as if you have gained fifty thousand dollars," George told Gordon, "and lost a father."

And yet, the fight had only begun.

When Gordon would return to San Francisco between his assignments with Getty Oil, he often stayed at a hotel on Lombard Street to be near the home of his high school friend Bill Newsom. At night he was invariably sitting at Mrs. Newsom's dinner table chewing the fat with Bill's father. During the day, with no one at home and having the whole house to himself, he composed on the Newsoms' piano. By the time of the dividend war, Bill Newsom himself was a freshly minted lawyer from Stanford Law School; Gordon engaged him for advice on his dealings with his father.

One evening, while they talked over dinner in a San Francisco restaurant, an acquaintance of both young men walked in with a breathtaking, brown-eyed redhead—five feet, eleven inches tall, with cover-girl beauty. "There's one attractive girl!" Bill exclaimed. He turned to Gordon, who was rapt.

"She's too tall for you," Gordon muttered.

Her mother was a wealthy fruit and nut farmer from the San Joaquin Valley, her father was deceased, and her name was Ann, just like Gordon's mother. Five days after Gordon's thirty-first birthday—thirty years to the day after his father and grandmother had hammered out the basic terms of the family trust—Gordon married Ann in Las Vegas on Christmas Day in 1964. Gordon's wish for the kind of big, happy *normal* family that Bill Newsom grew up in quickly came true; after thirty-six months of pregnancy over the next sixty months, he and Ann had four sons.

With a wife and growing family, Gordon continued looking for ways to supplement his trust income of about $50,000. Then one day in 1966, one of his stockbroker friends telephoned with some interesting news. While thumbing through a stock guide, the broker noticed that although Getty Oil had only recently begun paying a token dividend in cash, for years it had been paying a big dividend in *stock*.

Although the information had been routinely reported in Getty Oil's annual reports for years, Gordon had overlooked it. He and his lawyer, Bill Newsom, suddenly felt as if they had reached the end of the rainbow. If those dividends had been paid in cash, they would be equivalent to $100 million! Gordon's share alone would total about $7 million! If Gordon and Newsom could convince a court that Gordon and his brothers had the same right to the stock dividends as the cash dividends received by the trust, Gordon would be able to support his family and pursue his obsession with music unfettered by the need to grovel for a decent job with Getty Oil. "All of a sudden if you became aware of your entitlement to that much money," Bill Newsom would later explain, "you'd be a fool not to claim it." In 1966 Gordon filed suit—against his father, as trustee of the Sarah C. Getty Trust.

The Old Man reacted as if his life's work had been threatened—which in one sense it had. Liquidating years and years of stock dividends out of the trust would diminish its control of Getty Oil by one third—and thus reduce the iron-fisted domination of the company that the Old Man maintained as sole trustee over the trust. Getty Oil's most ferocious lawyers were put on the case.

For six years, until 1972, the messy case of *Getty v. Getty* was dragged through the courts of California and the newspapers of the world, and it was Gordon, not his father, on whom the humiliation was heaped. The company's lawyers called Gordon's suit "naive and childish," "reckless and shameful," and even an affront to the Fourth Commandment. They described him as an ungrateful, gold-digging son and repeatedly suggested that he was too inept to earn a living from his father. They said the law cited in Gordon's suit had long since been thrown on the "rubbish heap." And above all, they said, the lawsuits ignored "the financial realities of Getty Oil"—the reality being that unless the Old Man maintained his control, the company would die on the vine. George Getty, the company's number-two executive, jumped in, flinging his own attacks against his brother.

Because Gordon was suing over a trust matter, he technically had to name every other living beneficiary as a defendant, including his own born and unborn children—a fact that the Old Man's lawyers really took to town. "Gordon was, and is, arrayed against every other person" in the family, the Old Man's lawyers told the court. "Plaintiff would kill the goose that lays the golden eggs," the lawyers said, "imposing gross disadvantages on the trust, his father, his brothers, his nephews and nieces, and *his own children*." The lawyers demanded "no relief to plaintiff—other than a declaration that he *has no rights*." And that is all that Gordon got.

In the midst of watching her husband handed one defeat after another, Gordon's wife, Ann, decided to telephone her father-in-law at Sutton Place. "Your lawyers are killing my husband," she explained. Later, while talking to one of them, the Old Man urged no mercy. "Keep killing my son," he said.

But Gordon had staying power, and his three brothers did not. And the manner in which each fell would have an extraordinary bearing on events to come.

Ronald had never been a significant competitor for control of the family trust. Relegated to receiving $3000 from one of the world's largest family fortunes, he had little standing to assert a major role over the trust's affairs. And although he had begun his career at Getty Oil with such promise, it fizzled after several years. Ronald was prone to inflicting severe dressing-downs on subordinates, touching off brushfires that his half brother George would take upon himself to extinguish. Given the almost impossible assignment of marketing a weak Getty brand, Veedol Oil, in Europe, Ronald had also resorted to hiring away employees from competitors, which turned out to be a crime in France. Ronald received a suspended prison sentence and was plucked from his Europe assignment, eventually to leave the company altogether. "It's him or me," George eventually told his father, which was no contest. Ronald would go on to produce motion pictures and open a chain of Don the Beachcomber restaurants, but he certainly would be heard from again—when Pennzoil and Texaco would come along.

Yet Ronald's story was the least tragic by far.

Eugene Paul, who had gotten the plum assignment in Italy when Gordon had been sent to the Neutral Zone, had become quite the man about Rome. He changed his name legally to Jean Paul Getty Jr. and initially performed to the satisfaction of his father, but within several years he was no longer "setting the world on fire as a businessman," as the Old Man would write. Almost overnight, "everything started to go wrong."

Paul Jr.'s trim, Lincoln-style beard grew shaggy; his dark, carefully coiffed hair fell below his collar and his heavy eyeglass frames turned into Lennon-style wire-rim shades. The Italian *paparazzi* captured him, as one photo caption read, in "a tie-dyed velvet outfit that would make any genuine hippie green with envy." According to a variety of court documents, Paul Jr. became gripped with drug addiction; eventually he dropped out of the business and divorced his wife, Gail, after ten years of marriage and four children. Paul Jr.'s divorce decree was barely final by the time he married the breathtaking Talitha Pol, a Dutch actress with whom he had a beautiful, blond-headed boy whom they cursed with the name Tara Gabriel Galaxy Gramaphone Getty—a perfectly normal kid, as it turned out, who would unwittingly assume a center-stage role in the tragic downfall of Getty Oil. Talitha, in 1971, would die of a drug overdose; Paul Jr. would spend the next twelve years in London, struggling to overcome his grief. He too would play a central role in the corporate power play to come.

Remarkably, the chesty George Getty II—dubbed the "heir to his father's empire" by *Business Week,* fell the hardest of all. Even after becoming the company's number-two executive and after establishing an independent business reputation—as a director of Bank of America and Douglas Aircraft, for instance— George never escaped his father's humiliating heavy-handedness. Harold Berg, who would eventually succeed George, used to watch his boss hole up in his office for hour-long transatlantic conversations with his father—and then watch George take it out on everyone else. "It was obvious he'd been chewed out," Berg would recall.

When a complex merger was planned between Getty Oil, where the Old Man held the title of president, and Tide Water Oil, of which George was president, everyone, most of all George, assumed that he would become president of the combined enterprise, while his father would adopt a new role as chairman of the board. The Old Man, however, viewed this lofty title as one that befitted only elderly, out-to-pasture executives, which at age seventy-five he certainly did not consider himself. J. Paul thus appointed himself president of the merged companies, demoting George, then age forty-three, to executive vice president. George never got over it.

Things only became worse. George's executive assistant, Stuart Evey, watched year by year as the Old Man's attacks grew more mean-spirited, taking a progressively heavier toll. For while the Old Man was berating Gordon, he was also saving up plenty for George. Evey would read the letters from Sutton Place in horror, saving copies of as many of them as he could. "We might as well wind up the company and put the proceeds in profit-making enterprises," George was told. Reduce losses at the nuclear-fuel business, he was ordered, "or I will drop an H-bomb on it." The company based its drilling on "wishful thinking." Getty Oil should rename its profit centers as "loss centers."

"If a company's management can't equal the industry average,

I think the stockholders should replace it,'' he said. And on another occasion he wrote George: ''I think there is a place for a nine-year-old Hottentot in GOC's management. He couldn't feel less inspired and he might do better.''

Years later, when Gordon was asked in an interview about the abusive letters *he* had received from his father, he insisted, ''I laughed at them. My father was a mischievous man; he liked to ruffle people's feathers.''

If there was any joke intended—and it is hard to believe there was—George Getty never got it. George began stabbing the back of his hand with a letter opener. With other personal problems mounting, he became so concerned about his masculinity that his assistant Evey recalls him receiving injections of vitamin B12 in his office.

On June 5, 1973, George arrived at his home in the superswank Los Angeles suburb of Bel Air, pounded down several beers and a bottle of wine, grabbed a knife and stabbed himself in the stomach. The wound was superficial, but George's wife immediately summoned Evey, the company doctor and the Bel Air police. George panicked, locked himself in his bedroom and threatened to shoot everyone. Before long, he was quiet, and Evey kicked in the door to find him sleeping and drenched in blood. Evey and the company doctor admitted him to a local hospital under the false name George Davis—Evey had had lunch that day with former Army running back Glenn Davis—to spare George from any further grief from his father.

By the next day, when several empty or near-empty prescription bottles were discovered in the hamper and travel kit in George's bathroom, he was lying in the hospital, dead. ''Mr. Getty's death,'' the coroner's investigators wrote, ''was the result of slowly increasing personal stress which was triggered into an acute reaction by alcoholic intoxication and which manifested itself in violent and self-destructive behavior.'' Celebrity coroner Thomas T. Noguchi ruled it a probable suicide.

And suddenly, Gordon Getty was the Number-One Son.

Shortly after George had died, the Old Man reappointed Gordon to succeed him as trustee of the family estate. Gordon was not designated to serve alone, however. One of the very lawyers who had been ''killing'' Gordon in the war over the dividends was also appointed a successor trustee. In addition, the Old Man's longtime bank, Security Pacific National Bank of Los Angeles, would serve as a corporate co-trustee. This triumvirate would operate the trust under the same rules that had bound the Old Man, with one

difference: Except to avoid a catastrophe, the successor trustees
appeared to have no right under the trust instrument to sell the
trust's interests in Getty Oil Co.

In 1976, on the third anniversary of George's death, J. Paul
Getty went to his final resting place in Malibu, California, at age
eighty-three. He died contented that his company, if not his family, was securely intact, safe for the future.

"The tax men have no cause to poise themselves for a sprint,"
the Old Man wrote shortly before his death. "The Getty interests
are not about to sell off their shares or dissolve their companies.
Those companies have been built into a large and thriving *family*
business."

If any oilman ever measured up in the unforgiving eyes of J. Paul
Getty, it was J. Hugh Liedtke. In barely a year following his
seizure of control at the old South Penn Oil, Liedtke had succeeded in doubling the company's stock price—and thus doubling
the value of the Old Man's investment in that company to nearly
$10 million. On Christmas Eve in 1963, Liedtke was notified by
telegram that the Old Man wished to pocket his profits and sell out
his 10% stake in Pennzoil, a request to which Liedtke readily
assented.

Liedtke and the yellow can of Pennzoil were on a roll—one that
would continue for twenty years. Taking account of all the spin-
offs, refinancings and other creative deals along the way, a single
share of South Penn stock, worth $31.75 at the time Liedtke took
over in 1962, would multiply to $941 in value by 1983—a capital
gain, in round numbers, of 3000%. (By late 1986, this return
would rise to nearly 10,000%.)

He was hell-bent on becoming big from the beginning. Liedtke
took the modestly successful Pennzoil brand name and launched a
national advertising campaign, eventually with Arnold Palmer as
its spokesman. Even in the stagnant markets of the 1960s, as
automotive improvements extended the engine life of lube oils,
Liedtke increased sales of Pennzoil Motor Oil 45% to 145 million
quarts a year. A national brand was not Liedtke's main objective,
however. The cash flow from the yellow can was spent finding
more oil-in-the-ground—not Pennsylvania-grade oil, but crude oil
and natural gas in the burgeoning fields of South Texas, the Gulf
of Mexico, and eventually even the North Sea. More than a decade later, the Boston Consulting Group and other egghead outfits
revolutionized modern management with the "Cash Cow Investment Matrix," which favored milking cash from "mature" prod-

ucts and diverting it to growing businesses, just as Liedtke already had done.

Liedtke was ahead of his time in another respect as well—a more infamous one. Notwithstanding J. Paul Getty's takeover of Tide Water, hostile acquisitions were virtually unheard of until Hugh Liedtke came along. Essentially, the South Penn deal had been a hostile takeover, one that Liedtke won with the backing of a minority shareholder. Liedtke had established a pattern: identifying sleepy companies with weak management and stock prices so low they didn't begin to reflect the value of the underlying assets—just as J. Paul had done thirty years earlier in the Tide Water takeover.

Working from an office decorated with a poster of W. C. Fields as a top-hatted riverboat gambler, Liedtke found far more takeover deals than he had the time or money to pursue. One was an exploration and production company in Arizona in which Hugh and Bill Liedtke bought an interest as a prelude to a possible takeover. "It was dominated by a bunch of old farts that go to sleep," he would later say. But "I just didn't want to get into a big battle over that company. It just wasn't big enough." So the Liedtkes sold their stock interest to an up-and-coming oilman named T. Boone Pickens, a friend of theirs from the Midland years, who did the battle and won control.

Then, in 1965, came the deal that would forever give Liedtke a place in the annals of takeover history. Through business associates, he became aware of the United Gas Pipeline Company of Shreveport, Louisiana. The company operated the nation's busiest pipeline system, conveying three billion cubic feet of natural gas every day to the humming manufactories of the North and the petrochemical complexes of the Gulf Coast.

Like electric wires or household plumbing, pipelines are mere conduits—expensive to install, but cheap to operate, meaning that they generate prodigious quantities of cash. Owning one of the world's largest could provide the Liedtkes with a steady stream of dollars to finance their other ambitious growth plans. "We thought the pipeline was a cash cow," Bill Liedtke would recall— a source of money for finding more oil-in-the-ground. In fact, United itself already owned oil reserves aplenty, not to mention some of the world's largest deposits of copper, sulfur and other minerals—commodities which, like oil, the Liedtkes judged to be on the verge of short supply. None of the assets, in the view of the Liedtkes, were being properly utilized for growth. United's management, they determined, was way behind the times.

United was also *five times* the size of Pennzoil, but the Liedtkes were undeterred. They began slowly buying up shares in the stock market—but the word of a possible takeover leaked out, causing the stock price to soar. To make matters worse, another investment group emerged as a rival buyer.

As the Pennzoil board pondered its next move, a Scotsman serving as director suggested a new strategy: a cash tender offer, a takeover practice that was virtually unheard of in the U.S., but was widely used in Britain. Pennzoil could publicly announce an offering price to the public for only a portion of the shares; the shareholders, fearful that the stock price would tumble once the offer was closed, would ''tender'' as many shares as Pennzoil could afford to buy. The company's thunderstruck management resisted the offer in every way possible, but the shares flooded in and before long Pennzoil owned 42% of the company.

Pennzoil went so deeply into hock buying those $240 million worth of shares that it nearly went broke. ''My stomach churned over those loans,'' Bill Liedtke says. But Pennzoil achieved complete control of the United Gas board before the hapless management of the company knew what had hit. ''We had a hell of a fight,'' Liedtke now says. ''They lost and that was the end of it.'' Eventually the Liedtkes absorbed United into Pennzoil and began squeezing every dime of cash possible from the pipeline operations. Pennzoil had more than quintupled in size, mostly on borrowed money.

From the late 1960s onward, Hugh and Bill Liedtke proved time and again that Pennzoil's relatively small size did not prevent it from acting like Exxon, Mobil or any other major oil company.

More confident than ever in his conviction that most of the major onshore oil fields already had been discovered, Hugh Liedtke was anxious to get in on the huge discoveries being made in the distant and deep waters of the Gulf of Mexico, a playground that Big Oil had all to itself. There was no way Pennzoil could begin punching the water with the kind of multimillion-dollar wells you had to drill in the Gulf, until Hugh Liedtke came up with an imaginative financing arrangement. He set up new companies called Pennzoil Offshore Gas Operators—POGO, for short—and Pennzoil Louisiana & Texas Offshore Incorporated—PLATO, an apt acronym for the former philosophy major. POGO and PLATO sold stock directly to the public, thus relying mostly on shareholders' money, not Pennzoil's, to finance exploration. Investing public money directly in such rank, large-scale wildcat exploration was unheard of—practically tantamount to selling shares on the turn of a roulette wheel. But Liedtke gave investors special incentives and security, promising that if the ventures went bust, stockholders would eventually be repaid with stock in Pennzoil. Wall Street applauded the imaginativeness of this fund-raising method, and the shares were snapped up like tickets to a reunion concert of the Beatles.

The concept was a smashing success, permitting Pennzoil to outbid every other major oil company for some of the choicest leases in the Gulf. But the natural gas discovered by Pennzoil wasn't much of a bargain; at one point, federal energy regulators were permitting PLATO to charge the highest price ever received in the natural-gas industry—more than twice the rate received by virtually every other federally regulated gas producer.

In the toss-up between favoring consumers and enriching share-holders, the Liedtke brothers saw no contest.

Personal friendships, too, continued to rank high in the Liedtkes' eyes, and they had not forgotten theirs with George Bush.

In both presidential election campaigns of Richard Nixon, no regional finance chairman exceeded his fund-raising quota by a wider margin than Pennzoil president Bill Liedtke, who says he accepted the assignments as a personal favor to Bush. Liedtke spared little effort fulfilling the request of his old friend in 1972, raising $700,000 in anonymous contributions, including $100,000 that was cycled through a Mexican bank account to preserve the anonymity of the donor. Bill Liedtke proved himself adept not only at raising funds, but at arranging for their transportation. Two days before a new law was scheduled to begin making anonymous donations illegal, the $700,000 in cash, checks and securities was loaded into a briefcase at Pennzoil headquarters and picked up by a company vice president, who boarded a Washington-bound Pennzoil jet and delivered the funds to the Committee to Re-elect the President at ten o'clock that night.

The episode might never have received the slightest attention, except that the Mexico checks were turned over to campaign undercover man G. Gordon Liddy, who passed them to one Barnard Barker, who was found to be carrying some of the cash proceeds when he was arrested at the Watergate Hotel a few weeks later. Bill Liedtke recalls appearing before three grand juries investigating various elements of the Watergate scandal, but neither he nor Pennzoil was ever accused of wrongdoing.

Watergate investigators did establish, however, that the Nixon

White House valued Bill Liedtke's point of view. "He recalls several instances in which White House staff members contacted him seeking his recommendations for available government positions," the investigators wrote in an account of a 1973 interview with him. When the president had two positions to fill on the Federal Power Commission—which regulated Pennzoil's pipeline operation—Bill Liedtke recommended two lawyers from Midland who had previously worked at Pennzoil's Houston law firm of Baker & Botts, an office whose ties with Pennzoil were so close that one of its partners, Baine Kerr, left the firm to become Hugh Liedtke's top in-house lawyer, and eventually the president of Pennzoil.

According to a Louisiana appeals-court opinion written pertaining to the matter, the FPC approved a series of transactions that permitted Pennzoil to divert natural gas committed for low-priced sale within Louisiana into other markets, where the price happened to be much higher. Hugh Liedtke hotly contests that account, but there was little doubt that at the time Louisiana wound up with a severe shortage of natural gas for industrial use and home heating. "The health and physical safety of millions of Louisiana's citizens are gravely threatened," the state announced in the winter of 1972. Governor Edwin Edwards attacked the company's "absolute disregard of the public interest in this state." Layoffs hit the industrial plants, and a lawsuit claimed that Pennzoil-United had breached its contracts. Middle South Utilities estimated that it had to charge its electricity customers in the dirt-poor Deep South an additional $200 *million* in order to arrange alternate supplies of boiler fuels. (Many of the contracts, though, had been signed long before Pennzoil owned United Gas and were probably low-priced to begin with.)

In 1974 Liedtke faced a difficult choice, for Congress was stirring toward regulations that would prevent pipelines from owning offshore gas-in-the-ground. "We can't afford to be frozen out of the gas exploration business," Hugh Liedtke said, so United was set free as an independent company.

However, United was not spun off as the same company that the Liedtkes had acquired nearly a decade earlier. Pennzoil kept for itself United's oil-in-the-ground as well as its huge and growing minerals operation, Duval Mining. Pennzoil also kept $100 million in United's capital, a sort of fond farewell that severely weakened United's financial condition just as it was being set up as an independent utility intended to serve the public's need for

energy. Pennzoil's stock leaped when the terms of the spin-off
were announced.

"Love her and leave her" is how *Forbes* characterized the spin-off. "That, say the critics, is just what the Liedtke brothers did
with United Gas: acquiring it, deflowering it, then dumping it."
As a final act in the corporate reorganization, the Liedtkes' first
cousin once removed was named to head what remained one of the
world's largest and most important pipeline systems. He was
forty-two-year-old J. Hugh Roff Jr., the cousin with whose family
Hugh Liedtke spent the summer while working in his first oil
patch job, and who became a lawyer over a small staff of regulatory attorneys at AT&T.

Federal securities regulators jumped on the spin-off transaction
like roadrunners attacking a rattlesnake. They forced Pennzoil to
return United's $100 million. Moreover, they discovered that the
Liedtkes' private investment partnership had purchased 125,000
Pennzoil shares while the brothers "were in possession of material, nonpublic information" regarding the spin-off and the $100
million gift. Long before the world had ever heard of Ivan Boesky
and the other celebrated Wall Street figures felled by insider trading charges in 1986 and 1987, the Liedtkes were forced to disgorge $108,125 in allegedly ill-gotten gains and signed a consent
decree in which they neither admitted nor denied guilt, but in
which they promised never again to commit such an indiscretion.
Publicly, they asserted that the stock purchases had had nothing to
do with the spin-off.

With his lawyer's instincts and good luck, Hugh Liedtke guided
Pennzoil through the turbulent seas of litigation without taking on
a drop of water. United, not Pennzoil, was stuck defending the
breach-of-contract case. The political controversies were quickly
forgotten. And Pennzoil's expanding base of oil-in-the-ground
enabled it to borrow more money to find *more* oil-in-the-ground.
While most of Big Oil was consuming its reserves far faster than it
could replace them, Pennzoil was holding steady.

Pennzoil entered the 1980s as a $2-billion-a-year company with
a go-go performance record in the stock market and one of the
most impressive profit margins in business. Still, having reached
his late fifties, Hugh Liedtke remained restless. "He dreams great
dreams," his cousin Hugh Roff would say. "He wants to accomplish something major, a big step." Hugh Liedtke wanted to make
Pennzoil a world-scale oil company.

                               *   *   *

While Pennzoil and its yellow can flourished, the Big Red Texaco
star was fading fast. For between the late 1960s and late 1970s,
the unthinkable but inevitable finally occurred: In Louisiana and
elsewhere, Texaco's oil fields began running dry. And with insufficient lifeblood, the world's largest oil-refining and sales organization began to wither.

   The company that had once promoted itself as "Number One,
Second to None" was now ranked behind everyone. In 1968 it
lost its title as the nation's largest crude-oil producer to Standard
Oil of New Jersey (soon to become Exxon). In 1974 it lost its
long-coveted title as the nation's most profitable oil company, also
to Exxon. In 1976 it lost its position as the number-one U.S.
gasoline retailer to Shell. By 1977 its profitability had slipped
behind not only Exxon but Gulf, Chevron and Mobil. Once the
second most popular mutual-fund investment on Wall Street, behind IBM, in 1977 Texaco became the first major U.S. oil company to lose its triple-A bond rating from Moody's and Standard
& Poor's. In a decade's time, Texaco had become the ugliest of
the Seven Sisters.

   In addition to the problems of its own making, Texaco was
confounded by more than its share of outright bad luck. It was a
partner with Union Oil Company in the infamous 1969 Santa
Barbara blowout that gave birth to the environmentalist movement. Several years later, a Texaco drilling rig hunting for oil
under a lake in Louisiana managed to drill into a salt mine, draining the entire lake like a huge bathtub; in the whirlpool, even the
drilling rig and a barge were sucked underground.

   Overseas the company wasn't faring much better. Castro nationalized its Cuban operations when the company refused to refine Soviet crude. Five years later Texaco was nationalized in
Ceylon. Four years after that, after making a huge discovery in the
rain forests of the Amazon, two-thirds of the interest was nationalized by Ecuador. Libya, Trinidad, Angola, Saudi Arabia,
Iran—everything was going under in a tide of Third World assertiveness. When Muammar Kadaffay decided to make an international demonstration of his territorial claims in the Mediterranean,
he made a convenient target of a hapless Texaco drilling rig,
which he harassed and chased away.

   All major oil companies were running low on oil-in-the-ground,
of course, but none as severely as Texaco, a company whose
geologists, oil scouts, landmen and lawyers were all subject to the
iron-fisted domination of the company chieftains in New York.

When a group of Texaco explorationists became excited at the oil-bearing potential of a place called Prudhoe Bay, Alaska, top management determined that the operation would prove too costly and refused to approve the drilling, making Texaco the only major oil company to miss out on what became the biggest oil field in North America. Texaco was also a latecomer to the action in the British North Sea, the hottest foreign oil field of the 1970s.

In part, the inertia gripping Texaco was the product of its size. The fourth-largest corporation in America, Texaco ended the seventies bigger than Ford, Chrysler and B. F. Goodrich put together. It was five times larger than Getty Oil, twenty times larger than Pennzoil, doing more business in a year than the entire country of Finland could produce in goods and services. But while most other major oil companies had long ago decentralized their structures to throw more authority into the field, Texaco maintained such inbred and hidebound traditions that it became a curiosity throughout the industry. Other oilmen came to describe the enigmatic decision-making process at Texaco simply as "the black box." Dealing with Texaco, a federal energy regulator once marveled, "is like watching snails mate." Decisions that field executives could make at other oil companies had to go before the board of directors at Texaco.

After leaving the Chrysler Building in midtown Manhattan in 1977, the corporate headquarters even assumed the physical appearance of a Washington-style bureaucracy. The immaculate new office building in White Plains, New York, about an hour from Manhattan, was three stories tall, with each floor covering as much real estate as three football fields. Mail and memorandums were delivered by robotlike little vehicles that spent all day beeping through the hallways.

No one in Texaco was so acutely aware of the company's problems as John McKinley.

To the extent that such a ponderous organization could foster creativity, McKinley showed it. After nearly fifteen years in the research department, McKinley spent much of the 1960s in the company's petrochemical operations, where he pioneered the almost heretical practice of buying raw materials from outside the company if doing so could save money. In the late 1960s he became the head of the supply and distribution system that kept the company operating in some one hundred fifty nations—a job that made him the captain of one of the world's largest private armadas. His performance in these jobs—as well as a series of personal tragedies involving two of his superiors—suddenly combined in 1971 to thrust McKinley into the number-two position as company president.

Although McKinley believed that the company had carried its centralization to the extreme, a man of such military training wasn't about to upset the chain of command. Throughout the 1970s, while Chairman Maurice "Butch" Granville mostly carried on the policies that had built Texaco in the 1950s and 1960s, McKinley quietly discharged his duties to the extent his authority permitted, but like a good soldier he awaited his turn, hoping that by the time he had a shot at becoming the big boss it would not be too late.

Much to the delight of McKinley and everyone else at Texaco, there was one occasion in the late 1970s when it briefly appeared that the company's fortunes finally had begun to reverse. Along with several partners—including Getty Oil—Texaco had undertaken a Herculean drilling venture one hundred miles from Atlantic City in the turbulent waters of the Atlantic. Texaco, alas, lost an electronic testing device in a hole three miles deep, but still managed to establish that the well looked "encouraging," which in the understated language of the major oil companies meant that *we could have a monster oil field on our hands*. A second test confirmed the presence of "significant" gas reserves. So great was the rumored natural-gas find that the dollar rose and fell with the latest speculation. A great environmentalist debate erupted over the effects of turning the Atlantic Outer Continental Shelf into one big piece of Swiss cheese. Then, two more wells were drilled to establish that the prospect was of commercial size.

It wasn't. Texaco had come up empty-handed again.

When John McKinley finally was designated as chairman and chief executive in January 1980, Texaco, at its current rate of production, had but 8.2 years' of oil-in-the-ground remaining.

· 5 ·

After the hostile takeovers and the years of hoarding dividends, J. Paul Getty had left behind quite a company. It owned oil wells from the South China Sea to the North Sea, from the Arctic to the deserts of the Middle East, with the densest concentration in the politically stable U.S. It had refineries on the East and West coasts, and eventually some four thousand five hundred gasoline stations in between. It mined uranium in Australia and processed plutonium in upstate New York. It had a copper mine in Arizona, vineyards and almond farms in California and choice commercial real estate all over the country.

But all that paled next to its 1.5 billion barrels of oil-in-the-ground in the U.S. alone. Although in terms of annual sales Getty ranked only fourteenth in the U.S. oil industry, its holdings of domestic oil reserves ranked it sixth, ahead of Texaco, Chevron, Gulf Oil and many other much larger companies. And the oil-in-the-ground actually logged on the company's books didn't begin to reflect what was *really* there. The thick, heavy oil in Getty's Kern River Field, near Bakersfield, California, could be produced only with steam flooding, but new advances in so-called enhanced secondary recovery techniques were increasing the amount of oil that ultimately could be produced.

When J. Paul died in 1976, the mantle of leadership at Getty Oil had been passed to another of Liedtke's old Tulsa friends, Harold

Berg. The Old Man had once called Berg a "crack, veteran executive," and, in an even larger compliment, a "genuine *oilman*." Berg indeed was every inch an oilman. He smoked Luckies, drank Coke for breakfast, stood six-feet-two with a waistline that reached well below his belt, and he oiled his hair and parted it high. He had joined Tide Water in the midst of its takeover battle with J. Paul as an engineer trainee, a job that in those days involved working as a roughneck and labor-gang roustabout in boomtowns like Kilgore, Texas, and backwaters like Arkansas. Berg typified Getty Oil: plain and rough-hewn, a feisty independent in a field of oversized heavyweights like Texaco. An us-versus-them esprit de corps had come to permeate Getty Oil, generating such employee loyalty that fully half the company's employees continued working beyond age sixty-five when mandatory retirement policies were outlawed in 1978.

But Berg knew better than anyone that he was a caretaker between two eras. For in 1978, when the time came to designate his successor, the oil industry was being thrust from the good-old-boy era into a brave new world. Taking customers to hunting lodges and receiving fur coats from suppliers were quickly becoming bygone customs. Young oil executives were eating lunch at their desks instead of at the golf course clubhouse. The environmentalist movement was gathering momentum, with the oil industry as its main target. The Republicans had lost the White House. The quickening and deepening cycles of economic expansion and contraction were widening the gulf between the haves and the have-nots—and the oil industry had more than anybody. A tax was even enacted on those so-called windfall profits.

As he deliberated over his choice of a successor, Berg also took account of another disturbing factor: oil-in-the-ground was getting more difficult and costly to find. A spreading school of thought held that oil companies were best off to put some of their profits in other ventures. Exxon was about to begin making motors and electronic typewriters. Mobil had bought Montgomery Ward. Sun Oil was in the medical-equipment business. An oil company, it seemed, needed more than a mere oilman in charge.

Berg summoned his two candidates for the top job at Getty Oil into his office. One was Robert Miller, a lifelong oilman fifty years old. Berg loved Bob Miller. So, in fact, did everyone. Overweight and slightly frumpy, with heavy, dark eyeglass frames, he had demonstrated his loyalty on occasion after occasion, accepting every Godforsaken transfer that the company had imposed without ever a complaint. If Berg had a protégé in Getty Oil, it was Miller.

The other candidate was Sid Petersen, who besides being two years younger than Miller had never worked outside of the executive offices in California, had never worked on an oil well and was far too fastidious a fellow ever to get his hands dirty. Yet at the management level, Petersen had excelled in an unusual diversity of assignments, serving as a district engineering manager without ever having been an engineer and as the company's controller without ever having been an accountant.

Moreover, the feeling was widespread that J. Paul would have approved of Petersen—even though he had twice misspelled Petersen's name in his autobiography, *As I See It*. The influence of the company patriarch had continued to linger long after his death—almost as if Getty Oil had a ghost sitting at the directors' table—and everyone recognized that Petersen had always had a way with the Old Man. Older, higher-ranking executives were often pleased to let Petersen do the talking during the planning and budget meetings at Sutton Place, and Petersen did not shrink from the opportunity. The late George Getty thought the world of him. "Mr. Petersen is an outstanding young executive," George Getty had written the Old Man years earlier, "and throughout his years with the company he has always carried out his responsibilities in a first-class manner." It was upon the death of his mentor George Getty, in fact, that Petersen was first given a seat on the Getty Oil board.

"If I'd have let my personal prejudices decide, I'd have picked Bob to succeed me," Berg would recall. "But I picked Sid."

Miller, the runner-up, accepted his mentor's judgment with disappointment but without protest. Petersen, Berg recalls, "was tickled to death." At the relatively young age of forty-eight—and he looked even younger than that—Sid Petersen was the top man in one of the nation's largest companies. When a reporter came around for an interview, Petersen said, "After the flush wears off it's back to work"—but the flush definitely was there. No one, in fact, marveled more at Petersen's success than Petersen himself. "There are almost twenty thousand people in this organization," he once remarked to a lawyer friend, "and I'm the one who made it to the top."

Petersen was a thoroughly modern manager with a perfect situation. Getty Oil was rolling in cash. The company's North Sea

production was rapidly increasing. President Carter had de-
controlled the price of the heavy, sludgelike crude that Getty
produced in abundance in California—and heavy crude was ex-
empt from the federal windfall-profits tax on oil. Moreover, the
company's dividend, although slowly increasing, was still among
the lowest in the industry. In his first year as Berg's designated
heir apparent, Sid Petersen had $1 billion burning a hole in his
pocket. "The problem was what to do with all that money,"
Petersen would say.

Although Petersen remained committed to maintaining the oil
operations of Getty Oil Company—"oil is our middle name," he
was fond of proclaiming—neither did he harbor any romantic
attachment to them. The company entered into a real-estate partnership with Jack Nicklaus to develop suburban resort properties.
Another property venture followed with Blyth Eastman Dillon &
Company, an arrangement rendered all the more convenient by
the presence of Willard Boothby, a Blyth Eastman official, on the
Getty board.

Then there was ESPN—the Entertainment and Sports Program-
ming Network. Getty Oil would sink more than $60 million into
the network on the theory that cable-TV subscribers would jump
at the chance to view pre-recorded speedboat races and Third
World soccer matches. In fact, this theory proved correct: ESPN
rapidly became the nation's most-watched cable network. One
unhappy woman even named Getty Oil in her divorce petition,
claiming that ESPN had alienated the affections of her husband.
However, the proliferation of cable channels diluted advertising
dollars, and ESPN was the one network practically guaranteed not
to reach the market most coveted by advertisers: women aged
eighteen to forty-nine.

Getty tried its hand at other entertainment ventures, most nota-
bly a cable network called Premiere, to which Columbia Pictures,
Twentieth Century–Fox, Paramount Pictures and MCA agreed to
give first broadcast rights for new films—at prices mutually
agreed upon by the studios. HBO, however, called this a "pat-
ently illegal" scheme, and the Justice Department sued it as an
unlawful boycott and price-fixing arrangement. Attorneys general
in thirty-one states joined in the attack, and Premiere never pre-
miered.

But cash flow continued to build and Petersen continued to
spend: $60 million for a coal mine in Utah, $70 million for another coal mine in Colorado, $1.5 *billion* to dig a new copper

mine in Chile with General Electric. Petersen's deep pockets emboldened him to bust up a "definitive" takeover deal between two other companies: Reserve Oil & Gas of Denver had already agreed to be acquired by a Canadian firm, but assented to a takeover by Getty Oil when Petersen put $630 million on the table. It was a noteworthy signpost of the events to come.

Sid Petersen's masterstroke occurred when ERC Corporation, a major Kansas City-based insurance company, was fending off a hostile takeover by Connecticut General. Getty Oil stepped in as the white knight for ERC, which was delighted to be taken over by an oil company that didn't know the first thing about insurance and that therefore wouldn't be firing any of the top brass. At $570 million, Sid Petersen thought ERC was a bargain; events would eventually prove him right. *The New York Times* hailed the deal as "another bold move to diversify beyond the petroleum business, a strategy being carried forward by the company's new president and chief operating officer, Sidney R. Petersen."

The Old Man was certainly never opposed outright to corporate diversification—in principle, at least. Besides buying Sutton Place in the name of Getty Oil, he had purchased the Hotel Pierre and eventually much of the Fifth and Madison Avenue real estate around it. He had built the Pierre Marques Hotel in Acapulco and operated cattle and sheep ranches, a string of cafeterias, Spartan Aircraft and even a small insurance operation. But these were mere dalliances, not the cornerstone of a business strategy. Next to oil, "they're all of minor importance," he once told an interviewer with a wave of his hand. "I think management is difficult enough if you stay in the energy business," he said on another occasion.

Petersen rationalized the departure from tradition as a necessity of the times. "I suspect making a major investment outside of oil would not have found favor in Mr. Getty's eyes in *1970*," he said. "But in *1980* he might have recognized some of the changes taking place in the oil-and-gas business. I'm not sure he wouldn't have said, 'Yes, it's time.'"

But something strange happened when the ERC insurance-company deal had come before the Getty Oil board for perfunctory approval. The *new* Mr. Getty—Gordon Getty—voted no.

Harold Stuart, longtime friend of J. Paul Getty, member of the Getty Oil board and one of the largest nonfamily shareholders of Getty Oil, would watch Gordon and wonder, *What is going*

through that man's head? There he would be, seated at the directors' table, his head thrown back and his lips moving—looking
straight at the ceiling! Was he reciting poetry? A libretto, perhaps?
Sometimes Gordon wouldn't even move his lips; he just *sat there
with his eyes closed!* Stuart had seen all kinds in his nearly seventy
years—as a lawyer, judge and assistant Air Force Secretary under
Harry Truman—but never anyone quite like Gordon Getty.

Stuart wasn't alone in his marvel. Among men who had built
successful careers through purposefulness, consistency and con-
formity, Gordon was an enigma. "It's like he came to us from
Mars," Sid Petersen recalls.

Petersen, in fact, had been snickering over Gordon's stumbles
for years. As the executive assistant to Gordon's brother George,
Petersen had heard all the stories—about Gordon's imprisonment
in the Neutral Zone, the lost cars, the fight with the Old Man over
Getty Oil's dividend policies. *That's Gordon for you,* the executives would say. Petersen had even been one of the up-and-
coming executives whom George Getty had assigned to tear apart
Gordon's off-the-wall consulting recommendations. "Gordon
would take a trip, write a report and send us a bill," Petersen
would recall many years later. "We'd spend the next six months
trying to convince his father against it."

Some of Gordon's unpredictability made Petersen and the board
of directors downright uncomfortable, but for the most part they
viewed him as a harmless irritant. He sometimes asked questions
about the first item on the board's agenda when the last item was
under discussion. He once mistook a break in the meeting as an
adjournment, and left the building before the day's business had
been concluded. And cars, of course, continued to confound Gordon, such as the time Petersen had to help Gordon comb the
company parking garage for a vehicle whose whereabouts Gordon
had forgotten. Stories from Gordon's personal life even began
making the rounds, such as the one about Ann Getty sending
Gordon to pick up hamburgers at Miss Brown's on Lombard
Street; forgetting that he had driven there himself, Gordon waited
to be picked up out front while chowing down a burger. *That's
Gordon for you.*

Gordon rarely felt embarrassment at such episodes—and at
times even seemed to delight in his loony-tune image. "I guess
I'm just an absentminded professor," he would often tell people.
"Laughing at myself," he says today, "gives me the license to
laugh at others."

The reason that Petersen could not figure out Gordon may have been that Gordon had not quite figured himself out. By the time they both finally knew, it would be too late.

Gordon was a man who reveled in indecision, someone of such intellectual unpretension that he was unwilling to commit to abstract concepts that most people took for granted. Gordon was an agnostic: "It seems to me there aren't enough clues to know what's going on in that regard," he once explained. He eschewed ideology: "I have a certain fondness for democracy, but I wouldn't rid the world of everything else." Among the things that Gordon greatly admired was the naiveté and uncorrupted curiosity of children. "Any reflective child," he says, "will cover the ground that most philosophers cover."

The unwillingness to commit to real-world interests, as well, had driven Gordon from one avocation to another in adulthood. In the early 1970s he went through a period of passionately attempting to solve complex municipal-planning and traffic-flow problems on an early-vintage personal computer. After seeing a television program about the activities of the anthropologist Louis Leakey, he became obsessed with studying the origins of man and became a major backer of the Leakey Foundation, the Jane Goodall Institute and the San Francisco Zoo, where *People* magazine once photographed him with a baby gorilla named Bawang hanging from his neck. He boned up on Resnick and Halliday's *Physics* and took the final exam for freshman physics majors at Berkeley, purely "for the hell of it." (He passed.) In the late 1970s, when his passion for composing reawakened, Gordon began going to "work" each day in a soundproof music room decorated in a shade he described as "paperbag brown," enjoying a breathtaking view of San Francisco Bay, the Golden Gate Bridge to the left and Alcatraz to the right.

"Money," John Jacob Astor once said, "brings me nothing but a certain dull anxiety," and the same might have been said of Gordon. By the early 1980s, with his income from the trust totaling $28 million annually, Gordon had everything he wanted: a Jeep Wagoneer and an AMC Pacer, both parked in the street; a glistening black Yamaha grand piano, a ten-year-old reel-to-reel tape deck and a pair of Bose 901 speakers. A man whose richest tastes ran to compact discs could be quite content with a great deal less money than Gordon made; in the time it took to listen to one

$14.95 CD, Gordon earned enough to buy nearly two hundred of them.

For kids growing up with two butlers and a French chef in a five-story neoclassical mansion, the four sons of Gordon Getty led excruciatingly normal lives. He threw the football in the street and watched the San Francisco 49ers on television with them. They raised guinea pigs. He took them backpacking. "Almost any father would like to raise his sons to appreciate nature and the world and not be insulated from it by wealth," Gordon would say. "It's too easy to tune the world out and wait for the next dividend check. That's a miserable kind of life."

His wife, however, had different tastes.

Ann had become one of the country's most stunning socialites, a true "gold-plated beauty," as the *San Francisco Chronicle* called her. Ann's dinner guests might well include Princess Margaret, Mikhail Baryshnikov, Rudolf Nureyev, actor Christopher Lee and Prince Saud al Faisal, foreign minister of Saudi Arabia. She would enclose the courtyard of their home with a glass roof so that 120 guests could be accommodated at a time. ("I never come down until the party is starting," she once said.) She would co-host a floating soiree for nearly one hundred on the Nile, with the French wine—not to mention all the guests—flown in for the occasion at her expense. Ann would eventually expand her interests to include investments as well as invitations, and her appearance would change as well. Ann Getty, the hostess hailed by *Vogue* in 1977 for her "feminine and untailored" fashion look, would eventually appear on the glossy cover of *Manhattan, inc.*, a business magazine, in a severely tailored black suit, holding hands with her new business partner, the cigar-chomping British publisher Lord George Weidenfeld.

Ann apparently had no pretensions about the odd traits of her husband. A friend once asked her to describe Gordon's reaction when she drove home in a new turbo Porsche. "He hasn't noticed it yet," Ann replied. When Ann held a dinner party, Gordon, who loathed small talk, would often try to get up a group to repair to the music room. He was much happier when the dinner parties included artists and intellectuals; when he made the guest list, it was apt to include Nobel laureates from the Bay Area.

But it was always clear that Ann lavished her attention and her protectiveness on Gordon above all others. At the height of the battle to come, she would fly his friends to New York for Gordon's fiftieth birthday, help arrange for some of his musical compositions to be performed at Lincoln Center and secure Luciano Pavarotti to sing "Happy Birthday." As bad blood began to spill from Getty Oil, it was Ann—not the cook—who scrambled the eggs for Gordon's morning strategy sessions with his lawyers. It was Ann who, as a young wife in her mid-twenties, complained to her father-in-law that the company's lawyers were "killing my husband," and who had even traveled to Sutton Place to try to bring peace between the father and son.

"Ann is no dragon lady," says Bill Newsom, Gordon's life-long friend. "But nobody pushes Ann around."

However unsuited Gordon was to day-to-day management in the oil business, no one disputed that his intellect *was* considerable, and that he had every right to demand a voice in the company's affairs. He was, after all, not only a company director but also co-trustee over the family's 40% ownership of Getty Oil.

But in his first several years on the board, Gordon recognized that Getty Oil was performing exceptionally well without Gordon's help. Despite the industry's growing political problems and diminishing base of oil-in-the-ground, in the late seventies and early eighties Getty Oil was enjoying its most stunning profits ever, thanks mostly to the oil shock in Iran. In addition, Sid Petersen was on a roll. It didn't bother Gordon in the least that he needled Petersen when he abstained or voted against an important management proposal, such as the acquisition of the ERC insurance business. As Gordon would later explain, it was only his dissent, however meaningless, that prevented Petersen from having a board comprised entirely of yes-men. "I don't feel alone or threatened as a minority of one," he would say. But for the most part, Gordon saw little reason to make himself an obstacle to what Petersen and the board wanted.

However, the most important reason for Gordon's passivity was Lansing Hays.

Hays was an arrogant, abrasive and intimidating lawyer who served with Gordon as co-trustee of the Sarah Getty Trust. Heavy-set and white-haired, he had been the Old Man's lawyer for years, and could be every bit as ruthless as the Old Man required. Hays was also exceptionally articulate, though in a sardonic way; he was said to be one of the few people who could conjure a laugh from the Old Man. J. Paul Getty had variously referred to Hays as "my great friend and attorney *par excellence*," a "loyal friend" and "brilliant legal brain." If anyone carried J. Paul's proxy from the hereafter, it was Lansing Hays.

Hays, by all accounts, viewed his board membership and his role as co-trustee as a license to beat up on whomever he pleased. When staff lawyers dared to commit such unforgivable acts as splitting infinitives or dangling participles, Hays could drive them to the edge of tears. During directors' meetings, Hays would go out of his way to inflict one humiliation after another on the company's top in-house lawyer, Dave Copley, whose wording of board resolutions often failed to conform to Hays' specifications; when a resolution on the same subject came up the following year, Hays would attack the language that he himself had previously suggested. "He was a very nasty man, and I'm giving him the benefit of the doubt," recalls Jack Leone, a longtime Getty Oil PR man.

Nobody, however, knew the extent of Hays' ruthlessness better than Gordon. It was Hays who had helped depict Gordon as an ungrateful son and as a buffoon of a businessman during Gordon's dividend war with his father. Now, whatever Gordon's inclination to assert himself as a Getty Oil director or as a co-trustee, Hays stifled it. Hays openly boasted to people outside the company about the ease with which he "controlled" Gordon. "While Lansing was alive, my views as trustee weren't worth a hoot unless he joined in them," Gordon says.

Hays succeeded in intimidating Gordon to a degree that even the Old Man couldn't. Gordon once suggested, for instance, that Getty Oil should use some of its excess cash to buy stocks in a big way, in the manner of a mutual fund. Other directors, already practiced in humoring Gordon, politely pointed out that trading away the company's assets would eliminate the jobs of all those loyal, hardworking employees. Hays, however, immediately cut to the quick: "That doesn't make any sense," he barked. "Let's quit wasting our time and move on to something else."

As Sid Petersen would later maintain, "Lansing Hays had nothing but contempt for Gordon Getty." For his part, Petersen considered Hays a curmudgeon—a bully, even. "If his bluster caused you to back up a step, you were done," he recalls. Petersen, however, did not back down, and despite having to put up with Hays' insults, Petersen was left free to manage Getty Oil largely as he saw fit.

It was a small price to pay, for Petersen was reveling in the chairmanship of Getty Oil, and people had begun to notice it. He knocked down a wall in Harold Berg's old office to create a conference room. He traded in his company-owned Cadillac Seville for a Jaguar, and traveled extensively on a corporate jet whose expense, as a lower-level financial executive, he had determined to be steep. He seemed to be doing more and more deals with movie studios, and appeared so frequently in the social notices that his old boss Harold Berg began to wonder if Petersen hadn't "gone Hollywood." "I wouldn't say he flaunted it," Berg would recall. "But he made it pretty obvious around L.A. that he was in that position." Throughout Getty Oil, a name was adopted for the office of Sid Petersen. They called it the "Imperial Chairmanship."

A new maxim swept the oil patch in the summer of 1981: "Eat or be eaten." There had been a time when a hostile takeover of one oil company by another would have been unthinkable; the corporate chieftains of Big Oil hunted, fished, played golf and drank beer together, and they respected the sanctity of one another's turf. But oil was now dominated by men of high finance, who recognized that the price of oil-company stocks hadn't begun to keep pace with the increased value of oil-in-the-ground. Every oil company but Exxon was said to be vulnerable, and all of them, including Exxon, needed additional oil reserves.

Suddenly, everyone had to have a multibillion-dollar "line of credit"—a loan commitment ready to be drawn on a moment's notice. A line of credit enabled an oil company to act quickly as either a hostile raider or as a white knight, riding to the rescue of a company suddenly caught in the clutches of an unwanted takeover. Having a line of credit also sent a signal to the world that *your* company was not for sale; if someone tried to take away your oil-in-the-ground, you could always price yourself out of the market by purchasing someone else's. In a matter of weeks U.S. banks arranged loan commitments totaling at least $25 billion—an amount so large that some experts began fearing that other borrowers would be crowded out of the lending markets. Even Hugh Liedtke of the medium-sized Pennzoil—a man known for prizing his flexibility—went to Citibank, collateralized some of his oil-in-the-ground and walked away with a $2.5 billion loan commitment, just in case any ripe pickings happened along.

Texaco, however, didn't quite get with the program.

John McKinley had been chairman less than a year when the company was invited by an investment banker to step up as a white knight for Conoco Incorporated, which was seeking to avoid a hostile takeover by Seagram. But Texaco was still playing by the gentlemanly rules of yesteryear; when the chairman of Conoco Company told McKinley that he would prefer to avoid a takeover by another oil company, Texaco refused to follow through on its offer, and Du Pont signed the deal for the identical price that Texaco had bid. A few months later, Texaco was invited to make an offer for Marathon Oil Company, which was also seeking a white knight; this time U.S. Steel walked away with the prize. Nobody could believe that the nation's third-largest oil company—and the major most desperately in need of new reserves—was letting the merger wave roll by.

"Everybody else in this race is riding a thoroughbred," said a takeover professional on retainer by Texaco. "I've got a jackass."

While Texaco was one of the few companies unwilling to launch a hostile takeover, Getty Oil was one of the few that could feel secure from receiving one. The reason, of course, was that 40% of the company stock was held in a family trust that was legally prevented from selling *any* of its stock except under extraordinary circumstances. As the megamergers in oil began gathering momentum, Sid Petersen received a sales call from an imaginative young investment banker named Martin Siegel, who was marketing antitakeover defense strategies as "products" for Kidder, Peabody & Company. Petersen was intrigued, but Hays pointed out to Petersen that such defenses were unnecessary. As a trustee governing a virtually controlling interest that could not be sold, Hays said, "I am your best defense against a takeover."

But Sid Petersen knew that good things don't last forever.

What if Gordon Getty and Lansing Hays were flying to a directors' meeting somewhere and went down in a plane together? The trust would have no trustees, no one available to vote 40% of the company's stock. In the months it would take the courts to find new trustees, 40% of Getty Oil's stock would be in a deep freeze, leaving Getty Oil like a sitting duck for such corporation hunters as T. Boone Pickens.

There was another possibility that disturbed Petersen: What if Lansing Hays died? He was a seemingly healthy man only slightly over sixty, but as Petersen told him, *Lansing, you know what those cabdrivers in New York are like*. If Hays were out of the picture, Gordon Getty *alone* would control 40% of the stock. And although Gordon seemed safely co-opted for the time being, there was never any telling what he might do.

His refusal to vote in favor of Petersen's vaunted acquisition of ERC, though unimportant to completion of the deal, had been troubling; no director had ever dissented from a major move by management. Moreover, Gordon had explained his vote by saying that he simply felt somebody ought to be voting no every now and then—an explanation that had struck Petersen as nothing short of peculiar. If something happened to Hays, who would keep Gordon in line?

Just maybe, there was a solution.

Although the business of the Getty family technically was none of Petersen's business, the Sarah Getty Trust and Getty Oil had always functioned more or less as members of the same big, happy family. Everyone overlooked the conflict of interest inherent in the relationship; nobody tried to pretend that the arm's-length relationship customary between shareholders and managements existed in the case of Getty Oil. In J. Paul Getty's time, and later in Lansing Hays', the trust *was* the company, the company *was* the trust and both were instruments of the same patronage. Petersen felt no restraint from trying to keep himself abreast of the trust's business, at least to the extent that Lansing Hays would permit him.

Petersen knew that when the Old Man last amended the trust documents after George Getty died, he provided not only that Gordon and Hays should serve as co-trustees, but that his long-time bank, Security Pacific National, should serve with them as a *corporate* co-trustee. After all, the bank's chairman, Frederick "Fritz" Larkin, was a good friend of the Old Man and a director of Getty Oil. The bank was delighted at the chance to share in the lucrative trustees' salary—a total of roughly $3 million, under a fee-splitting formula. Banks love fees; unlike making loans, fee business involves minimal risk.

Well, most fee business, that is. After J. Paul Getty died and the bank was preparing to accept its appointment, it learned that the Old Man had never bothered to inform the tax authorities of the United Kingdom about the existence of the $3 billion worth of stock he had accumulated in the Sarah Getty Trust, "hopefully avoiding subjecting the trust or his personal assets to potential UK taxation," according to an internal report later prepared by the bank. Security Pacific grew even more concerned when it learned that a trustee could be held liable for the tax deficits of an estate— and that under similar circumstances, the United Kingdom actually had seized certain assets of Bankers Trust.

By the early 1980s, the law had been clarified in a way that removed the bank's tax liability; perhaps, now, Security Pacific would agree to become the third trustee, Petersen thought. He called Fritz Larkin, the banker and Getty Oil board member, urging him to reconsider. Lansing Hays also urged the bank to accept the appointment, pointing out that "Gordon would be around for a long time and would obviously run matters by himself at some point" unless the bank agreed to serve. Hays assured the bank that "Gordon would have little choice" but to capitulate to the bank in matters concerning management of the trust.

But Security Pacific *again* refused its appointment as co-trustee. Fritz Larkin explained why in a letter he sent directly to Sid Petersen's home, noting that "I didn't care to have it floating around your office." And it was easy to see why.

In light of the events that would occur in a Houston courtroom exactly five years later, Larkin's response may have been one of the most amazing acts of corporate clairvoyance ever. "The size of the trust is just too big for the bank, should there be a one-in-a-million serious development," Larkin wrote. The risk, he explained, was this: The trust document that J. Paul and Sarah Getty wrestled over that Christmas Day forty-five years earlier prevented any sale of Getty Oil except to avoid a major loss. If it ever came to that—of course, it almost surely never would, for selling the trust shares was unthinkable—but if it did, "litigation resulting from such questions would be difficult to defend and the results unpredictable, with potential judgments of unprecedented proportions," Larkin's letter said. In fact, the report concluded, "the potential magnitude of any loss resulting from acting as co-trustee of the $3.2 billion trust could jeopardize the very existence of the bank."

Alas, there would be no corporate co-trustee. Only Lansing Hays could prevent Gordon from exercising complete control over 40% of Getty Oil.

Years later, one of Sid Petersen's lawyers would call the bank's refusal to serve as co-trustee "the seed that sowed an $11 billion court judgment." But at the time, Petersen was not overly concerned about the matter. Any problems involving the trust were years, perhaps even decades, away, and Petersen was building for the future. Petersen had announced plans to erect "Getty Plaza," the tallest building in the San Fernando Valley—thirty-six stories sheathed in green granite, intended to consolidate the company's three dispersed locations along Wilshire Boulevard into a single structure, with Sid Petersen at the top. The $100 million building.

0% owned with MCA, the entertainment conglomerate, would
e first-class the whole way. Escalators would link the upper
xecutive floors, where each vice president was authorized to
pend $250,000 on decorating, fixtures and furnishings—and
here $750,000 was to be spent completing the office of Chair-
aan Petersen, according to Stuart Evey, the executive in charge of
ae project. Petersen even became involved in the selection of
olor schemes for executive washrooms, Evey says. Petersen calls
vey's recollection of the lofty budget figures "crazy," but
oesn't dispute that all the executive offices were to be costly.

Some of the oilmen in the company considered it all a supreme
aste of money by the "Imperial Chairmanship": Why build a
ky-high palace, they wondered, when you could look for more
il-in-the-ground? They did not know, of course, that Sid Petersen
ould never get to set foot in his new office.

n May 10, 1982, Lansing Hays died of heart disease at Monterey
ommunity Hospital in California. He was sixty-six.

Before long, Harold Stuart of the Getty Oil board would swear
aat Ann Getty was more carefully coiffed and beautifully dressed
aan ever. "She really began to blossom out," he would remark.

· 6 ·

Where is the proxy?"

Lansing Hays had been dead three days, the 1982 annual meeting was tomorrow and nobody could locate the crucial proxy that would cast the trust's thirty-two million shares to re-elect management's slate of directors. Petersen' aides rifled through Hays' office to no avail. Had he dropped it in a mailbox just before he died? Had he taken it home? Shareholders would be gathering in twenty-four hours and unless somebody found that proxy and desposited it with the corporate secretary, 40% of Getty Oil's stock would be legally absent from the meeting. There might not be a quorum. Getty Oil might not be able to re-elect its board of directors. Getty Oil would be the laughingstock of the corporate world.

When it occurred to Petersen to ask Gordon, the new sole trustee, whether *he* knew the whereabouts of the trust's proxy, Gordon replied that the proxy was safely in his possession. *May have it?* Petersen asked. *I'll bring it to the meeting tomorrow,* Gordon replied.

But the following morning, as shareholders filed into the Crystal Room of the Beverly Hills Hotel, neither Gordon nor the proxy showed up. Minutes before Petersen was scheduled to call the meeting to order, board member Harold Stuart anxiously phoned Gordon's suite and reached Ann.

"You put him on the elevator up there," Stuart commanded her, "and I'll get him at the bottom."

The matter of the errant proxy was initially passed off as another antic of the Absentminded Professor. And indeed that's all it may have been, for Gordon's recollection of the proxy incident is that there was no incident at all. "It seems to me that things passed along pretty normally, as near as I could tell," he would say. Gordon says he just as easily could have voted the shares by raising his hand at the meeting, making a proxy unnecessary. That company officials would circulate such stories "tells more about management's fretfulness," Gordon insists today, "than about me."

Indeed, the top managers of Getty Oil would later become convinced that Gordon had deliberately withheld the proxy until the last minute as a show of force, as a proclamation that *I alone am in charge now and everybody better realize it*. At a company the size of Getty Oil, annual meetings, after all, are carefully rehearsed events at which little is left to chance. With the benefit of hindsight, events they had attributed solely to Gordon's odd and unusual ways would seem like giant warning signs they had failed to observe.

"Looking back on Gordon Getty and his activities, it's amazing how naive we were," Petersen says.

Not everyone at Getty Oil was so naive. Stuart Evey had spent much of his career at Getty Oil as a gofer, which often meant attending to Getty family business. As George Getty's executive assistant he had looked after some of his boss's investments and helped maintain his social calendar, and says he once even came to the rescue when a Getty grandchild had a scrape with the law. Even now, as a company vice president in charge of ESPN and other far-flung investments, Evey was the company's top schmoozer, the expediter, a self-described manipulator, the guy who always knew where to get a ticket long after the game was sold out. Evey could cuddle up to anyone, and he liked to think he understood what made people tick. And when he looked at Gordon, Evey saw a powerful giant awakening from a long sleep.

"We should have someone who does nothing but massage Gordon," Evey recalls telling other executives after Lansing Hays had died. It only made sense, Evey thought, that he should handle the assignment himself. "Christ," Evey said, "I can keep in touch with the guy and take him to lunch every few weeks."

Of course, an occasional phone call and lunch could never have suppressed the assertiveness stirring within Gordon. But Evey was right about one thing. When Hays died, "Gordon felt he had a sense of responsibility," Evey says today. "He wanted to know what was going on. He was saying, 'Don't laugh at my ideas. Just tell me why not.'"

Management's failure to appreciate Gordon's goals and motives—and later its overreaction to them—would help cause the downfall of Getty Oil. And it would begin to explain why, in the eyes of some, the same company—Getty Oil—would be sold twice: first to Pennzoil, then to Texaco.

Gordon needed to turn no further than his father's own autobiographies for a lesson on the rights of ownership. In one book, the Old Man recalled taking the five dollars he had earned at the age of eleven from selling *The Saturday Evening Post* door-to-door and buying 100 shares of his father's stock at a nickel each. "There—now you're part owner of the company for which I work. You're one of my bosses," his father had remarked. J. Paul himself would once say: "A corporation's officers were bound to consider themselves employees of the shareholders, to whom they owed loyalty and for whose financial interests and welfare they were responsible."

In the case of Getty Oil, the company's management and its controlling shareholder had been the same person. But that was no longer so. Although Getty Oil was Gordon's birthright, it was never his career. He had refused to join in the cozy, all-in-the-family relationship that the trust and the company had maintained from the beginning. Gordon viewed himself as an outsider, as someone with a "*stockholder's* viewpoint." That, in effect, made him a potential *adversary* of Getty Oil. Sid Petersen and the Getty Oil board of directors owed a duty of allegiance and accountability not just to the twelve thousand members of the public who had purchased stock in Getty Oil, but to Gordon Getty, the largest shareholder by far.

To the same extent, Gordon himself now shouldered a heavier duty than ever to the other members of his family. By this time, the trust had two dozen beneficiaries, divided into two groups. One group consisted of "income beneficiaries": Gordon and Paul Jr. each received one-third of the trust's dividend income—about $28 million each by this time—with the remaining third divided among the three daughters of the late George Getty. The other group of beneficiaries—consisting of the Old Man's grandchildren and great-grandchildren—received no income, but stood in line to split up all of the trust's assets remaining after Gordon and his brothers had died. These beneficiaries, called "remaindermen," now numbered eighteen. Gordon alone now had the power—and the legal duty—to manage the trust's 40% shareholding to the maximum benefit of all beneficiaries.

This wasn't as easy as it may seem, for the 1934 Trust instrument practically guaranteed pitting the interests of one generation against another. Should the company increase the dividend? This would bring more cash to the income beneficiaries, such as Gordon, but it would reduce the capital ultimately available for remaindermen, including Gordon's own sons. Should Getty Oil increase oil drilling? This could reduce the company's profits in the short run, hurting income beneficiaries such as Gordon, but might enlarge the assets ultimately to be divided among his nieces, nephews, children and grandchildren.

Or should Getty Oil be sold? Selling the stock and buying a cool $3 billion or $4 billion in Treasury bills would hugely increase the current income while protecting future generations against the risk of the company's outright collapse, through mismanagement or other problems. This possibility, however, presented complications of its own. There were the nearly twenty thousand loyal, hardworking Getty Oil employees whose jobs would be endangered; didn't Gordon owe them the duty of loyalty, as well? And that arcane, anachronistic Christmas Day agreement between his father and grandmother—it categorically stated that trust shares could be sold *only to avoid economic loss*.

Gordon faced another problem as sole trustee—a problem that his father had never experienced. In effect, Gordon had only half a load of ammunition.

The rest resided in a charitable institution that was surely the oddest yet the most powerful of its kind in the world. In perhaps his quirkiest act ever, the Old Man had laid the groundwork for an *art museum*, of all things, to assume a center-stage role in a multibillion-dollar takeover contest—and later an even larger role in the multibillion-dollar lawsuit. It was the J. Paul Getty Museum of Malibu, California.

Nearly fifty years earlier, after abandoning Gordon's mother, the Old Man had established what he called a "home base" in New York. His landlady there, Mrs. Frederick Guest, was one of the most noted art patrons of her day, and the sprawling penthouse apartment she rented to him was outfitted not only with antique

French and English furnishings but with one art treasure after another. "The more of it I saw, the more of it I wanted to see," he would later write, "and the desire to see may well have by itself developed into the desire to own"—all the more so, as he would note on another occasion, when "I discovered I would be able to purchase some priceless treasures for almost shamefully low prices." Rembrandt's *Marten Looten,* for instance, would be his for the "bargain price of $65,000."

"One develops conscience pangs about keeping them to himself," the Old Man wrote, so in the early 1970s he commissioned the construction of a structure to house them—a garish, $17 million monstrosity full of proto-Classical geegaws, modeled after an unearthed villa at Herculaneum, Italy, said to have been owned by Julius Caesar's father-in-law. J. Paul, who openly admitted to thinking of himself as a modern-day Caesar, took such an interest in the project that he personally approved each paint sample. The ostentatiousness of the museum building reminded the *Los Angeles Times* of "a Beverly Hills nouveau-riche dining room" and *The New York Times* of a "gussied-up . . . Bel Air dining room." The collections hauled into it, though valued at $200 million, were eclectic and acknowledged even by the museum's managers as rather second-rate on a worldwide scale: fine Greco-Roman antiquities and 18th-century French furniture, but too few to rank either as world-class; an Old Masters collection that was unranked, medieval illuminated manuscripts ranked probably sixth or seventh in the world.

When the Old Man died, his last will and testament provided token bequeathments to eleven women; about $330,000 in stock to his son Ronald, who had been cut off from the trust; and "$500 and nothing else" to his sons Paul Jr. and Gordon, who were otherwise provided for in the trust. He willed the remaining $600 million or so of his personal fortune, mostly in Getty Oil stock, to support the dog-and-cat art collection and the tasteless structure he had built to house them amid the sycamores above Pacific Coast Highway in Malibu, which he died without ever visiting. By the time the will was probated, the value of the museum's inheritance had soared to $1.2 billion.

Besides leaving the museum more than it knew how to spend easily, the most striking effect of the bequeathment was the division of Getty Oil's majority control into two *minority* blocks. The Old Man had voted proxies for 64% of the company's stock, between the shares he controlled in the trust and those that were his alone. But after his death, and after the executors of his estate had unloaded a portion of the museum shares for cash, about 12% of the control over Getty Oil was left with the museum, 40% with the Sarah Getty Trust.

Yet Gordon had also not forgotten the strong position taken by his father and brother during the dividend war of the 1960s: *It was very important for the trust to have control of the corporation,* they had said. *Its principal strength is its ability to vote a large block of stock to control the destiny of a very large group of oil companies.* Without that control, *there is every possibility that the company will lose competitive position, be worth less, fade away. . . . The trustee should be buying Getty Oil shares, not selling them.*

Gordon vowed to study ways of restoring the family's interest in the company—an interest he personally managed—to a majority. With absolute control, Gordon could assure that the company was managed in the best interests of the family; he alone would be accountable for its performance. But for now, with only a minority interest under his control, Gordon had to rely on the judgment and management of Petersen and the board. And for the primary barometer by which he would judge their performance, Gordon chose the stock price of Getty Oil. As long as the stock price remained high, Gordon could satisfy himself that Getty Oil was being managed to the advantage of all the trust's beneficiaries, born and unborn alike.

Gordon did not like what he saw.

In a sense, the dilemma facing Gordon—how to satisfy the beneficiaries of today without hurting the beneficiaries of tomorrow—had come to afflict the entire oil industry. The "easy oil" had already been discovered. Except for drilling on Wall Street, the only places left to look were hostile and remote—the frozen tundra, the mile-deep waters in the Gulf of Mexico, the rain forests of the Amazon—places where a single well could easily cost $100 million and take twenty years to pay off, if it ever did. But all those billions of dollars poured into the ground for future returns was money that the oil companies were *not* paying to *today's* shareholders.

This began to concern Wall Street, which was undergoing some fundamental changes of its own. The buy-and-hold philosophy that ruled Wall Street for a century was replaced by the obsession for a quick killing. By the 1980s, the individual investors who had once controlled trading on the New York Stock Exchange had been shoved aside by pension programs, bank trust departments and mutual funds. The professional money-management firms that competed for business managing these multibillion-dollar portfolios were judged by their performance on a yearly basis, sometimes on a quarterly basis. The patience for long-term capital gains had all but evaporated from the stock market.

Worse still for Getty Oil, the company seemed to be losing its magic touch in the oil patch; only Texaco and a handful of other majors were now performing worse. Just as had Texaco overcentralized its operations, Getty Oil had *under*centralized, giving local managers unusually large authority in picking the best drilling projects in their jurisdictions. This was a fine strategy providing there was plenty of oil left to be discovered. But in the picked-over U.S. oil fields of the 1980s, the best drilling project in one location might be far from the best project available to the entire corporation.

Getty's oil-in-the-ground—not counting all its other assets— was still worth more than $100 a share. But from 1980 to the time of Lansing Hays' death, the company's stock had fallen to $50 from nearly $110. It was one of Wall Street's worst performances.

"It was kind of hard to ignore the sorry performance of the management of Getty Oil Company," Gordon would recall. "They were able men, yet they managed to do a wretched job."

Gordon decided he had better begin spending some time down at the office.

In June 1982, just as Gordon was awakening to the problems at Getty Oil, Texaco made its first entry in the record book of astonishing legal cases. It agreed to pay $1.7 billion to escape from long-term natural-gas supply contracts with Louisiana Power & Light Company.

Texaco's problems in Louisiana had reached unimaginable proportions—far worse than John McKinley could ever have feared. He discovered that the company's Louisiana officials had been overestimating the company's oil and gas reserves for years. He learned that the company was committed to deliver 700 billion cubic feet more natural gas than it really had—and at prices as low as one-tenth the current market rate. Once again, Louisiana was caught in a terrible natural-gas shortage—worse, in fact, than the one attributed to Pennzoil nearly a decade earlier. Chemical companies filed suit, claiming Texaco had fraudulently induced them to locate in Louisiana with promises of gas that didn't exist.

Layoffs swept the chemical plants. Utilities using Texaco natural gas to make electricity had to increase consumer rates by several hundred million dollars (although these rates were unusually low to start with).

McKinley's discovery of the problems forced him into the humiliating position of having to order "writedowns" in the company's energy reserves—in effect, admitting that it had never owned all the oil and gas it claimed to have owned. "It's just amazing how bad Texaco looks," a Paine Webber stock analyst said.

Texaco was experiencing other terrible embarrassments in the state that had made the company into what it was. A state investigation found that Texaco (as well as Exxon and Shell Oil) had been illegally buying oil field supplies from members of the Louisiana Mineral Board, which administered the one million acres of state leases owned by the company. Separately, a preliminary audit found that Texaco might have cheated the state out of anywhere between $100 million and $300 million in royalty payments; the audit, however, was canceled by the administration of Governor Edwin Edwards, who, it turned out, had performed legal work for Texaco and had other close ties to the company. *Gris Gris,* a spirited magazine in Baton Rouge, would eventually depict Edwards on the cover wearing the uniform of a Texaco gas-station attendant. "You can trust your state to the man who's on the take," the magazine cover read.

One good thing happened a few days after Texaco bought its way out of the contract with Louisiana Power & Light: Texaco hit oil off the coast of California in an area called the Hueso Field. It was a major discovery—far too little, unfortunately, to reverse Texaco's worsening shortage of oil-in-the-ground, but enough to boost morale, to convince the company that it was doing the best thing by redoubling its effort to find new "elephant" oil fields.

For one of Texaco's small-fry partners in the Hueso well, however, the discovery held far greater significance. Several months earlier, on barely twenty-four hours' notice, J. Hugh Liedtke of Pennzoil had agreed to take a 25% interest in the well.

"Other companies might have required *weeks* to make that decision," Liedtke crowed to securities analysts. Moving with alacrity was a Liedtke hallmark—one that was soon to serve him well in a deal of far greater size.

*   *   *

Among the Wall Street analysts who specialized in oil stocks, none shared Kurt Wulff's reputation for irreverence. Oil analysts generally went out of their way to avoid offending corporate managers, on whom they depended for the steady stream of minute financial detail whose analysis was their stock-in-trade. But at Donaldson, Lufkin & Jenrette, Wulff had created himself a niche identifying companies that were weakly managed—companies that *deserved* to be taken over in the name of improving their performance. A photographic portrait of T. Boone Pickens, the notorious oil-industry raider and liquidation artist, hung in Wulff's office overlooking the East River.

In the oil industry Wulff was viewed as a gadfly, but among the mutual funds and other institutional investors subscribing to his research bulletins, he was a hero. Wulff had correctly predicted the takeovers of Marathon Oil, Conoco and Cities Service—home runs for which he would be voted the top oil analyst on Wall Street by *Institutional Investor* magazine.

Now, Wulff was turning his attention to Getty Oil.

He had been preparing a negative report on the administration of Sid Petersen, whose oil-finding record had taken a turn for the worse and whose diversification into insurance and cable TV Wulff considered a waste of corporate assets. But a friend of Wulff's at a money-management firm urged him to reconsider issuing a negative recommendation on Getty Oil stock. The friend knew that Lansing Hays had recently died, and suggested that Gordon Getty, as sole trustee over 40% of the company's stock, might be open to fresh ideas to enhance the value of the trust's investment. Wulff began studying Getty Oil to the point of reading one of the Old Man's autobiographies.

In October 1982, while continuing his analysis, he issued a brief report noting almost offhandedly that Getty Oil might be a candidate for some sort of corporate reorganization. A brief time later, Wulff's secretary stuck her head into his office. "Hey, you'll never guess who just asked to be put on the mailing list," the assistant said. "Gordon Getty!"

Wulff was not alone in recognizing that Gordon Getty was searching for options.

A certain old-boy chauvinism pervades the offices of Mesa Petroleum Partnership in Amarillo, Texas. The secretaries are uniformly gorgeous. Self-promotion is rampant, from the

Styrofoam Mesa coffee cups to the Mesa napkins and Mesa sta-
plers. Western art is everywhere: an oil painting depicting drifters
cantering on horseback through the mud-drenched streets of an oil
boom town; a sculpture, *Angry West,* of two Indian chiefs firing a
rifle and an arrow from a gallop, and, in the office of T. Boone
Pickens, the painting of an Indian brave paddling a canoe through
a marsh at dusk, in search of prey. The painting's title, *Silent
Hunter,* would be apt, except that sandy-haired, fifty-eight-year-
old Pickens is anything but silent when he goes on the warpath.
Pickens in the 1980s became one of the most forceful agents of
change in the history of the oil industry—and it was inevitable
that Getty Oil would rise to the top of his hit list.

Boone Pickens became convinced early in his career that the
major oil companies had a limited future. After spending three
years as a geologist at Phillips Petroleum—a company created in
the same oil fields where J. Paul Getty and his father made their
first millions—he packed in his career with Big Oil, took the
$1300 in his company savings account, made a down payment on
a 1955 Ford station wagon, outfitted it with well-site equipment
and struck out on his own. But Pickens discovered what J. Paul
Getty had also eventually learned: that buying oil-in-the-ground
was less costly than actually going out and drilling for it. And in
this pursuit, Pickens received his first big break from Hugh
Liedtke.

When the Liedtke brothers and George Bush formed Zapata
Petroleum, one of the founding investors was the Clark Family
Estate, which owned the Singer Sewing Machine fortune. Apart
from its investment with the Liedtkes and Bush, the estate had
come to own an interest in the Hugoton field of Kansas, one of the
world's great onshore deposits of natural gas. Pickens and Hugh
Liedtke, meanwhile, had become friends doing a few small deals
in the Midland area. So when Pickens wanted a chance to buy
Hugoton Production Company, he turned to Hugh Liedtke, who
saw that Pickens received the necessary introductions to the trustees of the Singer Sewing fortune. On that deal, following a tough
battle with management, Mesa Petroleum joined the big leagues.

But Pickens would not achieve his greatest notoriety until the
1970s, when he began promoting a concept that would be variously hailed and maligned—but that no one in the oil industry
would ignore. It was called the royalty trust.

The philosophical underpinnings of Pickens' notion was that
because the world was running out of oil, it was a waste of the

shareholders' money to reinvest profits in a futile search for new oil-in-the-ground. The royalty trusts worked like this: Through a series of bookkeeping entries and stock registrations, an oil company took some of its oil fields and transferred ownership to a trust. Anyone owning stock in the oil company itself also received shares of the trust. Thus, the profits, or "royalties," earned by the oil field flowed to the shareholders through the trust, whose income was tax-exempt, rather than through a corporation, whose income was taxable. Moreover, the shareholders' direct ownership of the oil field enabled them to take the kind of big tax-deductions on their personal tax statements usually reserved for big corporations that own "depleting assets," such as oil.

This was a notoriously unpopular concept among the empire builders of the oil industry, among executives who remained convinced that there *was* oil left to be discovered and who wanted to dedicate their cash flow to that pursuit. So Pickens began trying to sell his idea to major shareholders instead of executives.

One of Pickens' first efforts to promote the idea involved a company that bore a striking resemblance to Getty Oil. It was Superior Oil Company of Houston, a $2-billion-a-year company whose ownership, like Getty Oil, resided partly in a group of trusts benefiting the descendants of the company's founder. That founder was William Keck, "the greatest wildcatter of them all," a man who tested for oil by licking core samples lifted from deep underground. Pickens traveled to Santa Barbara to meet with Keck's daughter, Willametta Keck Day, a fiery old woman who had always resented never having had the opportunity to take an active role in her father's oil company. Management of the company had gone instead to her brother Howard. Willametta and Howard had been feuding nearly sixty years, ever since her pet ostrich had gotten its neck stuck in a fence hole and Howard had stuck an orange in its mouth, strangling the bird. Now Willametta was convinced that her brother, as chairman of Superior Oil, was trying to entrench himself against a possible takeover—one that might provide her and other shareholders with a big profit on their investment.

Pickens met with Willametta by the side of her pool and tried to convince her of the virtues of establishing a royalty trust at Superior Oil. "He told me there's no oil left to be found!" she would recall before her death in 1985. "I damn near flipped into the pool. I said, 'Wait a goddamn minute!' " The world is full of great oilmen, she said—not her brother, perhaps, but many others with oil aplenty left to discover. "I nearly threw that Learjet cowboy into the pool," she would say.

But Pickens' visit to Willametta turned out to be only a dry run of sorts. Pickens received a decidedly more cordial reception from another oil heir, Gordon Getty, who had agreed to meet with Pickens in a suite at the Beverly Wilshire Hotel in Los Angeles.

"You've got a dying duck on your hands," Pickens told Gordon.

"Well, what can I do about it?"

It would be best, of course, to get management to submit voluntarily to a restructuring plan, Pickens said. "You've got to get Sidney Petersen on board."

"How do I do that?" Gordon asked.

In Pickens' view, the lust for money and privilege made the world go around. "If we put fifty points into Getty Oil stock, how much would Petersen make?" Pickens asked. He believed that it would take $10 million—$20 million, even—to give Petersen the incentive to submit to a massive restructuring of the company that kept an airplane at his beck and call, that bought him whatever club memberships he wished and that gave him such prestige. Petersen, alas, owned far too few shares. The meeting ended inconclusively, but the royalty trust idea remained alive.

Pickens was far from the only takeover maven to cuddle up to Getty Oil's 40% shareholder.

Within several months of the time that Pickens came calling, so did Sid Bass, one of the ridiculously wealthy Bass brothers of Fort Worth, Texas, who were themselves oil heirs. The Basses had added to their family fortune through an adroit and aggressive investment program, including speculation in the stocks of take-over candidates. Sometimes, when no takeover materialized, the Basses threatened to launch one themselves, which caused companies to buy out their stock holdings at a rich premium over the price available to other shareholders—a practice that had achieved infamy as "greenmail."

The Bass brothers owned 1% of Getty Oil. A hostile takeover was unlikely, but Sid Bass knew that if Getty Oil would simply buy back a huge chunk of its own shares—20% of them, say— that would reduce the number of shares in public hands, thereby increasing the price of the individual shares remaining; each share, in short, would control a larger slice of the company. Bass pointed out the obvious advantage to Gordon: The trust's "slice" of the company would increase above 50%, and Gordon alone would have unassailable control. This was *very* intriguing, Gordon thought. "I was pleased and honored to have his advice," he would later say.

By September 1982, the persistent "value gap" in the stock price of Getty Oil moved Gordon to assume some initiative of his own. He decided to launch a fact-finding mission on Wall Street.

It was an extraordinarily brash undertaking. For a 40% shareholder in one of the nation's major corporations to traipse through the canyons of Wall Street soliciting advice was tantamount to hoisting a "for sale" sign over Getty Oil. It defied the protocol in shareholder-management relationships. Gordon, however, wanted answers and advice, and would not discriminate against any method simply because it was unconventional.

Gordon undertook the trip to Wall Street at the urging of his friend Alexander Papamarkou. A fifty-three-year-old stockbroker and onetime correspondent for the British Broadcasting Company, Papamarkou had spent years developing contacts in the upper echelons of European and New York society. He had met Gordon and Ann during one of their visits to Sutton Place, after Papamarkou had become a friend of the Old Man through the Duke and Duchess of Bedford. Papamarkou was already helping to introduce Ann into New York society life, escorting her to a variety of high-society soirees, counseling her on her business interests and even recommending schools for the Getty boys. Ann would publicly identify him as "sort of a nanny" to her children.

Papamarkou arranged for Gordon to see not only a variety of money managers and other financial experts, but to chat over lunch with William Tavoulareas, the president of Mobil Corporation. Seeking advice about "value gaps" and corporate restructuring from a fellow like Tavoulareas was not quite like sitting down with a Dutch uncle for some friendly advice about girls. Mobil had only recently engaged in a pitched battle for control of Marathon Oil, ultimately losing out to U.S. Steel, and would eventually succeed in taking over Superior Oil, where at the moment Boone Pickens was known to be causing trouble. Tavoulareas briefed Gordon on the antitrust obstacles standing in the way of Getty Oil's acquisition by a major oil company, and suggested that selling the company's refining and marketing operations might enable a company such as Mobil to buy Getty Oil's oil-in-the-ground. Gordon listened with interest.

Boone Pickens and his royalty trust, Sid Bass and his share-repurchase concept, Bill Tavoulareas and his takeover briefing.

Yet another oilman came calling with an idea that Gordon found particularly intriguing. Corbin Robertson Jr. of Houston, an heir to the billion-dollar Cullen family fortune, suggested joining forces with Gordon and taking Getty Oil private—buying out the museum and all the public shareholders at $80 a share with borrowed money—leaving himself and Gordon in control. Robertson, however, would have wound up with majority control, an idea that didn't appeal to Gordon in the slightest. But the concept of going private—now there was a notion worth thinking about.

The evening following their meeting, Gordon took Robertson to a performance of Rossini's *La Cenerentola,* an opera based on the fairy-tale story of Cinderella. Gordon hadn't the slightest idea that his footsteps through the takeover community would quickly touch off an earthquake at Getty Oil.

In early January 1983, Gordon read the latest "action recommendation" from Kurt Wulff of Donaldson, Lufkin & Jenrette:

> Getty Oil is the best statistical choice for major restructuring. . . . Only six months ago, Gordon Getty became the sole trustee of the Getty Trust and, as such, must now become more active and might well question why investors hold Getty Oil in such low esteem. . . . Investors are acting as though they have a very low level of confidence in the management of Getty Oil. . . . By any measure, Getty Oil is a cheap stock. Now, let's hope something favorable happens to it.

Sid Petersen was beside himself. After months of trying to accommodate Gordon's requests for information, Petersen believed that his largest shareholder was spinning out of control.

In July 1982, at the time of a board meeting on the campus at Texas A&M, Gordon had mentioned almost offhandedly that he had met with Sid Bass, the much-feared takeover artist, to discuss ways of enhancing the value of Getty Oil. Gordon, Petersen decided, was no longer merely absentminded. Gordon was bizarre.

Gordon had moved into a beach house in Malibu—no doubt, Petersen figured, so that he could meddle in the company's affairs without bothering himself with a regular commute from San Francisco. (Actually, Gordon's family simply wanted a beach house for the summer.)

Then, in the fall, Petersen had met Gordon at the Phoenix Airport so they could together interview a new candidate for the

Getty Oil board: John Teets, the chairman of Greyhound. Gordon had flown into Phoenix from New York, where, he told Petersen, he had spent several days seeking advice from oil experts and Wall Street professionals. *What kind of Quixotic adventure am I hearing about now?* Petersen asked himself. In a taxi on the way to the Greyhound offices, Gordon even began recounting a meeting with Bill Tavoulareas, the president of Mobil Corporation—another known aggressor in the takeover game—and sharing with Petersen what he had learned about the virtues of royalty trusts. Petersen was upset not only to learn about Gordon's New York mission, but that he would feel free to discuss it in front of a cabdriver. "A lot of these cabdrivers are pretty smart," Petersen would later explain. Gordon, however, recalls talking only in "harmless generalities."

Now, in early January 1983 it was really hitting the fan.

Chauncey Medberry, an outside director who had served as chairman of Bank of America, excitedly informed Petersen of a rumor that Gordon was passing out a confidential study about royalty trusts, which Getty Oil had conducted, to Corby Robertson, the Houston oilman. Medberry had also heard that Robertson was lining up an $8 billion line of credit for a possible takeover of Getty Oil.

Petersen claims that when he confronted Gordon with the issue, Gordon not only confirmed that he had distributed an internal company report to Robinson, but that he intended to give him the follow-up report as well that the company was preparing.

This sharing of information with outsiders is not only idiotic, Petersen thought, *it's illegal!* He imagined being called before a federal grand jury to explain why he as chairman had done nothing to prevent a company director from leaking information that could potentially give the recipient an edge over other investors.

Petersen decided that the time for politely humoring Gordon had passed. *Gordon is not only bizarre,* Petersen now thought. *Gordon is dangerous.*

On January 12, 1983, Petersen and Gordon held a summit meeting at the Bonaventure Hotel with all their sidemen present. One objective was to get Gordon to discontinue his demand for studies of royalty trusts and other schemes to close the value gap. But as Petersen would recall, the main agenda item was inside information. "Our principal purpose at the meeting was to get Gordon from flopping around in the market-place."

Petersen indicated that he understood Gordon's desire for the trust to acquire control of Getty Oil. *Just be patient,* Petersen said. It had taken J. Paul Getty from the early 1930s to 1954 to consolidate his control of Tide Water. *Your father was a very patient man, Gordon.* As long as Gordon insisted on more studies of royalty trusts or other ways of closing the value gap, the company could never legally buy back shares from the public to help bring Gordon to a controlling position, he explained. That would be buying shares on the basis of inside, nonpublic information. If Gordon could see his way clear to drop the studies, then Getty Oil could slowly resume repurchasing its shares—something it had done from time to time in the past—and eventually Gordon would achieve the majority control he desired.

Gordon was unrelenting. The studies must go on.

Petersen's lawyer Bart Winokur tried to point out the dangers of dealing with outsiders—not just to Getty Oil, but to the trust. Winokur was a master at the use of metaphors, and had plenty at his command for this situation. Knowledge of turmoil within Getty Oil would spread "blood on the water," which as everyone knows draws sharks, Winokur explained. And if Corby Robertson or anyone else managed to snap up every share of Getty Oil *except* those held by the trust, then Gordon would be left as a locked-out minority shareholder, subject to the whims of a majority partner. "Somebody could get squeezed," Winokur told Gordon. "And you could be the juice."

Petersen's people then gingerly addressed themselves to the legality and ethics of Gordon's handling of inside information—and the mild-mannered trustee, a man who usually came to anger slowly if at all, blew up. Gordon had been patronized once too often.

He told Petersen that he was perfectly capable of deciding by himself what material could be shared with outsiders; he was a director of the company, and no one, other than the shareholders, outranked the directors. Petersen would later swear that Gordon came unhinged at the attack on his morality. *You are questioning my ethics?* he recalls Gordon demanding. *I feel I have a higher sense of ethics than anyone in this room.* When Gordon's lawyer Tim Cohler entered the discussion, Petersen sensed that he too was siding with the company. As Petersen would recall it, Gordon announced that he considered his ethical standards to be higher than those of his own lawyer!

Gordon would not recall making such sanctimonious statements, but he did consider the whole episode to be an affront to his integrity. "I thought it was insolent and ridiculous," he would say.

Petersen, for his part, was utterly floored. He couldn't believe his ears. He had gone into the meeting convinced that Gordon could be reasoned with over the insider-trading issue, that he would readily agree to discontinue talking outside the company once he recognized the potentially dangerous implications of doing so.

"If there was ever a turning point, that was it," Petersen would recall more than three years later. "Gordon had been testing his wings for a year. But I came away from that meeting convinced we couldn't just react defensively to every one of his wild initiatives." It was time to *do* something about Gordon Getty.

Petersen reached a further conclusion at the Bonaventure meeting. "It turned out to be a mistake to remind Mr. Getty of anything that his father did."

The guardian of the nation's greatest family fortune had been accused of leaking inside information, and even his own lawyers now recognized the need to call Gordon's attention to the risks of meeting with outsiders. Four days after the ill-fated meeting at the Bonaventure, Gordon's lawyer Moses Lasky had a letter hand-delivered to Gordon's home, noting that "for obvious reasons, I am not willing to entrust this letter to the mails."

If Ward Cleaver had ever lectured the Beaver about insider trading, he couldn't have done a better job.

One of the most sensitive and trickiest aspects of the law has to do with inside information and insider trading. . . . The business world, particularly the world of investors, is full of people seeking "tips" or even bits of information. . . . I think it is imprudent to have *any* communications with securities analysts *at all*. . . . It is imprudent to talk to investment bankers, unless they have been engaged. . . . It takes a great deal of foresight to thread one's way through these thickets.

Lasky had a further piece of advice: Keep Petersen's people fully abreast of any discussions you have with outsiders.

Despite getting taken into the woodshed, two weeks later Gordon graciously welcomed Kurt Wulff into the music room, where they had a lengthy discussion about the future of Getty Oil. Wulff found Gordon to be "quite sensitive" about discussing certain company information. But otherwise Wulff was taken aback at Gordon's familiarity with Boone Pickens, and at his acquaintance with such arcane topics as royalty trusts. "Anybody who knows about it has to like the idea," Wulff believed. *Gordon is not afraid to shake things up.*

On March 23, 1983, Wulff circulated another report.

> Any portion of Getty Oil converted to a royalty trust or limited partnership would be worth about twice as much in the stock market. . . . There is no J. Paul Getty today who combines a blood relationship to the beneficiaries of the trust with an intense personal involvement in the Getty Oil Company. . . . Management should see that voluntary changes are better than an involuntary takeover.

Seventeen years earlier, while working as a consultant for Getty Oil, Gordon went to the unusual length of soliciting advice from Loeb, Rhoades & Company about the possible sale of some company-owned gasoline stations. Getty Oil, however, had no retainer with Loeb, Rhoades. George Getty was startled to learn that his kid brother was discussing internal corporate matters with an outside investment banking firm. "Please nicely conclude your discussions with Loeb, Rhoades," George instructed his brother in a memo.

History repeated itself in February 1983, when Petersen was told that Gordon was considering the engagement of an investment-banking firm for an independent, in-depth evaluation of the value gap at Getty Oil. *Another harebrained scheme,* Petersen thought. The engagement could easily put Getty Oil "in play." And besides, Getty Oil would *never* divulge the kind of confidential inside information necessary to the successful completion of such a study to somebody else's investment banker—even one on retainer by the company's largest shareholder.

Gordon's group and Petersen's people came to an understanding: Both sides would submit a list of major investment-banking companies to conduct a *company*-sponsored study.

Goldman, Sachs & Company was on both lists. The choice seemed proper, for Goldman was one of Wall Street's oldest and most distinguished investment-banking partnerships. It had underwritten the stock offering that took Sears, Roebuck public in 1906 and had been the leader in short-term corporate financing ever since Marcus Goldman walked the streets of lower Manhattan carrying IOUs in his top hat. But Goldman, Sachs in modern times had come to distinguish itself as Wall Street's busiest take-over manufactory, representing either the buyer or the seller in more than half the billion-dollar mergers that had ever occurred. Above all, Goldman was known as the first choice among invest-ment bankers when the decision was made to put a company on the auction block; nobody could sell a company better than Gold-man.

Although Goldman was hired only to conduct a "valuation analysis" of Getty Oil, Sid Petersen knew that the engagement could still put Getty Oil "in play" for a hostile takeover. If word spread further than it already had that the company's largest share-holder was unhappy—that he did not seem to trust management's own studies—the sharks and barracudas of Wall Street were sure to smell blood and swim in from the open sea. They might try to make a direct attack on Getty Oil in its entirety or they might try cozying up to Gordon, just as apparently Sid Bass, Boone Pickens and Corby Robertson already had. In their nickel psychoanalysis of Gordon Getty, Petersen's people had become convinced that Gordon—the man with the terrible father—was vulnerable to the offers of friendship and respect held out by any strong man.

In its effort to maintain confidentiality, Goldman assigned a code name to its study of Getty Oil. The choice, Project Plutus, could not have been more apt. Plutus was the god of wealth, but was blind, signifying that he distributed riches indiscriminately. And although Plutus was depicted as a lame god, because he came slowly, he also had wings—because he fled with such speed.

Practically by the day, Sid Petersen was growing more concerned that Gordon's unstructured activities could expose the investing public—itself a minority block of stock—to the risk of an unfair takeover. But Petersen was also concerned about Petersen. In May 1983 the Getty Oil board outfitted him with a golden parachute—his first-ever employment contract with the company, which would perpetuate his annual salary of $460,000 plus bonus long after a change occurred in the control of Getty Oil.

At about that time, Boone Pickens called Petersen asking for a meeting.

Pickens had already presented Gordon with a book outlining a plan to merge his Mesa Petroleum with Getty Oil, companies that he code-named "blue company" and "gray company." Gordon, apparently wishing to avoid the appearance that he was making any side deals with outsiders, had politely urged Pickens to present his plan to Sid Petersen. Petersen agreed to the meeting, but only after Pickens consented to signing an agreement that he would not make a tender offer for Getty Oil.

When Pickens and two associates met with Petersen's people at the Century Plaza Hotel, Pickens was a more fearsome figure in the oil patch than ever. He had only recently thrust Cities Services into the arms of Occidental Petroleum, collecting more than $40 million in profits by selling his hostile investment in Cities Service stock. He had also purchased 2.5% of Superior Oil, shares which only a month earlier he had voted against management when Willametta Keck Day conducted a successful proxy fight that imposed certain limitations on the management discretion of her brother Howard; before long, Superior Oil would pay $31.8 million of greenmail to buy out Pickens.

And Pickens reveled in his reputation for ferocity. When he reached Petersen's suite with two aides, they noticed a pile of breakfast dishes. "You guys didn't order any breakfast for us?" he jokingly inquired.

The Getty people were startled—"nervous, uptight," Pickens would recall. "Oh, well! We'll get you some."

"That's all right," Pickens said. "We ate *hours* ago." When the "Blue-Gray" merger proposal was distributed around a conference table, Petersen asked the first question: "Does Gordon Getty have this book?"

"Yes," Pickens said.

"What is the date on Gordon's book?" Petersen asked stonily.

When Pickens informed him that Gordon had seen the proposal the prior month, he watched Petersen's jaw tighten and the color drain from his face. The company had spent days preparing for the meeting with Pickens by trying to anticipate the subject matter— and all the while one of the company's own directors had had a copy of the report!

The proposal was explained by a tax specialist from Baker & Botts in Houston—the law firm that had cut its teeth in the takeover game through its work for Hugh Liedtke's Pennzoil. Under the proposal, Mesa and Getty Oil would merge, with many of their assets eventually to be spun off in a royalty trust. Petersen's people were convinced the Internal Revenue Service would never approve the plan because it contained a tax dodge that one of them would later describe as "borderline, to put it mildly." The meeting ended inconclusively, with what Pickens took as a don't-call-us brush-off from the Getty people.

Pickens would later recall being impressed with the thorough knowledge that Petersen's aides had exhibited about complex, tax-related restructuring proposals. But he considered the resentment that Petersen appeared to exhibit toward Gordon, the company's largest shareholder, to be unbecoming and unwise.

"They won't be around very long," Pickens remarked to his compatriots as they walked through the hotel lobby. "This company is long gone."

The Getty Oil limousine that had dropped off Pickens and his people hadn't yet returned to the hotel. "Do we want to wait for it?" someone asked.

"Hell no," Pickens said. "Let's take a cab. They'll figure it out around noon tomorrow."

· 7 ·

Another split had erupted within Getty Oil. The cadre of lawyers who had spent years in the shadow of the mean-spirited Lansing Hays began to choose sides between Gordon and Petersen.

There was, for instance, Moses Lasky of San Francisco, the outside lawyer whom Hays had hired in the 1960s to help defeat Gordon in the dividend war with the Old Man—one of the lawyers whom the Old Man had instructed, "Keep killing my son." Lasky went on to become one of the most irascible and expensive lawyers on the West Coast, extracting a $1 million fee, for example, by writing a single legal brief for Telex Corporation in its ultimately unsuccessful antitrust suit against IBM. Now, Lasky's clients included the Oakland Raiders, the meanest team in professional football. When the mayor of Oakland took the witness stand to prevent the Raiders from moving to Los Angeles, the white-haired Lasky—by this time long past retirement age—left the mayor speechless. "Isn't it a fact," Lasky demanded, "that sports fans are fickle brutes?"

In the fight for control of Getty Oil, Lasky lined up with Gordon. "Now Lasky is Gordon's son of a bitch," one of Lasky's adversaries would remark.

A protégé of Lasky's, Charles "Tim" Cohler, an equally intense litigator with steel-blue eyes and curly black hair, also be-

came part of Gordon's group. Cohler himself had represented
Getty Oil management for years, forging a deep friendship not
only with Sid Petersen but with Dave Copley, a gentlemanly and
soft-spoken in-house lawyer. Cohler and Copley used to travel the
country together as litigators for Getty Oil. Once, Cohler and his
wife joined Copley, Petersen and their wives for a vacation in
Greece. When Hays died, Copley, after nearly twenty-five years
with the company, was finally promoted to general counsel; his
allegiances were now clearly cast with Petersen's people, just as
Cohler's were with Gordon. One of Cohler's friends from
Amherst and later Harvard Law School—a bookish and mus-
tached securities lawyer named Thomas Woodhouse—had joined
Lasky's firm and was now also a member of Gordon's group.

A young partner in Hays' Philadelphia-based law firm, Barton
Winokur, had also moved up in the world when Hays died, assuming command of the Getty Oil account for the firm. Winokur,
a workaholic who specialized in merger and other securities cases,
had graduated from Harvard Law School only a year ahead of
Cohler and Woodhouse, although they were not acquainted there.

Lasky, Cohler and Woodhouse for Gordon; Copley and
Winokur for Petersen: They were five outstanding lawyers, many
of them friends, but aligned on different sides—and ready to
prove themselves worthy as successors of the indominable Lansing Hays.

While Goldman, Sachs was undertaking Project Plutus, Petersen's
people launched a plan that Gordon might have been forgiven for
calling Project Brutus. Secretly, Petersen's people were plotting
to put Gordon on ice.

After Gordon's forays into the takeover community and his
holier-than-thou rejection of Petersen's lecture about inside information, Petersen had convinced himself that something had to be
done to neutralize his unilateral control over 40% of the company's stock. Members of management were persuaded that if
Gordon wasn't motivated by greed, then he was acting out of
some deep psychological need to prove himself to his late father.
They speculated that after being denied a career in his father's
company, Gordon was now obsessed with dealing it away or
taking control of it himself—somehow getting the last laugh on
the Old Man.

"Gordon didn't want to run the company; he didn't want to
work that hard," Petersen would later insist. "Gordon wanted

absolute power. His father had absolute power and everyone looked up to *him.*''

Board members shared the mistrust of Gordon's motives; Harold Stuart, for one, believed that becoming chairman of Getty Oil was Gordon's ''long-cherished desire.'' Even Harold Berg, the retired chairman, who remained a director and who had a personal liking for Gordon, thought Gordon had gotten carried away. ''He was all pie in the sky as far as I was concerned,'' Berg would say. Berg, Stuart and others—all men of a generation in which women did not take the interest in business that Gordon's wife did—also began to sense that Ann was stirring her husband's assertiveness, perhaps so that she could be ''introduced as the wife of the chairman of Getty Oil rather than the wife of a director,'' as Berg would later speculate.

The course of action set by Petersen's people would later be described in the Houston courtroom as a case of corporate betrayal and deceit of the most arrogant and high-handed form. The incumbents of Getty Oil, however, would defend it as an effort to protect the public and fulfill the wishes of J. Paul Getty. Whatever it was, Project Brutus would certainly go down as a stunning case of corporate miscalculation. If any action would be singly responsible for the destruction of Getty Oil and the incredible court case that would rise from the rubble, it would be this.

The legal documents governing the Sarah Getty Trust required the unanimity of any *two* trustees in the deployment of their 40% influence over the company; if the trustees were stalemated, the 40% might as well be zero. Now that Hays was gone, there was no one available to exert a positive influence on Gordon, Petersen thought. There was no one around to *stalemate* him.

J. Paul Getty had appointed Security Pacific Bank to serve alongside Gordon and Hays, but time and time again the bank had refused the appointment, fearful of the multibillion-dollar liabilities that could befall any trustees if the terrible and unforeseen occurred—a legal judgment, for instance.

Within barely a month of the January 1983 Bonaventure meeting over the inside-information controversy, Petersen's people had made contact with Bank of America—the nation's largest bank, which Gordon's big brother George had served as a director before his suicide and where Chauncey Medberry of the Getty Oil board had recently served as chairman. Before long, Bank of America agreed it would accept an appointment as Gordon's corporate co-trustee.

But bringing in a new bank presented its own complication. A court would have to approve Bank of America's appointment as co-trustee, and court approval was far more likely if someone with "standing"—such as a trust beneficiary—came forward requesting the bank's appointment.

A solution was soon found.

In 1949, California's title companies and bank-trust departments lobbied for passage of a law permitting court appointment of a guardian ad litem—meaning "for the purposes of the suit"—someone, usually a lawyer, who could represent the interests of unborn generations in land disputes and other inheritance matters. The law was later broadened to include guardians for young children, incapacitated adults and others. Petersen's people recognized that the courts wouldn't confer authority over one of the world's greatest family fortunes to just anyone. A distinguished member of the bar would have to "volunteer" to act as guardian for some member of the Getty family. And surely *some* member of the Getty family would be happy to participate in an effort to dilute the power of Gordon, the family kingpin.

Getty Oil's lawyers initially aroused some interest at the white-shoe Los Angeles law firm of Gibson, Dunn & Crutcher, the nation's fifth-largest law partnership. But the firm quickly got cold feet over the idea of jumping into the affairs of the Getty family, particularly as part of a secret plan being engineered by the management of Getty Oil. According to a participant in the discussions, Gibson, Dunn decided it would feel compromised or embarrassed if it came out that the nonfamily management of Getty Oil had recruited the firm to serve as a Getty *family* guardian. "People would allege a *conspiracy* and deviousness and so forth." Undeterred, Petersen's people took their search for a guardian to O'Melveny & Myers of Los Angeles, the nation's seventh-largest law firm. O'Melveny, however, also declined to have any of its partners serve as guardian for any member of the Getty family.

Then someone thought of Seth Hufstedler, a lawyer who enjoyed one of the most stellar reputations in the Los Angeles bar. Hufstedler expressed interest in receiving an engagement as a Getty family guardian, but he made it clear to Petersen's people that he would have absolutely no part in going out and soliciting a client—youngster or adult—to sue a relative. Petersen's people said they would take care of that; indeed, lawyers representing some of Gordon's relatives had already made contact with management, expressing concern about Gordon's activities as family trustee.

Petersen's people met twice with Hufstedler without receiving his commitment. Finally, Hufstedler told them that "If I was nominated by an appropriate family member, I was prepared to proceed."

After working through the entire spring and into the summer of 1983, Petersen's people had at last made progress in the effort to neutralize Gordon. The co-trustee, Bank of America, had already been lined up. So now was the guardian—Hufstedler—who would sue in the name of a Getty family member to secure the bank's appointment.

It was two down, one to go. All that Petersen's people needed was the family member.

Two days after his lawyers began setting up meetings with dissident relatives of Gordon, Sid Petersen convened the July 1983 meeting of the Getty Oil board of directors. The main item on the agenda was the long-awaited presentation by Goldman, Sachs on Project Plutus, more than four months in the making.

The report initially seemed made to measure for management. Any effort to increase the stock price should have "affirmative *long-term* impact," the investment bankers said, rather than merely the "one-shot impact" that Gordon seemed to be seeking. The company had obligations to its employees and the communities where it operated. Liquidating the company—turning it into a royalty trust or limited partnership—was impractical. *Very good so far.*

Then Goldman dropped its bombshell. If the company really wanted to elevate its stock price, the best alternative was to buy back a significant number of the company's own shares—something that could deliver majority control of Getty Oil to Gordon! Petersen moved quickly to head off immediate implementation. He explained that his people needed time to regroup; the company had to study the study.

Finally, someone solicited Gordon's view.

Despite his delight at Goldman's report, Gordon was now holding his cards closer than ever. After his consultations on Wall Street and his meetings with the sharks, Gordon had formulated an altogether new plan for closing the value gap while putting him in control, and Gordon did not consider the timing ripe to disclose his plan to his fellow directors.

What do you think, Mr. Getty?

"What I really want," some directors recall Gordon replying, "is to find the optimum way to optimize value."

The directors looked at each other silently for a moment. The first to speak was the newly recruited director John Teets of Greyhound—a man who could bench-press 350 pounds and who climbed nineteen flights to his office every day.

"Gordon," said Teets, "you may know what you just said, but nobody else in the room does."

Gordon's secret plan involved rejoining the two blocks of stock once held by his father—the trust's, and what was now the museum's—to reimpose unassailable Getty-family control over Getty Oil.

The J. Paul Getty Museum board of trustees—chaired by Harold Berg, the Old Man's successor at Getty Oil—had all but panicked when J. Paul's will was opened and they discovered they were in charge of a billion-dollar charitable endowment. Upon catching their breath they began an exhaustive search for a president, someone whose job description would emphasize investment management far above the curatorial aspects of the position.

Then a fortuitous event occurred: Jimmy Carter was defeated by Ronald Reagan, meaning that Harold Williams, Carter's chairman of the Securities and Exchange Commission, would soon be in the job market. Williams was a Harvard lawyer who had once specialized in tax issues, of which the museum faced many as a tax-exempt institution. A gregarious and jowly man with a thick mane of white hair, Williams had reached a top position in Hunt Foods in the late 1960s. But when the company was merged with Canada Dry Corporation and McCall Corporation to form Norton Simon Incorporated, Williams lost out in a management shuffle involving another executive, and left to become dean of the Graduate School of Management at UCLA. It was from there that he was recruited to head the SEC in Washington.

With careers in industry, academia and government, Williams' qualifications to serve as museum president seemed unsurpassable. Williams happily accepted the position, entirely unaware that he would soon be thrust into a chaotic contest for corporate control unlike any he had ever seen in his job as the nation's highest-ranking corporate policeman. Williams controlled the wild card at Getty Oil—the shares that could decide whether Gordon controlled a majority or a minority of the stock. Williams had become the kingmaker.

Shortly before the Getty Oil board had pondered Goldman's Project Plutus report, Gordon had conceived of a variation on the popular "leveraged buyout" scheme, or LBO, that would enable the Sarah Getty Trust and the Getty Museum to own *all* of Getty Oil for themselves without posing undue risk to either: They could pledge the 48% of the company not owned by them as collateral to borrow enough money to buy the same 48% held by the public. Gordon dubbed this concept an "LBO with a fence." Before long, Goldman got a look at the proposal and judged it to be impractical. Williams, although intrigued by the plan, politely declined to participate.

But Williams recognized that he had not heard the last from Gordon Getty. Williams did not relish dirtying his hands in a messy battle for control, but he liked Gordon and considered him bright—if a bit out of his element in the world of high finance. And neither could Williams ignore the arithmetic: By itself, the museum with its 12% stock interest was a minority shareholder, subject to having almost its entire hoard of assets controlled by others, perpetually exposed to the risk that the remaining 88% would fall into unfriendly hands. But the presence of Gordon's 40% gave the museum a much louder voice in the affairs of Getty Oil.

Williams would keep an open mind. It wouldn't hurt if he and Gordon flirted a bit.

Petersen would look with fear and wonderment on the willingness of a man like Williams to listen to Gordon, to treat him seriously. "Harold Williams," Petersen would become convinced, "had a desire to be the power behind the throne."

Although Gordon's LBO with a fence concept had died, it had, in effect, branded him as a raider. Whatever community of interests had existed between himself and the other directors representing the public shareholders had all but disappeared. Thus, Petersen's people expanded their number by bringing in someone else from Goldman, Sachs—someone who knew how to defend against takeovers.

As a child, Geoffrey Boisi's father would enliven the dinner table conversation with tales of big-time dealmaking in the New York real estate market. The excitement and intrigue of the negotiating process captured the imagination of the young Boisi, who in not too many years would achieve partnership status as one of the top dealmakers at Goldman, Sachs. In 1980 Boisi (pronounced BWA-zee), at age thirty-three, would become the head of the firm's takeover group, helping to make it the industry's most prolific merger-and-acquisitions department.

When Boone Pickens attacked a big independent oil company called General American Oil, Boisi engineered the company's rescue by Phillips Petroleum. When Pickens made a run at Cities Service, Boisi ushered in Occidental Petroleum as the white knight. After Mobil made a hostile bid for Marathon Oil, U.S. Steel called Boisi at home on a Saturday and had his wife, Rene, track him down on a jogging course; the next day Boisi had a team of takeover specialists at U.S. Steel headquarters, and before long Big Steel was into Big Oil. Basketball-tall, with overbroad shoulders, enormous green eyes and boyish good looks, Boisi would describe himself as "a naturally aggressive, intense personality." On another occasion he would enumerate the other attributes necessary to success in the takeover-counseling business: "We're a psychiatrist, a father confessor, a coach, as well as a financial architect."

When Boisi entered the power struggle at Getty Oil, his psychiatrist persona initially played an unduly prominent role. He recognized the difficulty—perhaps the impossibility—of ever defeating an alliance between Gordon and the museum if one were forged. So, like Scarlett O'Hara plotting to upset the engagement announcement of Melanie Hamilton to Ashley Wilkes, Boisi moved to see that the marriage never happened. He moved to plant mistrust, warning Gordon's group that the museum could sell its shares to a raider, who could also buy up all the public shares and leave Gordon out in the cold as a "frozen-out minority shareholder." The raider and the museum could then do whatever they liked to Gordon—including forcibly buying out his shares at whatever unfairly low price they chose.

"Maybe Harold Williams already has a shark in his hip pocket," Boisi at one point whispered to a member of Gordon's group.

There was possibly a way to take care of Williams, Boisi said. As one of Gordon's advisors would recall the proposal, Getty Oil could approach Williams on a Friday and inform him that on the following Monday the company intended to announce an offer to buy back some of Getty Oil's shares. *Sell us all your shares now for cash*, the museum would be told, *or take your chances that the deal won't be as good on Monday*. To Gordon's group this became known as the "three-day museum squeeze."

But Gordon's group wasn't about to choose sides between the

company and the museum without first settling some business of its own.

Gordon remained as committed as ever to re-establishing absolute family control over the ownership of Getty Oil. The trust, however, could not directly buy its way into that position because it was legally barred from borrowing money. The only realistic alternative was to have the company buy back many of its own shares from the public—to reduce the total number of shares outstanding and thereby shrink the size of the pie—so that Gordon's slice would become larger. And indeed, the company's own investment bankers—Boisi's own firm, Goldman—had recommended buying back shares as the most strategic way to increase the value of the shares that would remain in investors' hands.

Gordon's bargaining power was the implicit threat that he could grab control of the company by another method: combining with the museum or with a raider like Pickens. Petersen's people began jumping through hoops trying to reach an accommodation—but only to a point. Placating Gordon's wish for majority *ownership* was one thing. Using the company's money to buy back shares and put Gordon into majority *control* was something else altogether. Getty Oil, in short, would give Gordon his cake without letting him eat it.

Throughout the summer Petersen's people met regularly with Gordon's group to try to reach a compromise. But in their efforts to plan for every possible contingency, the legion of Harvard lawyers quickly became bogged down in minutiae. Petersen's people agreed to have the company buy back shares, leaving Gordon with more than 50% of the stock, but only if Gordon agreed to "safeguards" limiting his voting power to *less than* 50%. It wasn't the trust that management feared, but Gordon as trustee; under one proposal, the *trust* could have majority control but only if the *trustee* consented to the appointment of a counterweight—two additional trustees, each with one-half of a vote to Gordon's full vote.

Gordon was offended at the suggestion of a need for "safeguards." "The trustee is not interested in accepting any handcuffs," one of his representatives said at one point.

Gordon's group offered an arcane proposal under which a "supermajority" of the board—80%—would have to approve certain actions, but a supermajority of *shareholders*—chiefly Gordon—could approve others. Petersen's people heaped even more complexity into the discussions with a plan involving staggering directors' terms on the board and installing a sliding scale of voting power in each group of directors. ''Bart,'' said Gordon's lawyer Moses Lasky to Petersen's lawyer Bart Winokur, ''if we took your proposal and put it in *The Wall Street Journal,* you'd be the laughingstock of the nation.''

Lasky, fresh from the Oakland Raider battles, made a direct approach to Petersen. ''Sid, why don't you and I sit down and talk this thing through?'' he asked. Petersen was always willing to meet with Gordon, principal to principal, and to have *his* lawyers meet with *Gordon's* lawyers. But in so contentious a situation, a principal-to-lawyer meeting was ruled out.

''So, Sid, it has come to that,'' Lasky replied sadly.

The point was, these were no longer friends dealing as friends.

In the course of the ''handcuffs-safeguards'' negotiations, Getty Oil even offered Gordon the inducement of a title as chairman of the board. Petersen, whose magnanimity had its limits, would remain chief executive officer and would take the title of president away from his number-two executive, Bob Miller, whom he had passed over five years earlier. But Getty Oil was mistaken if it thought Gordon solely wanted a big-league title; time and again, Gordon would reject plans to make him chairman if they did not also accord him all the *rights* of a chairman who happened to own the majority of the company.

Communication breakdowns flourished. At one point Petersen's people breathlessly rushed from Los Angeles to San Francisco expecting to hear a breakthrough proposal; it turned out to be more of the same. On Yom Kippur, Lasky, a Jew, flew off to Philadelphia to meet with Winokur, also a Jew, expecting that one of Gordon's proposals would be accepted; instead, Winokur put forward an altogether new proposal.

And through it all, Gordon was losing his patience.

He turned to Bill Newsom, the friend who functioned as his lawyer during the dividend wars with the Old Man. Newsom, now a California appeals court judge, could easily tune in to Gordon's wavelength; the judge belonged to Friends of the Sea Otter, sat on the Leakey Foundation board with Gordon and would accompany him to the opera, occasionally stopping afterward for a few bottles of medium-priced Italian wine at a restaurant called Harry's. Newsom urged Gordon to abdicate the whole mess. *Just walk away from it, let someone else run the trust and sit on the company board,* Newsom said. Petersen will ''cut your throat,'' he warned. Gordon, however, refused to abdicate.

* * *

By the late summer of 1983, Harold Williams of the museum recognized that things were getting serious. He continued receiving overtures from Gordon and he wasn't exactly hitting it off with Sid Petersen—he'd occasionally call him for a lunch date but never had the message returned. For all its value, that 12% block of stock he managed was a powder keg. Unless the museum kept clear of the crossfire between Gordon's group and Petersen's people, it could explode.

Williams knew exactly where to turn for advice. Although Wall Street had hundreds of takeover lawyers, there were really only two at the top of anyone's list when the stakes reached into the billion-dollar range. One was Joseph Flom of Skadden, Arps, Slate, Meagher & Flom, the nation's fourth-largest law firm, which often took the lead in representing corporate raiders. The other was Martin Lipton of Wachtell, Lipton, Rosen & Katz, a much smaller firm that tended to specialize in defending companies *against* raiders. Lipton also happened to be a close friend of Williams'. In the end, few characters would have so great a bearing on the outcome of the battle for Getty Oil, nor on the result of the trial it led to.

Despite his Coke-bottle glasses and his middle-age paunch, the fifty-two-year-old Lipton projected a rather hip image, and had served with *Rolling Stone* publisher Jann Wenner as the top New York fund-raiser for the antinuke presidential campaign of Senator Alan Cranston. But it was aggression in the world of business law rather than pacifism in the world of politics on which Lipton had made his reputation. With merger mania sweeping Wall Street in the early 1980s, Lipton became so well known in New York that a small, unrelated advertising agency called Martin/Lipton had to alter its name; Martin Lipton, the lawyer, kept getting calls intended for Martin/Lipton, the ad agency. "Martin Lipton: The Big Deal Lawyer," *M* magazine would crown him.

Embattled managements-on-the-defensive were so eager for Lipton's counsel that they gladly paid fees starting at $350 an hour per lawyer, and sometimes three times that amount when the action really got hot and heavy. Lipton's Park Avenue firm could be counted on to pull out all the stops where the situation required, to "turn the firm upside down on a full-time basis," as Lipton would once explain. Although with about eighty lawyers the firm ranked only seventy-fifth in size in the U.S., its estimated

$880,000 in annual profits per partner (in 1986) ranked it number one, according to a survey by *The American Lawyer*.

Part of the incentive to paying such rich fees was gaining access to the brain of Marty Lipton, for if anyone could find the loophole through which to escape a takeover it was he. Lipton had concocted the "poison pill"—a hastily issued class of stock that in one variation permitted shareholders of a target company to buy stock in a raider's company at a discount. *Fortune* at one point would call the poison pills "a dastardly device that makes hostile takeovers deathly expensive," so much so that by 1986, well over one hundred companies would have poison pills in reserve to keep raiders at bay.

Lipton had been particularly busy defending corporate America in the weeks and months prior to his call from the museum's Harold Williams. He helped defend General American from Liedtke's friend Pickens long enough to permit Geoff Boisi's client, Phillips Petroleum, to come in as a white knight. He helped prevent the acquisition of Texas Gas Corporation by Houston oilman Oscar Wyatt. He put a poison pill in the medicine cabinet of Superior Oil. In addition, he helped General Felt attempt to take control of Sotheby's, the art-auction house, where Ann Getty would soon join the board of directors. He represented Diamond Shamrock in its merger with Natomas, an energy company where Tom Woodhouse, now a member of Gordon's group, had recently worked as a lawyer. Small world, this takeover community.

Lipton's engagement by Williams was desultory to begin with, purely to act as a sounding board for his old friend. As Lipton saw it, a classic battle for control had erupted: It was not unusual, he thought, that the son of the company founder wanted control, nor that the professional managers wanted to prevent him from attaining it. But within weeks Marty Lipton, the veteran of more than one thousand takeover deals, would watch this textbook case turn into something that he could only describe as "bizarre."

What was supposed to be the Getty Oil board's annual sun-and-fun meeting in Pebble Beach, in September 1983, instead became the fire-and-brimstone meeting.

Petersen informed the board that among other things the company was trying to placate Gordon with the title of chairman—chairman of this very board. *What?* Gordon as *chairman?* One longtime director who had been an old friend of the Old Man said he would resign from the board the day that Gordon became chairman. And the idea of having the company actually *help* Gordon attain majority control? To several directors, the notion was pure folly.

One of the more recent additions to the board, Henry Wendt, did not share in the institutional memories of the Absentminded Professor, and thus wasn't quite so consumed with emotion. But Wendt—the chairman of the $3-billion-a-year SmithKline Beckman Corporation, which made Contac, Tagamet and Allergan products—also had not had time to build up any emotional loyalty to the company. Putting the public shareholders at the risk of control by Gordon was an unacceptable alternative. *Let's just sell the whole thing,* Wendt said. *Yes,* some directors said. *And let's do it while we still have control of this company.*

Petersen was shaken. "There was clearly a sense that I hadn't been tough enough on Gordon," he said. Nevertheless, he prevailed on the directors to permit him one more attempt at a negotiated solution.

Neither Harold Berg, the retired company chairman who now headed the museum board of trustees, nor Gordon, of course, was present for the discussion. Because matters involving both the trust and the museum were at issue, they had voluntarily excused themselves. As one of Petersen's people would explain, "Gordon was no different than Boone Pickens at that point."

Getting up and leaving the meeting was a practice that Gordon would quickly wish he had never begun.

Gordon refused to believe the secondhand accounts he heard of the outrage directed at him in the boardroom in Pebble Beach. So he decided to find out for himself what had really happened, to "persuade them that I didn't have horns, and a tail"—and to begin recruiting support away from management. "Let's face it," Gordon says today. "I was also trying to go over Sid's head, get a little power play going."

One of Gordon's lawyers would dub it "The Medberry Affair." Gordon had called his father's old friend Chauncey Medberry of the Bank of America, who agreed to meet him for lunch at the Beverly Wilshire. Gordon thought he could count on Medberry for some straight talk, and perhaps for some support; although Medberry did not know it, Gordon had once saved his seat on the board when a personal dispute had erupted between Hays and the banker. (Gordon had refused to go along with Hays' wish

to vote the trust's proxy against Medberry's re-election to the
Getty Oil board.)

Gordon was correct about the straight-talk part, but not about
the support.

After making some unflattering comparisons between Gordon
and Walter Mitty, Medberry hit close to home. *You are not your
father,* he said. Gordon could not entirely disagree: "That was
true—as far as being CEO of an oil company goes."

But Medberry cut even closer to the quick. *I will not support
anything that you propose,* he declared.

That's it, Gordon decided. He stuck Medberry with the
check—and began planning to do worse.

"Instead of coming away crying, he came away thinking," one
of Gordon's lawyers would recall. And what Gordon had thought
up would have been unthinkable to almost anyone else. It was
brilliant—and it bordered on reckless. Gordon would present the
museum with an incredible inducement—and together they would
throw out the entire Getty Oil board of directors. "I thought
something should be done about them," Gordon would later say
of the directors, "before they did something to me."

Under Delaware corporation law, shareholders representing the
majority of a company's stock have the right to impose new
bylaws over their company simply by filing a Form of Consent of
Majority Shareholders with the Delaware secretary of state. The
mere act of filing such a statement gave the majority the right to
impose its will in whatever manner it chose.

Gordon had his lawyers draw up a form that simply read:

The undersigned, holders of an aggregate of 41,125,428 shares
constituting 51.9705% of the outstanding common stock of
Getty Oil Company, a Delaware corporation (the "Company")
hereby consent pursuant to Sections 141(k) and 228 of the Delaware Corporation Law, to, and take, the following action:

1. Each member of the Board of Directors of the Company other
   than Gordon P. Getty is hereby removed.

2. The following new directors are hereby elected:
                    Harold Williams
                    Moses Lasky
                    _____
                    _____

In the blank spaces, additional designees of Gordon and the museum would be named.

To enlist the museum's cooperation in so daring an act, Gordon conceived an ingenious if unusual proposal. The museum could name its price for selling its interest in Getty Oil back to the new Getty Oil, with Gordon in charge. As a check against enabling the museum to demand too high a price, Gordon would have the option of offering to *sell* Getty Oil to the museum at the *same* price. Such buy-sell toss-ups aren't all that unusual, although this was perhaps the first occasion in history where one was proposed in a transaction potentially worth $10 billion. The whole thing was vaguely reminiscent of the consulting recommendation that Gordon had made to his big brother George nearly twenty years earlier. "In view of your basic conclusions on selling service stations," his brother had responded, "why do you want to buy service stations?"

Most amazing of all, Gordon's lawyers estimated that the proposal could, potentially, bring the museum a windfall of a half-billion dollars by enabling it to buy Getty Oil and promptly resell it. A half-billion-dollar inducement: a small price to pay for control of Getty Oil. Gordon would present "the museum proposal" at the October 1983 meeting of the museum board, scheduled for the next week in London.

When Petersen's people were told that Gordon was preparing a plan that "the museum could not refuse," they immediately recognized the gravity of the words. *Gordon is going to throw us out,* they thought, leaving the public shareholders at the risk of his whimsy.

Three years later, after lawyers, jurors, judges and journalists had picked over the events leading to the downfall of Getty Oil, a great mystery would remain. Why was Gordon Getty never notified in time to attend the special Getty Oil board meeting on Sunday, October 2, 1983, in Philadelphia? It must have been Western Union at fault, Petersen's people would say. Unaware of the "special" meeting called on one day's notice, Gordon had already boarded a plane to London. But his presence was academic anyway, for the directors held a rump meeting in Wendt's office at SmithKline-Beckman before regrouping at the official meeting place at Getty Oil's Philadelphia law firm—Dechert, Price & Rhoads, where Bart Winokur worked.

The directors were overcome with fear and loathing. "They were convinced that Gordon would destroy the company," recalls Herbert Galant, a noted Wall Street takeover specialist who had recently been hired as special counsel to Getty Oil. *Certainly, Gordon must want Sid Petersen's job; isn't that what all this is about?* the directors wondered. And by this point, the directors were convinced that Harold Williams of the museum just might go along with whatever outlandish scheme Gordon might be planning. Williams actually seemed to take Gordon seriously! Williams seemed coy; his intentions were unclear. Some people thought that after losing the power play at Norton Simon thirteen years earlier, perhaps *Williams* was angling for Petersen's job.

Suddenly Sid Petersen had the most popular job around.

Under Plan A adopted at the emergency board meeting in Philadelphia, Gordon would become chairman of Getty Oil and the trust would become the majority shareholder. And although some "handcuffs" would remain on Gordon's voting control, these would be unlocked within five years, possibly sooner.

If Gordon turned his back on the proposal for five years of peace, the company would fall back to Plan B—its antidote to any alliance between Gordon and the museum. It would issue more shares—out of thin air, if necessary—and dilute the company's entire base of stockholdings, so that Gordon and Harold Williams' combined interest would be reduced below 50%. It would be an extraordinary action; for one thing it was considered unlawful to use "corporate machinery" solely to entrench management, and for another, issuing more issues would dilute not only Gordon but the museum and every other shareholder, rendering each individual share worth so much less.

Immediately after the meeting, Petersen and four aides flew to London.

It was on October 4, 1983, at Claridge's Hotel in London, that Gordon Getty's lawyers lost control of their client. They warned him that on the one hand, the "museum proposal"—his plan to pitch out the board—would be viewed by the world as a coup, as a sign of irreversible upheaval within Getty Oil. By contrast, they said, the compromise proposed by Petersen's people was the "best solution."

But Gordon would not approve any proposal that would perpetuate the power of the board. Gordon told his lawyers the directors were a "bunch of snakes" who would "loot" the company if given the chance, according to a memorandum that Gordon's lawyer Lasky wrote. For nearly two hours the lawyers implored him to back off his plan to deep-six the directors. "In every possible way short of using the word 'mistake,'" Lasky would recall, "we told Gordon that his decision was a mistake." Gordon says today that Lasky must have misunderstood him—that even if he used the words he never really thought that the directors were snakes or looters. But Gordon does not dispute that he was more resolute than ever. *Present the museum proposal!* he commanded his lawyers.

Gordon, alas, quickly learned that he had completely miscalculated the museum's reaction to the "proposal the museum could not refuse." Harold Williams and his lawyer Marty Lipton were outraged at the very suggestion that the former chairman of the SEC would conspire to oust the directors of a huge, publicly held corporation. "It would have invited suit after suit," Lipton says. "If you had dreamed up a law school exam question on how much trouble you could have in a corporate transaction, this was it."

As they departed from the museum's suite at Claridge's, Gordon's group passed Petersen's people in the hallway.

The museum learned that it wasn't only Gordon who had arrived in London with a bold scheme. Getty Oil—Lipton and Williams discovered—was ready to fall back on Plan B, the plan to open the floodgates with newly issued company shares, diluting Gordon and the museum to a position below 50%. Once Getty Oil had neutralized Gordon and the museum, management would begin trying to sell the company outright, for its investment banker, Geoff Boisi, was chomping at the bit to undertake the search for a white knight to rescue the company from Gordon.

Lipton and Williams were astonished at the intensity level of the turmoil, that the civil war had escalated into a dangerous game of brinksmanship. "The situation was so bizarre; it needed time, and cool heads," Lipton recalls.

It was thus that the concept of a standstill agreement was born. Lipton grabbed a Claridge's note pad and began outlining the elements of a truce; under it, *nobody* would do *anything* to *anybody* for eighteen months. Petersen's people agreed to sign— if Gordon agreed to sign within the hour. Gordon, however, refused, repeating a maxim he had once heard from his father: "Whenever someone gives an ultimatum that something has to be done immediately, the thing to do is refuse to act immediately." *I'll think about it,* Gordon said.

Members of all three groups—Gordon's, Petersen's, and the Museum, then left London with frayed nerves and deeper suspicions than ever about the motives of the other two. But at the behest of the museum's Marty Lipton, the two war camps were at least thinking about consenting to a truce.

Petersen's people, however, had not forgotten Project Brutus—the plan to neutralize Gordon through a lawsuit filed by some other family member. Thus, before leaving London, they attended to another piece of business while riding in a limousine to Heathrow Airport.

They rode with Vanni Treves, the London solicitor who represented Gordon's full brother, J. Paul Getty Jr. Over the years, Paul Jr., a collector of antiquarian books, had maintained a certain affection for Gordon's aesthetic streak—but when it came to family financial affairs Paul Jr.'s attitude toward his brother bordered on contempt. "He admires his talents," Treves would say, "but he has long been of the view that Gordon is *not* a businessman, that Gordon's view of important business matters is very often distorted or myopic." As Paul Jr. saw it, Gordon suffered from a "very, very deep streak of stubbornness. And of course," Treves says, "when that quality is allied with Gordon's other characteristics—which include intelligence—he is extremely self-righteous."

At the time, however, Paul Jr. had a few problems of his own. Still anguishing over the death a decade earlier of his second wife Talitha Pol, Paul Jr. remained in the grip of narcotics dependency. Treves himself was concerned enough about representing a man of such affliction that he got one noted physician after another to swear that Paul Jr. had full possession of his faculties. "I didn't want to be personally charged with the possibility that when Paul signed documents or gave me instructions, he didn't fully understand what he might be doing," Treves would later say.

Petersen's people found an especially receptive ear when Treves joined them in the limo ride to Heathrow. Treves was *very* interested in their report on what had occurred in London the preceding two days. He said he would discuss the matter with Paul Jr. and talk to them again soon.

"It was never my intention to take anything away from the trust," Petersen would later insist. Tattling on Gordon to other beneficiaries was part of a strategy "to bring some stability and business experience to the trust"—and, Petersen would insist, to fulfill the wishes of J. Paul Getty, who had designated a bank to serve alongside his son at the helm of the family trust.

"Was it evil to go to these people and say, 'Here is what this nitwit is doing?'"

Three can play at this game, Gordon's group decided. Petersen's people had hired not only Boisi of Goldman, Sachs, but now the noted takeover lawyer Herb Galant. Everyone but Gordon had takeover counsel, an "M&A specialist," as in "mergers and acquisitions."

En route from London to San Francisco, Gordon's lawyers Tom Woodhouse and Tim Cohler stopped in New York, where they had arranged an appointment to see Martin Siegel of Kidder, Peabody & Company, a strikingly handsome, dark-haired investment banker of thirty-five years who happened to be a close friend of Marty Lipton's—"Marty & Marty," "the two Martys" they were called. Siegel had assisted Lipton in the popularization of the poison pill takeover defense. Only a few months earlier they had worked side by side around the clock in the circuslike takeover drama between Bendix and Martin Marietta, a battle that involved the "Pac Man" takeover defense, in which two warring companies buy out a piece of the other. In recognition of Siegel's "meritorious achievement in eating the dots," Lipton had even awarded him his "lifetime membership in the Royal Order of the Golden Joystick," with a "Pac-Maniac Certificate" encased in Lucite that Siegel proudly displayed on his credenza.

When Gordon's lawyers showed up for their appointment with Siegel, they were shown instead to a Kidder, Peabody partner, Peter Goodson, who explained that Mr. Siegel had become tied up with a major deal. "How can I help you?" Goodson asked.

Woodhouse explained that they represented Gordon Getty, trustee over 40% of Getty Oil Company.

Goodson was stunned.

"Will you excuse me a moment?"

Goodson sprinted down the hall to Siegel's office.

"You'll never guess who these guys represent," Goodson said. "Gordon Getty!" Siegel accepted the engagement.

Getty Oil wasn't entirely unfamiliar to Siegel. Several years earlier, when he originated the concept of selling takeover defenses as products, he had tried to land a sale with Sid Petersen. Siegel was also the investment banker who three years earlier had arranged the sale of ERC, the insurance company, to Getty Oil. Siegel had not forgotten Petersen's puzzlement over Gordon's refusal to vote in favor of the ERC deal. Now, Siegel was representing the selfsame Gordon.

Although they were representing different parties—the trust and the museum—the personal acquaintance of the two Martys quickly proved important. Within two weeks of the action in London, Lipton's back-of-the-envelope truce had become bogged down in legal complexity in the hands of Getty Oil's lawyers. Lipton's loosely written agreement reminded some of Petersen's people of SALT II—an agreement that either superpower could walk away from when it chose. But the two Martys had become convinced that Petersen's people were so gripped with paranoia that they were incapable of drafting the kind of plainly worded, layman's-language document necessary to "sell" Gordon.

Flying together on Pan Am from New York to San Francisco on October 19, 1983, they resurrected and modified the standstill agreement that Lipton had scrawled on a hotel note pad in London. Under this deal, Harold Williams of the museum would join the board, along with four additional members designated by Gordon. Other than that, everything would remain frozen for a year; nobody could do anything to anybody.

Petersen's people were summoned from Los Angeles to the San Francisco offices of Lasky, Haas, Cohler & Munter, where Gordon's group maintained its command post. They were led into a small conference room decorated with a pencil drawing of horrified-looking, zombielike figures. Lipton entered with the proposal he had written by hand on the airplane. "This is as far as I think Gordon will go," Lipton announced, "and I believe that it is 'take it or leave it.'"

Take it or leave it. So this was peace with honor.

One overriding question troubled the Getty Oil group when Lipton thrust the truce onto the table: whether the standstill would prevent the company from carrying on its campaign to find a relative willing to sue Gordon, to have him neutralized as trustee. Indeed, the company's only incentive to sign the standstill was that it bought time to bring Project Brutus to fruition.

Another of the issues left unanswered after the self-destruction of Getty Oil would be just how much Marty Lipton knew about the company's plan. Petersen would insist that Lipton was fully aware. "Sign this and you've got time to neutralize Gordon," Petersen would recall Lipton assuring the group. Lipton, however, would insist that all he ever heard was "a general statement in London that they were seeking to neutralize Gordon's power." All he recalls stating to Petersen's people in the take-it-or-leave-it discussion of the standstill was: "Sign this and it will give the company time to work things out."

This much was certain: Neither of the warring parties—Gordon nor Petersen—was overjoyed to sign Lipton's truce.

Shortly before Gordon affixed his signature to the one-year agreement, he and his advisors ate a carry-out Chinese dinner. Later, after Gordon finally had relented to signing the document, his gut began to churn. Marty Siegel watched Gordon's face turn shades of green.

Gordon quickly excused himself from the room.

"To this day," Gordon would say, "I don't know whether it was my apprehension over the standstill, or the food."

The nearly twenty thousand employees of Getty Oil company were going about their business unaware of the tension gripping the executive suite. Thus it came as a surprise when the company's investor-relations and media-relations officials got some unusual orders from upper management: *Go through the files— newspaper clippings, Wall Street analysts reports, anything—and pick out material showing Gordon Getty's lack of business understanding.* Jack Leone, the media-relations man, would recall that at that moment, "I knew there was real trouble in paradise."

When the standstill was publicized a few days later, the rest of the world also sensed there was trouble—and in any company, trouble at the top often means takeover. During the fourth quarter of 1983, Bankers Trust picked up roughly $45 million worth of stock; Harvard College, the New York State Teachers Retirement Fund and the national YMCA each about $4 million. Takeover speculators began snatching up shares; Ivan Boesky, the most noted such professional on the Street, amassed a huge interest in Getty Oil shares even *before* the standstill had been announced. He would later attribute his stunning foresight to the psychological research he had undertaken into Gordon's motivations: "I thought the son wanted very badly to get back at the father." As it turned out, according to the U.S. government, Boesky also had the benefit of a little inside information about the goings-on in the Getty group—information that he obtained in exchange for suitcases full of cash.

There was an informant in the midst of the battle for Getty Oil. For more than three years, until long after Getty Oil was felled, no one would know.

The company's stock price, which Gordon had made a barometer of management's performance, was now a barometer of corporate tension; in one week it soared $5.87 a share to $75.37. Said *The Wall Street Journal*'s "Heard on the Street" column:

Standstill agreements between corporate managements and big shareholders usually cool takeover speculation in the company's stock. Not so with Getty Oil, which recently signed such a pact with its largest holders and has seen its stock soar.

The reason? Merger speculators and others who have carefully read the pact believe that it isn't really a "standstill" at all.

The pop psychoanalysis also quickly spread from one water cooler to another at Getty Oil headquarters on Wilshire Boulevard. *Gordon is causing all the trouble. . . . He was raised without a father image, his brothers didn't love him, he was reared in a household where he always got his way. . . . Have you heard? Harold Williams wants Petersen's job! . . . Marty Lipton is going to deal this company away for sure!*

One morning someone in the headquarters building inquired, "Well, who's taking us over today?"

"Xerox!" some wag piped up.

That afternoon someone from the Bakersfield office called headquarters. "Hey, have you heard?" he breathlessly demanded. "Xerox is taking us over!"

· 8 ·

In boom-town Houston, the skyscrapers erupted from the ground in only a decade's time, giving downtown the appearance that it was built overnight. Houston became the most wholly modern major city in America, full of chrome, reflective glass, granite and marble. Each building might have leaped from the pages of Ayn Rand's *The Fountainhead*.

None of these buildings surpassed the Pennzoil Towers.

Designed by Philip Johnson and John Burgee and completed in 1975, about the time that Pennzoil had disgorged United Gas Pipeline, the buildings are solid black thirty-six-story trapezoidal boxes lying ten feet apart at odd angles to one another. At the twenty-ninth floor, the outer wall of each building is cut away to form inward-leaning diagonal planes that face the sky. The buildings are *trompe l'oeil*, looking square from some vantages and radically angular from others, appearing to have flat tops from one street corner and sloped tops from another.

In the peak of the North Tower, the Darth Vader of Houston plotted the takeover of Getty Oil.

Images of heaviness abounded. Six-feet-one, with the thrusting chest and midriff of a professional wrestler, Hugh Liedtke sat at the head of a sixteen-foot conference table made of six-inch planks from an old San Francisco shipwreck. His brawny forearms rested in grooves worn into the hardwood from years of heavy leaning. The floor was travertine marble, the same material used to pave the streets of Rome. Behind him, in an illuminated case, rested a collection of basketball-sized geodes and mineral specimens on shelves made of Austin limestone, a porous rock noted for its oil-bearing properties. However, Liedtke was often anxious to appear less ostentatious than his surroundings. "All they did was squirt the dirt off," he told visitors who marveled at the beauty of the rock collection.

He could not help leaving any visitor with an impression, and it was one that could either enrapture or intimidate. He had bovine eyes and a tubalike voice, his Oklahoma accent softened by an Eastern education. He projected a certain humility—an aw-shucks kind of charm—that bore the manner of a John Wayne character. Once, during an interview, his recollection of boyhood family memories sent a tear careening down his cheek.

But when the situation required, Liedtke could quickly shed his modest, Oklahoma-bred style. His own associates would describe how he could look them dead in the eye, his voice booming, imprinting his will on their nervous systems.

As a dealmaker, Liedtke had been uncharacteristically silent for some time; in late 1983, Kurt Wulff, Gordon's contact at Donaldson, Lufkin & Jenrette, was even preparing a research report for investors remarking on Liedtke's notable absence from the big deals sweeping the oil patch. The headline would be: IS PENNZOIL OVER THE HILL?

Fortunately for Wulff, Liedtke would not leave him time to publish the report. Liedtke might take a long time cultivating his opportunities, but when they ripened he was always quick to harvest them. And Getty Oil was almost ready to drop from the vine.

Liedtke did not look at Getty Oil by himself. Ever since his brother Bill had left Pennzoil in 1977 to run POGO Producing, Liedtke's chief confidant and ever-present Number Two was Baine P. Kerr, a former partner in the huge Houston law firm of Baker & Botts. Kerr had counseled Liedtke and George Bush in their Midland days and later through several of Pennzoil's most complex deals, before leaving the law firm to become Pennzoil's general counsel and eventual president. Kerr was a classic "Mr. Inside"—a green-eyeshade type whose caution and thoroughness complemented the daring of his boss. "Baine and Hugh are really a perfect team," says Liedtke's cousin Hugh Roff. "Baine is very careful, whereas Hugh is imaginative and bold."

From afar, Liedtke and Kerr probed Getty's defenses, searching for signs of weakness. They learned of Gordon's interest in roy-

alty trusts and share-buyback programs—and of how management
had balked at them. Kerr investigated the company's directors,
making note of their scant stock holdings in Getty Oil and of some
interlocking directorships: Two Getty directors also sat on the
board of Rockwell International Corporation. It struck Kerr that
Sid Petersen had a board filled with friends and loyalists—a fact
that would make a hostile approach against the company all the
more difficult to complete.

Liedtke studied several takeover prospects in the summer and
fall of 1983. Superior Oil was on the block, but Liedtke was put
off by its significant presence in heavily regulated Canada. He
looked at Gulf Oil, where his old friend Boone Pickens was threatening a takeover, but Gulf was probably too costly a target even
for someone as practiced as Liedtke at stretching his financial
resources. True, Getty also was a gigantic company by Pennzoil's
standards—at $12 billion a year, six times larger in fact. But
Liedtke had overcome nearly the same size difference almost two
decades earlier in his minnow-swallows-whale acquisition of
United Gas.

And Liedtke, who knew as much about Getty Oil as anyone
outside Getty itself, had known for years that hidden inside the
company was a half-slumbering giant.

It was the Kern River field—the nation's fourth-largest oil
field, nearly a century old but still very much alive, with a strong
heart. Kern River was a "huff and puff" oil field, one whose
heavy, sludgelike oil required a battery of gigantic steam engines
to pump it to the surface. Liedtke had remained obsessed with the
technology of enhanced recovery, which he had applied so successfully with George Bush back in the Midland days and had
improved upon in the oil fields owned by South Penn. Only a
small increase in the recovery rate at Kern River, Liedtke reckoned, could dramatically increase the amount of oil-in-the-ground
carried on Getty Oil's books—a bonus of as much as seven hundred million barrels, he estimated. Controlling Kern River would
be like getting seven hundred million barrels of oil without really
even paying for it!

Liedtke also thought back to his conversations with one of his
old Getty Oil friends, Earle Gray, a senior geologist. Gray used to
tell him about the "rock oil" fields owned by the company near
McKittrick, California. Special technology was required to separate the oil from the rock, but Liedtke knew that once the technology was developed, McKittrick too would become an
extraordinary oil field. "Earle Gray has told me that when he was

a young man on a hot day he would go down there and the oil would be running over the highway," Liedtke would say.

Getty Oil was a once-in-a-lifetime opportunity—just the kind of company whose acquisition could cap Liedtke's brilliant career narrowly within his retirement. "I hoped for years to be part of building a substantial, first-class oil company," he would say, and Getty Oil lay available for the fulfillment of his dream.

Liedtke could not move hastily, however. The power struggle at Getty Oil was difficult to comprehend from the outside, like watching three cats wrestling under a blanket. But Liedtke and Kerr knew that once they understood it more fully, perhaps an affordable means of entry could be discerned. They would spend October and November continuing to watch the situation closely, and to make whatever preparations they could.

Although the newly signed standstill prevented them from establishing a shareholders' alliance, Gordon and Harold Williams were furthering their friendship. Shortly before Halloween they jointly accepted an award that the Los Angeles Cultural Commission was posthumously presenting to J. Paul Getty. The black-tie evening was a welcome distraction from the fight over the Old Man's fortune. Another old man—Dr. Armand Hammer, the chairman of Occidental Petroleum—regaled the audience with tales of J. Paul. Gordon mingled with the likes of Peter and Kirk Douglas, Mr. and Mrs. Cary Grant, and Mr. and Mrs. Irving Stone. The guests thrilled at the special, one-time-only exhibition of drawings by Rembrandt, Raphael, Rubens, Cézanne and Goya.

But the restful moments never seemed to last for long. *Forbes* heaped on a new load of unwanted publicity, ranking Gordon, with $2.2 billion, as the nation's wealthiest man—even though virtually all of that money ultimately belonged to Gordon's relatives. "He has spent most of the day in the music room," Gordon's secretary told an inquiring reporter. "He is not happy at all about the article."

Gordon's mother, Ann, also considered the *Forbes* ranking unfair. "Why should they zero in on poor old Gordo?" she asked. "After all, his brother has the same inheritance he does."

Another disturbing thing was happening: Gordon was beginning to receive strange phone calls from that brother. Out of the blue, it seemed, Paul Jr. was calling from London and insisting that Gordon give up his sole control over the family trust; in one conversation Paul Jr. even burst into tears. Gordon didn't know

quite what to think. "My brother at that time was a drug addict,"
he would say.

In October a typewritten letter arrived in San Francisco:

Dear Gordon,
   Your refusal to agree to the appointment of an additional—
and of course neutral—Trustee of the 1934 Trust is unfortunate,
provocative and, in my view, wrong. I hope you will change
your mind very soon, because I'm saddened by the prospect of
another round of public dissension within the family. . . . There
should be no misunderstanding. I don't want to threaten you, or
even to appear to; but I'm afraid that litigation will be inevitable
if you don't quickly agree to another trustee and I'm sad to think
that I too would be sucked into it.

Love,
Paul

By what seemed a remarkable coincidence, another letter came
from twenty-nine-year-old Claire Getty, one of the three "beautiful
little daughters" of the late George Getty, as the Old Man called
them. Like Gordon, they were income beneficiaries of the trust, but
they were emotional beneficiaries as well. It was their father who
gave up his life to Getty Oil—and yet who had never attained the
position of unilateral control in which their Uncle Gordon now
found himself. Claire wrote with a beautiful cursive hand and a
flourishing "y" that looped all the way under the word "Getty."

Dear Uncle Gordon,
   I am writing to discuss some concerns I have regarding my
interest in the Sarah Getty Trust. In this connection I have consulted with my lawyer regarding my rights and expecta-
tions. . . . It seems clear that grandfather intended for there to be
three trustees—one a family member, one a business person,
and one a financial institution. I do not mean to be critical of
you, but I think you would admit that your business management
background is limited. . . . There are a number of provisions of
the trust that are important to me. For example, the trust clearly
prohibits the sale of Getty stock. . . . I have heard suggestions
that you are trying to acquire control over the board and over the
business of the company. If this is true, it would seem to me that
this is a conflict-of-interest problem. . . .

Best regards,
Claire

About the time that Claire was writing her letter, Gordon was replying to his brother's. Having signed the standstill agreement in San Francisco, Gordon wrote:

> Dear Paul,
>   . . . The relationship between the Trust and the Company is now amicable, perhaps more so than at any time since Father's death. Management and I are cooperating to the common good of everyone. I hope you will not try to upset this amity.
>
>   Love,
>   Gordon

Getty Oil was about to explode.

I wonder what Pennzoil has cooking? Herb Galant wondered.

In the midst of his retainer as Getty Oil's "special counsel" in the battle with Gordon, Galant received a call from Moulton Goodrum, a friend who worked at Baker & Botts in Houston. Goodrum wanted to set up an interview so Pennzoil could consider hiring Galant's firm for a possible takeover that was under study. Conveniently, Galant was scheduled to attend a meeting of the Getty Oil board in Houston; he would attend the meeting at Pennzoil on November 10, the day before the Getty Oil board was scheduled to meet.

Naturally, Pennzoil withdrew Galant from consideration when he mentioned that he was doing some work for Getty Oil at the moment. Galant, however, remained unaware that Pennzoil's possible target was the very company whose board meeting he had flown to Houston to attend.

Exactly two years later, the Getty Oil board meeting of Friday, November 11, 1983, would be reconstructed in vivid detail for a dozen workaday Houstonians chosen for the jury in a case called *Pennzoil v. Texaco.* They would marvel at what they heard, for it was at this meeting that Project Brutus came to fruition.

A faint trace of South Texas autumn was in the air—a temperature in the fifties and a merciful humidity reading of about 50%—as a lineup of Yellow cabs dropped off the Getty Oil directors at the company's newly completed research center on the West Side of Houston at 9 A.M. As they poured out of the cabs, the distinguished businessmen who belonged to the Getty Oil board were more committed than ever in their support of Petersen—and more upset than ever with Gordon Getty. Henry Wendt of SmithKline Beckman, for one, was all but apoplectic after hearing about Gordon's abortive effort in London to recruit the museum to a plan to oust the Getty Oil board. Gordon had "pulled the trigger of a gun pointed at the head of the directors," Wendt would tell the Houston jury two years later. "The gun wasn't loaded," he would add, "but nevertheless he pulled the trigger."

Some of Petersen's people were a few minutes late for the meeting; Herb Galant had a headache, and held up their cab at the hotel to return to his room for some Excedrin. But no drugstore analgesic could cure the legal headache that Getty Oil was about to inflict on itself.

The main item on the agenda was the board's formal ratification of the standstill agreement written by museum lawyer Marty Lipton and signed in San Francisco a month earlier. Because Gordon, as the family trustee, and Harold Berg, as chairman of the museum board, were also parties to the agreement, they excused themselves from the meeting, as they had done at earlier meetings whenever matters involving either institution came up for discussion.

While seated with Gordon in an adjacent office decorated in Getty Oil's trademark orange, the retired chairman used the opportunity to probe the son of his old boss.

"Gordon, what are you trying to do?" Berg asked. "Do you want to take over the company and run it?"

"No, I just want to be contacted on major decisions."

"Management matters? Policy? Or what?"

"I just want to be consulted on major matters."

"What you're talking about, Gordon, is veto power," Berg said.

"Well, I guess," Gordon replied. "The last thing I want is to be in at eight o'clock in the morning and stay till five. *I just want to be consulted.*"

As Berg and Gordon talked, three of Petersen's people—Geoff Boisi, the investment banker; lawyer Bart Winokur; and Herb Galant, the special counsel with the Excedrin headache—finally straggled into the meeting through the main entrance. From the office where Gordon and Berg were having their tête-à-tête, neither man could observe the arrival of the advisors. Once the board had decided to ratify the standstill, Berg was called back into the meeting. Gordon, however, was not.

Petersen and his advisors announced they had been in touch

with lawyers representing various of Gordon's relatives. The daughters of George Getty—a man who once sat on the board with several of the same directors serving now—wished for the appointment of a co-trustee. But this could be accomplished only through a lawsuit, and none of George's girls, as Petersen's people called them, had yet come forward to sue their Uncle Gordon.

But Petersen's people had also been meeting with Vanni Treves, the solicitor of Paul Getty Jr., Gordon's hermit brother in London. *Would you like to consider what we have to say about Gordon?* Treves had been asked. *Would you like to act on it?* The response had been affirmative: Paul Jr. had agreed to have a suit filed against Gordon. Only the previous day, in fact, Paul Jr. had signed a statement consenting to the lawsuit during a meeting with his solicitor, Treves.

This suit, however, would not be filed in Paul Jr.'s own name, but in that of his estranged, fifteen-year-old son, Tara Gabriel Galaxy Gramaphone Getty.

Meanwhile, Gordon had begun stirring in the suite of offices outside the boardroom, wandering through the executive-reception area and studying the paintings of Rocky Mountain scenes. The secretaries on the floor were thrilled at his presence—Gordon Getty, in the flesh!—and felt obliged to offer him some magazines from the building library. "I thought it was a little strange that he spent so much time *out* of the board meeting," Pat Williams, one of the secretaries, says today.

Back in the boardroom, the directors were told that the distinguished Los Angeles lawyer Seth Hufstedler would file the suit as Tara's guardian ad litem, seeking appointment of the Bank of America as Gordon's co-trustee. (One of the company's directors, Chauncey Medberry, was the retired chairman and remained a director of the bank; he, however, was not excused from the discussion.)

Getty Oil itself would somehow have to come clean about its role in instigating this family dispute, Petersen's people said. Hufstedler had insisted on the company's support in Tara's lawsuit against Gordon, and in any case, the board was told, it would be improper to continue to hide the company's involvement in the affairs of the Getty family. Getty Oil could file an affidavit admitting to its role in the legal action. Or the company could simply intervene in Hufstedler's petition—essentially filing its own suit against Gordon.

Berg could not believe his ears. "Here you've just signed a

standstill agreement and now you come up with this brainstorm! I think it's terrible!'' *It is legally proper,* Petersen's people insisted. "I don't care if it's legally proper!" Berg boomed. "It doesn't seem ethical." But most of the directors supported the move. By consensus, rather than by a formal vote, they agreed that Getty Oil too would file suit—against its own board member, the manager of 40% of its stock—seeking the appointment of a co-trustee.

Petersen's people—Galant, Winokur and Boisi, who had entered the room in Gordon's absence—now departed, and Gordon was readmitted, unaware of the discussion that had occurred in his absence.

Gordon had been bamboozled.

It had taken Petersen's people nearly a year to put the plan in motion. A troubled man, identified in court filings as a heroin addict, secretly consents to the appointment of a distinguished attorney as guardian for a son he has barely seen in twelve years— and who has never asked to get involved. The guardian files a suit in the name of the man's son, demanding appointment of the nation's largest bank as a co-trustee to stalemate management of the family jewels.

When the final preparations were complete, two suits would be filed in the Superior Court for Los Angeles County. Tara's suit would claim that Gordon, as an income beneficiary, had taken actions "which enhance income at the expense of principal . . . at the expense of company reserves," namely, proposing to liquidate the company through a royalty trust. Gordon had even once *sued his own father* to extract more cash from the trust assets, Tara's suit would say. Old Man Getty never intended for Gordon to serve alone, it would add, pointing out that Bank of America "is willing to fill the vacancy."

Getty Oil's companion suit, being prepared in Los Angeles, would cite the company's vital interest in the matter. A corporate co-trustee was needed for the Getty family trust, the company would argue, to impose "a cautious and conservative attitude, in order to promote growth and to avoid turmoil and upset in the business affairs of Getty Oil."

As the lawyers undertook their final preparations, the soul-searching was heavy on two continents. "We realized that what we were doing was something of the greatest seriousness," Paul Jr.'s London lawyer, Treves, would say, "something that could have deep and wounding repercussions for many people." And at Getty Oil headquarters in Los Angeles, a distraught look overcame Dave Copley, the retiring general counsel of the company.

"There's no turning back now," he remarked to another executive.

On Thursday, November 10, the day before the Getty Oil board had gathered in Houston, twenty-three-year-old Mark Getty visited his father in London and had a disturbing conversation. Mark, an Oxford graduate, was one of four children from J. Paul Getty Jr.'s first marriage, but he had also grown close to his Uncle Gordon. In fact it was Gordon who helped defray the medical expenses of Mark's brother J. Paul Getty III, a kidnapping victim who eventually experienced severe brain damage from a lethal combination of barbiturates; Paul Jr. had balked at paying those expenses himself.

Mark was troubled to learn that his father seemed to be taking sides in the civil war within Getty Oil—and that he had chosen Sid Petersen's side. Worse still, Mark's father said something about having a lawsuit filed against Gordon, partly at the behest of Getty Oil. Mark was gravely concerned at the prospect of another round of terrible litigation in the family.

On Saturday, the day after the Getty Oil board had met in Houston, Mark and his wife enjoyed the welcome distraction of a visit to Mark's half brother, Tara, at Tara's boarding school in Dorset County, England. Although the children of different marriages, Mark and Tara had become close over the years. Mark's mother, Gail Getty, had taken in Tara after Tara's mother—Paul's second wife, the beautiful Talitha Pol—had died of a heroin overdose. Mark and Tara had lived together as full brothers until 1973, when Tara moved away to live with Talitha's parents near St. Tropez.

During their visit together on Saturday, Mark and Tara toured the school, ate lunch and attended a rugby match. Tara was one member of the Getty family who seemed completely removed from the tension of the moment; Mark learned that Tara had not even seen their father for six months.

But Mark could still see the storm clouds of litigation gathering over the Gettys. So on Monday he paid a visit to Vanni Treves, his father's solicitor, expressing concern over his father's possible involvement in a suit. Treves did not want to lie to Mark, but neither did he want to blow the cover of the suits against Gordon to a young man who, however well intentioned, might "speak out of turn." There was no telling what action Gordon might take against Getty Oil if he knew what was about to hit. So Treves decided he could only be "economical with the truth," as he would later put it.

"Mark," he finally said, "you must realize that I hate litigation, everybody on this floor hates litigation and everybody in this building hates litigation. We do everything possible to avoid litigation."

Mark later learned that the California lawsuit was filed in Tara's name that very day.

Hugh Liedtke harbored a certain vague mistrust of the New York-based professional advisors in the takeover game. Moreover, Liedtke didn't really need investment bankers and takeover lawyers to conceive his dealmaking strategy; he did that as creatively as any takeover professional ever could. In addition, he liked to do business through connections; unlike many corporate chairmen, Liedtke would never dream of turning over leadership in a critical deal to a complete stranger from Wall Street.

But Liedtke also knew that it would be foolhardy to undertake a major corporate transaction without hiring professional advisors—even if only for a nominal role. Takeover advisors have minute knowledge of the latest twists and turns in the continually shifting specialty of securities law. Also, you needed a takeover advisor of your own to talk to the takeover advisors on the *other* side of the deal; all these guys knew each other and dealt most comfortably with members of their own professional fraternity.

For years Liedtke had relied on investment bankers with whom he had a personal connection. For the takeover of United Gas Pipeline and for the creation of POGO Producing, he had hired his roommate from Amherst, Robert Green of Merrill Lynch. But now, with Getty Oil in play, Liedtke would rely on his longtime friend James Glanville of Lazard Frères & Company, whose personal background made him somewhat of an oddity among New York investment bankers.

Glanville was a Texan, raised in Dallas and educated in chemical engineering at Rice University in Houston. Eventually he became a petroleum engineer for Texaco when it was still the Texas Company. But Glanville was an investment banker in oilman's clothing. He conducted valuation studies of oil-in-the-ground and other economic analyses. He had a glib personality, and he greatly enjoyed participating in major corporate transactions. In 1959 he left the oil patch for Wall Street.

Now, at age sixty, Glanville was uncharacteristically old for a

profession with so high a burnout rate. But he was a survivor. Five
years earlier, at a time when defections from major investment
banking houses remained a rarity, Glanville caused a stir on Wall
Street by leading an exodus from Lehman Brothers Kuhn Loeb,
where he had reviled at the diversification of the firm beyond its
traditional investment-banking roots. About the time of his resignation, questions had also reportedly arisen about Glanville's possible personal financial interest in a property deal involving a
client. But thanks to the size of his reputation as a dealmaker,
Glanville had had no trouble landing another major-league position on Wall Street. Lazard Frères was so enthusiastic to recruit
Glanville that some partners offered to reduce their shares of the
firm's earnings to make him the most attractive offer possible.

Glanville and Liedtke had been friends for twenty years, establishing their acquaintance, coincidentally enough, through a mutual friend who had served on the boards of Tide Water Oil and
Getty Oil. Besides working at Liedtke's side on a number of
Pennzoil's deals, Glanville had been fishing with Liedtke as far
away as Panama and was a frequent overnight guest at Liedtke's
home in Houston. "We have always had a very warm, effective
working relationship," Glanville says. Their wives had also become friends over the years.

Glanville, however, had absolutely no pretensions about who
was the boss. Glanville regularly referred to his friend as "Chairman Mao"—or sometimes "The Chairman" for short. Liedtke,
unmistakably, was the one giving the orders whenever it came
time to deal.

Glanville would not be Liedtke's only lieutenant on the Getty
Oil deal. In all megamergers, the investment bankers also hire
takeover *lawyers*, who do not work directly for the client but
whose fees nevertheless are paid by the client. After Liedtke
brought Glanville into his analysis of the Getty situation, Glanville, in turn, engaged Arthur Liman of the firm Paul, Weiss,
Rifkind, Wharton & Garrison. Liman was a garrulous lawyer who
had the rare combination of experience of working in corporate
law as well as trial law. He could alternately apply a velvet glove
or an iron fist to a takeover situation, and he was squeamish about
nothing. Lying on his credenza was a beatup hacksaw, a memento
of his service as general counsel to the commission that investigated the inmate uprising at Attica Prison.

Liedtke rounded out his group of lieutenants with several lawyers from the Houston law firm of Baker & Botts, a megafirm
whose 250 lawyers ranked it among the nation's twenty-five larg-

est—and whose annual fees per lawyer—more than $300,000—ranked it in the top ten. For twenty-five years the internal legal department at Pennzoil had been almost indistinguishable from Baker & Botts, which had supplied Pennzoil not only with Baine Kerr but also with Perry Barber, who had succeeded Kerr as general counsel when Kerr became president. Its involvement in so many of Liedtke's creative deals had helped to build a great repository of takeover-related expertise at the firm. At the moment, as some partners began working with Liedtke on his analysis of Getty Oil, others were busily helping Boone Pickens plot the takeover of Gulf Oil. It was an article of pride among the lawyers at Baker & Botts that in any takeover situation they could compete on an equal footing with Wall Street's best. Soon they would be put to their biggest challenge ever.

While entrusting his lieutenants with the spadework, Liedtke took some time in November for personal pursuits. He and his wife, Betty Lyn, flew to New York on a Pennzoil jet with ex-governor John Connally and his wife, Nellie, who by this time had forgiven Liedtke for the raucous law-school parties he and his brother had held when they'd lived with the Connallys on the Lyndon Johnson estate. A few days later he was hunting on the ranch of Texas oil heir Pat Rutherford. He and his longtime friend Joe Jamail, one of the nation's most successful and colorful personal-injury lawyers, attended the Dads' Day football game at the University of Texas, where they both had children in school. Liedtke also spent two days hunting in the Laurel Highlands of western Pennsylvania with his friend David Roderick, the chairman of U.S. Steel Corporation.

But throughout the month, Liedtke remained in almost daily contact with his takeover lieutenants, monitoring the progress of the corporate reconnaissance he had set into motion. His advisors assembled a set of Getty Oil's recent disclosure statements with the SEC. They searched through an electronic data base of media coverage for everything they could find on Getty Oil. They dispatched a paralegal to the California courts to obtain copies of the documents governing the Sarah Getty Trust. They studied the feasibility of creating a royalty trust out of Getty Oil's reserves.

Liedtke closely examined each new scrap of information as if he were turning over pieces of a jigsaw puzzle one by one. A picture was beginning to emerge in his mind—the picture of a takeover deal that was complex and risky, but a deal that could go down in history as one of the most ingenious of its kind.

*   *   *

In November, while Getty Oil plotted against Gordon and while Liedtke eyeballed them both, the rest of the oil industry drew its attention to a man-made gravel island in the icy waters off the coast of Deadhorse, Alaska. They called it Mukluk, and nobody was watching more closely than Texaco.

Now more desperate than ever for oil-in-the-ground, Texaco was one of several companies that had paid a total of $1.6 billion just for the rights to explore for oil in the area. It was history's most expensive exploratory-drilling venture, and if the wildcat came in as a producer, another $15 billion would be necessary to bring the field into commercial development.

But every dime would be well spent, for the geologists who explored the area estimated that the underground "structure" of the Mukluk field might easily hold five billion barrels of oil. Mukluk was no panacea for Texaco; the company held only 10.7% of the well. But even at that Texaco could duplicate many times over what it had discovered with Pennzoil a year earlier off the California coast. Many companies, such as Pennzoil, had decided to sit this one out, worrying that the huge geological structure contained nothing more than water—or at best natural gas, which would be uneconomical to transport from such a great distance. But executives at some oil companies began speaking of Mukluk as one of those epoch-making oil fields, every bit as significant to the industry as the Spindletop discovery that gave birth to Texaco eighty years earlier.

Back at Texaco headquarters, just after the drill bit had broken ground at Mukluk, John McKinley was being accorded the highest honor any corporate chieftain can receive. On November 14 it was announced that the Texaco board had asked McKinley to postpone his retirement beyond his sixty-fifth birthday in March 1985. There were several reasons for the request. It remained a toss-up whether Alfred DeCrane, Texaco's president, or James Kinnear, its vice chairman, should succeed McKinley at the top. Moreover, McKinley had undertaken Draconian restructuring actions in his three years at the helm—drastic, certainly, by Texaco standards. He had mothballed high-cost refineries and gasoline stations, recognizing the futility of maintaining a distribution network that had become far larger than the company could ever keep profitably primed with oil. He'd overhauled Texaco's hiring, salary and bonus policies, feverishly attempting to infuse fresh blood and entrepreneurial spirit into the organization. McKinley had slowly begun to "turn the ship," an analogy he used frequently to describe the painstaking process of overcoming decades of what many generously described as corporate inertia.

McKinley was receiving recognition from other quarters as well. He was inducted into the Alabama Academy of Honor and was accorded other distinctions. He had gone nearly as far as a businessman could go from a grease lab in Port Arthur, Texas.

If Mukluk came in as everyone hoped, he would go a whole lot further.

Mnemosyne, mother of the muses, was also the goddess of memory. But she was fickle, and besides bestowing memory she could take it away or confuse it. Mnemosyne would make many appearances throughout this drama.

For instance, Petersen's people would swear that they had given Gordon's group and the museum every reason to believe they were investigating ways to neutralize Gordon's sole say-so over 40% of the company's shares. Company officials would distinctly recall discussing the matter with the museum's Marty Lipton in London as well as in San Francisco on the evening the standstill agreement was signed. In addition, Petersen would recall telephoning Gordon to inform him that the company was joining in Tara's lawsuit, although not until just before the company's action was filed.

Giving the situation the benefit of the doubt, an extraordinary communication breakdown had occurred.

In mid-November, Gordon's investment banker, Marty Siegel, was attending a banquet at the Sheraton Center in New York when he learned from one of Petersen's New York-based people of the action taken in Gordon's absence during the Getty Oil board meeting in Houston. He immediately ran to a pay phone to call his client.

"Did you ever leave the meeting?" Siegel asked.

Gordon didn't think so.

"Not even to the bathroom?"

Gordon thought for a second and recalled being excused for the board's hour-long discussion of the standstill agreement. "Well, come to think of it, I was," Gordon said.

Siegel spilled the beans about the action taken in his absence. "He felt totally betrayed," Siegel says. "He was shocked."

Marty Lipton, too, would claim that the action came as a complete surprise. Lipton, after all, had made himself the mediator in the war between Gordon and Petersen—the architect of the truce signed less than a month earlier in San Francisco.

"You snookered me!" Lipton said in a phone conversation with Herb Galant and Geoff Boisi of Getty Oil.

"Marty, you shouldn't feel snookered," Galant replied casually. "If there's anyone who should feel upset, it's Gordon."

"Upset," could not begin to describe the reaction of Gordon's lawyers to the company's involvement in the Tara suit.

In a response filed with the Superior Court in Los Angeles, they called it "a drastic instance of cynical corporate maneuvering . . . a plot of manipulation and deception of unprecedented dimension and boldness. . . ." The court filing even stooped to suggest that a narcotics fix was withheld from J. Paul Jr. until he consented to have the suit against his brother filed. "Whether he was under the influence of drugs, or under the influence of a *need* for drugs . . . all such matters must be fully determined," Gordon's lawyers wrote.

For his part, Gordon would insist that by this time he was beyond feeling resentment. Instead, he would claim, he was struck simply by the *stupidity* of Petersen's people. "I felt the company had lost its mind," he would say. "It was comic book. I thought, 'What kind of childishness is this?'"

Snookered . . . lost its mind . . . a need for drugs—all on top of the company's accusation that its 40% shareholder had been tilting at windmills. In late 1983, Getty Oil seemed more like an asylum than one of the nation's largest corporations.

On December 2, Gordon Getty was in New York fulfilling a longtime dream: attending the first public performance, at Alice Tully Hall, of his composition "The White Election," a cycle of thirty-two songs based on the poetry of Emily Dickinson. Then, on Sunday December 5, he read the review by Bernard Holland of *The New York Times:* "Mr. Getty's settings are terse to the point of emptiness. There is, in other words, great deference to the text, but hardly any music at all. What there is is expressed in popular 19th Century American harmonies, colored by occasional Schubertian devices and interrupted with an occasional dissonant chord or two."

To make things worse, the vocalist was flat, the review said.

The next day Gordon was in Marty Lipton's office, engaged in a far more historic performance.

In the history of Delaware-incorporated companies—which is to say in the history of corporate America—no two shareholders, in anyone's memory, had ever joined together to alter the bylaws of a company in a manner that suited them. But in their outrage over the company's involvement in the Tara suit, the Sarah Getty Trust and the J. Paul Getty Museum did just that, jamming new bylaws down the throats of the entire Getty board and all of Petersen's people by the simple stroke of a pen. Lipton recognized it as an "extreme action." Gordon himself was delighted at the formal statement of unity against the company—and at the emasculation of the board. "We could no longer rely on them to behave rationally," he later explained.

The trust and museum had the power, if they chose, to throw out the entire board, as Gordon had attempted two months earlier in London. They could have greedily bought out the remaining shareholders at whatever price they chose and kept the company for themselves. They could have named Gordon as chairman, Williams as president and Sid Petersen as custodian of waste pits. But this was the former chairman of the Securities and Exchange Commission signing the documents here. There was no use in getting carried away.

Under the new bylaws, fourteen of the sixteen members of a newly expanded board would have to approve any major corporate action. Gordon had the power to appoint four additional board members—the same right he had received under the standstill agreement reached less than two months earlier in the office of his lawyer Moses Lasky. Williams, too, would join the board. The new "supermajority" voting requirement, in effect, gave Gordon, Williams and any *one* additional director veto power over the entire board.

As for the company's lawsuit against Gordon, the new bylaws specifically required its withdrawal. And for good measure, in what was surely one of the most bizarre corporate bylaws ever, a new Section 9 was added to Article II, stating that the board couldn't discuss any issue after any board member had been excused from a meeting, "without notifying such director of each and every matter to be discussed while he is out of the room."

Gordon and the museum issued a lengthy joint statement describing their reasons for the action by consent. Lipton, whose writing talents were considerable, felt little restraint from using the most colorful language possible.

Item: The trustee's personal absence from the November 11 board meeting was procured by fraudulent means. . . . These means included (1) inducing the trustee to leave the meeting while the proposed intervention (in the family litigation) was discussed without telling him that it was to be discussed and (2)

using the "back-door" maneuver of secretly introducing into the
room, through another entrance, counsel to the corporation to
advise on the proposed intervention. The result was a "rump
meeting" of the board, of a kind unquestionably prohibited by
the standstill agreement.

Item: In a subsequent telephone conversation, counsel to the
corporation admitted in so many words that the purpose of these
maneuvers was to "snooker" the trustee.

In fact, there was no "back-door" maneuver, literally or figur-
atively. Petersen's people had simply arrived late, and entered
through the front door. Gordon had left through a different exit,
one leading to the executive offices, which was the only place to
sit down. Petersen's people had then left through the same door
used by Gordon. The *discussion* occurring in Gordon's absence
certainly *was* questionable. But Lipton and Gordon's lawyers had
no idea that their slightly exaggerated use of the two little words
back door would set into concrete a revisionist view of what really
happened, and that through the extraordinary chain of legal events
to come, the words would help heap billions of dollars in liability—on Texaco.

The rift between Gordon and Petersen was now beyond repair,
and for Getty Oil the timing could not have been worse. Public
disclosure of the consent action would make it plain to the lords of
the takeover game and everyone in the oil industry that Getty Oil
was irrefutably "in play." And a fresh new takeover candidate
was just what the oil industry needed. A week before Christmas
Day in 1983, *The Wall Street Journal*'s widely read "Heard on
the Street" column noted: "Acquisition specialists are busier than
they've been in two years, mapping out takeover strategies and
Byzantine defenses. . . . The takeover game is definitely taking
off again. . . . People are dusting off their studies because they
think the oil business is bottoming out."

The writer was unaware that the day he wrote the column, Hugh
Liedtke of Pennzoil had convened a special meeting of his board
of directors. The three-way battle for control of Getty Oil was set
at last to become a four-way contest.

· 9 ·

A hostile takeover is little different from a military invasion. Wars are governed by the Geneva Convention, takeovers by the Securities and Exchange Commission; in each case aggressors often look for discreet ways to sidestep the rules without actually flouting them. Aggressors seek to catch their targets with their defenses down: In 1982 the chairman of Brunswick, the bowling-ball company, was watching the Super Bowl when his company came under attack, just as the Israelis were celebrating the High Holidays when the Arabs attacked in 1973. Knowing the whereabouts of the enemy leader is also essential: Oscar Wyatt of Houston's Coastal Corporation hired detectives to stalk the chairman of Texas Gas Corporation before launching a takeover bid. When the enemy leadership refuses to surrender, propaganda is aimed at the citizenry: Boone Pickens was now taking his battle for Gulf Oil directly to shareholders, recruiting their backing in a pitched proxy fight with Gulf's management.

Getty Oil was a divided opponent—and it was this fact that made possible an attack by a guerrilla army such as Pennzoil. By casting his lot with one or more competing factions, Liedtke could establish a provisional government—a ruling junta—with power far in excess of his own strength. At the same time, however, Liedtke knew from his boyhood experience re-enacting battles with tin soldiers that risk was present in attacking *any* enemy of greater numbers. Liedtke could engage Petersen's people in a battle to acquire the publicly held shares, just as Napoleon went toe-to-toe with the Duke of Wellington. But if Liedtke weren't cautious, Gordon Getty or the Getty Museum could launch a blindside attack against Pennzoil, perhaps even taking over Pennzoil, just as the Prussian forces of General Blücher attacked Napoleon's right flank. His method of approach would have to be perfect.

Liedtke had recognized by early December that his attack window would soon be closing, for the flurry of media coverage about the action by Gordon and the museum might easily have awakened other aggressors to the same opportunity that he saw. "One of the most high-stakes corporate battles in recent years is shaping up as Getty Oil Co. challenges a son of the late billionaire J. Paul Getty over control of $2.3 billion of Getty stock," *The Wall Street Journal* had reported. Moses Lasky had been quoted in *The New York Times:* "It's like a game of stud poker played with five cards," he had said. "Everyone has four cards on the table and what is in the fifth card, no one knows." Liedtke had long ago taken down the old portrait of W. C. Fields from his office, but he had not forgotten his card tricks.

The timing had become ripe in another respect, for Christmas was approaching. Liedtke would later deny any intention of capitalizing on the holidays as so many raiders make a point of doing. But there was no denying that even if the warring Getty parties could overcome their emotional obstacles to unity, the holidays would present logistical obstacles. The week between Christmas and New Year's Day was one of the slowest of the year, a time to travel to family get-togethers, to watch football games, to head for the beach. Gordon Getty would be playing "celebrity Santa" for the Little Sisters of the poor at Neiman Marcus in San Francisco. His investment banker, Marty Siegel, would be in a phoneless bungalow in the Virgin Islands. Bart Winokur would be visiting friends in Virginia. Liedtke himself had been planning a Christmas vacation in Hawaii, but his wife, Betty, would have to go by herself. Liedtke's lieutenants at the Baker & Botts law firm were already drafting takeover materials—offering circulars, SEC filings and the like—under the code name Project Sunshine.

By December 13 the paperwork was well under way at Baker & Botts, and the Sarah Getty Trust documents had been analyzed. By the next day a new company called Holdings Incorporated had been chartered to begin making open-market purchases of Getty stock, to establish a beachhead for the battle to come. That same day Baker & Botts began studying the securities laws of all fifty states to make sure that none presented an obstacle to any takeover action. Two days later a similar investigation was undertaken to rule out any chance that the acquisition of ERC, the Getty insurance company, would violate state insurance statutes. By the 18th, a group at Pennzoil was assembling pro forma income statements to analyze the effect on Pennzoil of paying anywhere from $90 to $120 a share for Getty stock; even at $120 a share, one of the studies suggested, Getty Oil was a "bargain," as one Pennzoil executive would later say. On December 19 the Pennzoil board met in Houston, hearing Liedtke's report on the warring camps within Getty Oil, discussing the best means of approach and giving Liedtke broad authority to pursue "a substantial equity interest" in the company. On Christmas Day the Baker & Botts lawyers continued preparing tender offer documents. On December 27 the lawyers made plans to camp for the night at the Charles P. Young Printing Company on Walker Street in Houston, where the offering documents would be printed and meticulously checked for accuracy as soon as the price and final terms of the offer were decided and phoned in.

The strategy was this:

Liedtke would offer to purchase 20% of Getty Oil's shares at $100 each—far less than the "bargain" price of $120, but a sufficient enough premium over the recent $80 market price that the public was bound to flood Pennzoil with shares.

The tender offer would give Liedtke clout—"a seat at the table," he called it, alongside Gordon's group, the museum and Getty Oil. But to what end?

Liedtke had personally worked on only one paragraph of Pennzoil's twenty-three-page offering circular. It was the critical "purpose clause," in which Pennzoil was obligated to state the reason for offering to buy shares held by the public: "Pennzoil's purpose in making this offer is to acquire a substantial equity interest in the company, with a view to participating in a constructive way in the formulation and implementation of a restructuring of the company."

In the oil industry of the 1980s, *restructuring* was a euphemism that could mean anything from "change in ownership" to "liquidation." As it turned out in this case, the word would mean both—change in ownership, followed by a liquidation.

"We're not in cahoots with anybody," Liedtke would declare when the offer was made public. But he knew of course that the only strategy that made sense was to approach the party most anxious to deal: Gordon Getty, the biggest shareholder by far, the most disaffected faction—and the son of Hugh Liedtke's own mentor.

Before the offer was made public the following morning, "courtesy calls" would be placed to each of the three warring parties at Getty Oil.

All that Gordon Getty knew about Pennzoil Company was that it sold motor oil in a yellow can. He had no idea that his father once owned 10% of Pennzoil, much less that his father had arranged to put Hugh Liedtke in charge of it. Gordon nevertheless was delighted to accept Liedtke's phone call on December 28, 1983. And when Liedtke proposed that they meet face-to-face at the earliest opportunity, Gordon considered the idea a splendid one.

But Gordon's investment banker, Marty Siegel, had something to say about that.

Siegel had assumed that the Christmas holidays would provide him, as well as his client, with a respite from the civil war within Getty Oil. He had flown with his wife and baby daughter to Caneel Bay, a Rockefeller-owned retreat in the Virgin Islands, for their first vacation in well over a year. A few days after Christmas the baby nurse would be flying down so that Siegel and his wife could have more time to themselves.

Then, a woman from the office knocked on the bungalow door. "The front desk has your office on the phone, sir," she said.

Although Siegel made plans to take a commercial flight back to the mainland later in the week to meet Gordon in New York, he saw no reason to react hastily. The situation would remain stable for a few days, after which "We'll talk to everybody—all sides," Siegel told Gordon. "Then we'll figure out what to do." This meant working both sides of the street—seeing what kind of deal could be made with Petersen's people, while also seeing what Liedtke had in mind. All the groups connected with Getty Oil, in fact, would be well practiced in working both sides of the street by the time Texaco came along.

Not until the next day, when the woman from the resort office was again pounding on the bungalow door, did Siegel feel any cause for immediate concern. His office in New York had passed along a message to call Jim Glanville, one of Liedtke's lieuten-

ants. Glanville was a friend, and he assured Siegel that there was
no reason for him to cut short his vacation. Immediately, a red flag
popped up in Siegel's mind: *Liedtke is going to try to make an end
run around the advisors and deal directly with Gordon*. There was
no way that Siegel was going to remain in the Caribbean while a
schmoozer like Liedtke was trying to make a direct approach
toward Siegel's impressionable client. Siegel decided to return
immediately.

The next few commercial flights from the Virgin Islands were
fully booked, so Siegel arranged for a chartered plane. Within a
few hours he was winging his way to New York, just as the baby
nurse was flying to the islands. Siegel's wife would spend the rest
of her vacation with the nanny.

The Getty Museum's 12% block of stock had been the decisive
factor so far in the battle for Getty Oil, enabling Gordon's group
to forge a majority and to establish itself as the real power at the
company. The museum was equally necessary to Pennzoil's effort
to grab a piece of Getty Oil for itself.

It made sense that among Liedtke's lieutenants, Jim Glanville
should break the news of Pennzoil's tender offer to Marty Lipton,
the museum's outside lawyer. Glanville and Lipton had been
friends ever since they had collaborated five years earlier in a
successful effort to defend Houston-based Rowan Companies, one
of the world's largest oil-drilling contractors, from a hostile takeover attempt by Schlumberger Ltd. Lipton was also extremely
close to Glanville's boss, Felix Rohatyn, the investment banker
who had risen to national prominence by leading the effort that
saved New York City from bankruptcy in the mid-1970s. In the
past few weeks, Lipton and Glanville had been involved on opposite sides in the $500 million takeover of Dr Pepper by Forst-
man Little & Associates. Lipton had no way of knowing that
besides working on the Dr Pepper deal, Glanville had also been
helping to plan Hugh Liedtke's raid on Getty Oil.

Lipton's cook had the night off, and his wife, Susan, was
whipping up some pasta when Glanville arrived for dinner the
evening of December 27. Glanville handed Lipton the printer's
proofs of the tender-offer documents that Pennzoil had prepared
under the code name Sunshine, with the price of the offer and the
identity of the target company left blank. He then handed Lipton a
Telex conveying the information intended to fill in the blanks: an
offer by Pennzoil to purchase 16 million shares of Getty Oil

Company at $100 a share. The offer would be announced to the press the next morning.

The terms of the offer smacked of a power play: Pennzoil was seeking to buy just 20% of Getty Oil's stock, which would put Pennzoil in a position to combine with Gordon's 40% and "squeeze out" the remaining shares—including the museum's—at whatever price they chose. But Lipton's experience told him that an offer like Pennzoil's was usually only a prelude to something else. Liedtke simply wanted a seat at the table, a way to obtain the upper hand in any private negotiations to purchase *more* shares. Surely, Lipton thought, Pennzoil would be willing to buy *all* of the museum's shares.

Lipton put on his poker face to suppress his delight at this turn of events. He knew that his client and friend Harold Williams had been anxious to sell the museum's shares ever since becoming chief executive of the museum. Over the long term, the museum—a "self-liquidating" trust—was required by law to spend its entire inheritance. In the short term, for a charitable institution like the museum to maintain almost all its assets in a single investment hardly seemed prudent. Yet the museum could never unload its 9.3 million shares in the open market without depressing the stock price. Only through a takeover could the museum obtain a premium price for its shares. Even at Pennzoil's nickel-biting price of $100 a share, the museum could sell out for $930 million and put the money in Treasury bills or whatever else it desired.

Moreover, unloading the stock would provide the museum a graceful means of escape from an unseemly power struggle. Getty Oil had become a tar baby. In his efforts to preserve the "control premium" of the museum shares, Williams had touched off fear within the company's management. In his efforts to bring peace, Lipton had furthered the divisiveness. Remaining a major shareholder of Getty Oil was an unacceptable risk—not just to the value of the museum's investment, but to the reputation of Williams, a former SEC chief, and of Lipton, one of the nation's preeminent securities lawyers.

If Williams and Lipton played their cards right, they could rid themselves of the problem forever while making the world's richest museum all the richer.

Lipton called Williams in Los Angeles and handed the phone to Glanville. Yes, Williams told Liedtke's lieutenant, the museum would be delighted to sell to Pennzoil, but only on the right terms and at the right price—say, $120 a share, which would bring

$1.11 *billion* to the museum. But the museum certainly wasn't about to commit at the lowball price Pennzoil was offering right now.

After Glanville left, Lipton called Williams a second time to confer in private. "We were overjoyed," Lipton would later recall.

Lipton knew that several days of hard bargaining lay ahead; if Pennzoil refused to pay top dollar for the museum shares, then perhaps another bidder would enter the fray. Lipton did not know, of course, that eventually a jury would sit in judgment on his efforts to obtain the best price possible for the museum's shares—nor could he have imagined how loathsome the jury would consider those actions.

On Wednesday, the day that the Pennzoil offer appeared in the press, Getty Oil's stock rose an astonishing $19.625, closing at $100 a share. Stock traders knew that Pennzoil was bidding for only a fraction of the shares; that *all* of the shares of stock should reach the $100 level was clearly a sign that the market expected more money to reach the table—or that it expected someone to bid for the entire pot.

Although Gordon had been blocked from meeting with Liedtke, his protectors did not prevent him from meeting with his friend and newfound ally, Harold Williams. On Thursday, before flying to New York with his phalanx of advisors, Gordon met with Williams in the music room to discuss the Pennzoil offer, which Gordon thought could provide the catalyst to at last reunite the Getty family legacies: the Sarah Getty Trust and the J. Paul Getty Museum. Together, Gordon told Williams, they could now remove the board and take the company private—leaving it finally and firmly in the control of the Getty name. Petersen, of course, would have to go.

But the museum, alas, wanted out. For good.

Gordon was a painfully precise person who believed strongly in the whimsical power of Mnemosyne. "Memory can be tricky," he would recall when questioned about the meeting during a deposition. "We remember some things that may not be the most important, but for some reason we remember them. I do recall him"—Harold Williams—"making the point that at one hundred dollars a share or higher, the museum was a seller and not a buyer—which was a rhyme.

"Put me among a circle of poets for that," Gordon went on to say. "That's the lowest circle in hell, by the way."

While the museum viewed Pennzoil's tender offer as a way to make a graceful retreat from its ownership of Getty Oil, Petersen's people approached it with all the enthusiasm of an underequipped army on its way into battle.

Boisi, Petersen's investment banker from Goldman, Sachs, had wondered whether these holidays would be the first in five years—including every Christmas, Thanksgiving and Easter—that would remain free of interruption from a major deal; he was just about to take his family out for a post-Christmas dinner, in fact, when he was notified of the tender offer. The phone was ringing in Bart Winokur's office when he walked in after a seven-hour drive from his vacation in Roanoke; immediately after receiving the news, Winokur began making arrangements for a six-way conference call to get all of Petersen's people on the line together.

Their first suspicion was that Gordon was somehow behind the Pennzoil offer. They quickly ruled out any conspiracy, but their dismay was unabated.

Liedtke had made himself the master of a multibillion-dollar shell game, and the permutations were mind-numbing. Despite an offering price they considered incredibly chintzy, Petersen's people recognized that Pennzoil was certain to receive all the shares it had bid for, because a partial tender offer like Pennzoil's typically causes a take-the-money-and-run panic among shareholders. If Pennzoil got its 20%, Liedtke could combine with Gordon's 40%, oust the entire board of directors and establish majority control of a company owned by a voiceless group of public shareholders. The museum, too, could be left as a frozen-out minority.

Or, if Liedtke were unable to make a deal with Gordon, he could expand his offer to buy even more shares—perhaps the entire 48% in public hands—and take majority control by combining with the *museum*. Under this scenario, *Gordon* would be left as the frozen-out minority.

Everybody connected with Getty Oil, in fact, was potentially a loser.

The best response was clearly a united front—one that permitted all three groups to maintain control of Getty Oil's destiny. Now was the time to put aside the differences of the past year. If everybody stuck together—if no one broke ranks—Pennzoil could not win.

It was a naive view.

For the last four days of 1983, representatives of the trust, the museum and the company made their final attempt at peace under the pressure of a common threat. Meeting in Marty Lipton's offices at Wachtell, Lipton, they discussed a plan to have the company announce its own, competing tender offer for its shares at a higher price, thus preventing Pennzoil from obtaining *any* shares. Then, for ninety days, Petersen's people would attempt to sell the company outright for the highest possible price.

The acrimony of the prior year was nearly intractable, and the skeletons continued rattling. By repurchasing its own shares, the company would again be confronted with the problem of Gordon obtaining majority control—if only for the ninety days while the company was on the block. Gordon, meanwhile, continued wishing Petersen's banishment from the executive suite, even for those final ninety days. It had taken weeks to reach the ill-fated one-year standstill agreement. Now, even a pact of only three months' duration was proving every bit as difficult to reach.

Ultimately, a possible arrangement was laid out. Getty Oil would announce a competing tender offer at $110 a share— enough to prevent anyone from selling to Pennzoil. By reducing its number of shares outstanding, Getty Oil would at last make Gordon the majority shareholder. For ninety days, he would operate with handcuffs, permitting management to attempt to sell the remaining shares in the company for more than $110 each, a task they considered a cinch; Petersen's people were convinced that even $120 could be easily obtained. If the company succeeded in the sale, fine: Gordon, the museum and the public would all receive a fair price for their shares. If the company failed to find a buyer at more than $110, then Petersen's people would walk away. Gordon's handcuffs would be removed. He would control Getty Oil.

Give us ninety days to find a buyer at a fair price, Gordon, Petersen's people were saying. *If we fail, Getty Oil is yours.*

By Saturday night—New Year's Eve—the plan seemed set. A vacationing Getty Oil lawyer would be summoned from Florida to assist in the preparations; a variety of major banks would be approached for the billions of dollars in loans necessary to finance the defensive self-tender. Geoff Boisi of Goldman was ready to preside over the "controlled sale" of the company.

Petersen's people meandered their way through the crowds heading for the midnight merriment at Times Square. But after

completing the two-block walk from Lipton's Park Avenue offices
to the Helmsley Palace at Madison Avenue and Fiftieth Street,
they went to bed. None felt much like celebrating. For weeks they
had been resigned to the eventual sale of the company; now, at
least, they had the consolation of knowing they had devised a way
to obtain top dollar for all shareholders. But it was hardly the
outcome that they had sought from the beginning. As soon as
Gordon signed the ninety-day "controlled auction" agreement,
Getty Oil would be as good as gone—if not to another oil company, then to Gordon himself.

Happy new year, indeed.

Alas, the lawyers and managers of Getty Oil had no idea that
even their consolation prize would never be collected. For while
negotiating with Petersen's people and the museum, Gordon's
group was also seeing what kind of deal *Pennzoil* had to offer.
"Gordon was in the catbird seat," Marty Siegel would recall.
"Our job was to fully develop the alternatives—and let him pick
which one he wanted."

Would Gordon decide to sell Getty Oil, as Petersen's people
now wished? Or would he decide to *buy* it, with Hugh Liedtke as
his partner?

Siegel had been stalling for two days over Pennzoil's demand for a
face-to-face meeting between Liedtke and Gordon, but Liedtke's
lieutenants had been unrelenting in their request. So on Friday,
after leaving one of the negotiating sessions with Petersen's people, Siegel made his way through heavy, pre-holiday midtown
traffic to a meeting with Liedtke's lieutenants at the offices of
Lazard Frères. The kitchen at Lazard churned out some stale tuna
sandwiches for the half-dozen lawyers and investment bankers in
attendance.

Expecting Gordon to negotiate a multibillion merger with Hugh
Liedtke would be like showing a Girl Scout into a biker bar. As
one participant in the meeting would recall, Siegel told the Pennzoil representatives that Gordon was "incapable of making a
prompt decision," although Siegel insists he said only that Gordon had *no reason* to make a prompt decision. Moreover, delay-
ing a meeting would give Gordon's group more time to see if they
could work out a favorable deal with Petersen's people on the
proposal for the controlled sale of the company. Siegel urged that
all the lawyers and investment bankers get together for *more*
meetings, just as secretaries of state meet in advance to work out

the agenda in advance of a superpower summit meeting. Once Pennzoil and the trust had come to terms on a merger arrangement, then it could be presented to Gordon for final approval.

Glanville called his client Liedtke in Houston, repeating the proposal for a series of "structured meetings."

"What on earth is that?" Liedtke demanded.

"Well, they want to present to you an outline of what to talk to Mr. Getty about, and then they'll go up and review it with Mr. Getty."

"Like hell they will!" Liedtke bellowed. "I'm going to deal directly with Mr. Getty and I'm not going to deal with third parties. I don't want them telling Mr. Getty what *I* think of what *I'm* offering. I want to sit right across the table from him and tell him myself so there won't be any questions about who said what."

On the same day that Gordon's group opened negotiations with Liedtke's lieutenants, after a series of earlier disappointing reports the world learned for good that Texaco and its Mukluk partners had drilled the most expensive dry hole in history. Instead of finding one of the last great "frontier" oil fields, they had poured well over $100 million into the earth and come up with a duster. Texaco had become the incredible shrinking oil company, ending 1983 with 27% fewer domestic reserves than it had five years earlier. It would begin 1984 with 7.8 years' worth of oil-in-the-ground, the shortest potential lifespan of any major oil company.

At the time, John McKinley was spending the holidays at his newly built, four-story second home on a bluff above Lake Tuscaloosa in his native Alabama. Fortunately, he had the welcome distraction of Bob Baldwin and his wife as houseguests for the New Year's holiday. Baldwin was managing director of Morgan Stanley & Company, one of Wall Street's premier investment-banking houses, and a golfer and tennis player besides. Baldwin was delighted to spend a few days with McKinley talking business on the sunny links and courts along the lake.

Just before leaving New York to join McKinley, Baldwin had read the news of Pennzoil's tender offer for Getty Oil on the Dow Jones broad tape. Pennzoil was really lowballing it, he told McKinley. Baldwin reckoned that the partial tender offer would only encourage other oil companies to make their own bids for Getty Oil—and there was no reason, he said, why Texaco couldn't get in on the action. As the two couples toasted the New

Year, McKinley was wondering just how Getty Oil might fit into Texaco. He would have to have his management team back at the headquarters in White Plains look into it.

When Ann Getty found out that her husband might be going into business with Pennzoil, she called a well-placed social acquaintance to find out what kind of people ran Pennzoil. He was Fayez Sarofim of Houston, whose money-management firm was enjoying remarkable success from savvy betting on oil stocks—and whose clients happened to own not just a huge share of Pennzoil's stock, but also about 1.3 million shares of Getty Oil. If Pennzoil were to buy those shares for, say, $110, Sarofim would make a profit on the stock of a tidy $50 *million* or so. Thus, when Sarofim, in turn, talked to his friend Liedtke, Liedtke had an acutely interested listener on the line. Liedtke complained that Marty Siegel was stalling the meeting with Gordon. Before long, Sarofim called Liedtke back. The meeting would be held the next day—New Year's Day at 4 P.M.—in Gordon's suite at the Waldorf.

Late on New Year's Eve, Liedtke climbed aboard the Pennzoil jet for the four-hour trip to New York. His number-two executive, Baine Kerr, and Pennzoil's general counsel, Perry Barber, would follow in another plane the next morning in deference to a company policy that prohibited the top officers from flying on the same aircraft. Pennzoil owned a suite at the Waldorf-Astoria where Liedtke regularly stayed, often with his wife. This time Betty was in Hawaii, and he reached the hotel alone. He dumped his bags and took the ten-block stroll through the brisk, midwinter air to Times Square.

It was after midnight, and Liedtke found himself kicking his way through the broken glass and crumbled party hats littering Broadway. Liedtke himself would soon be celebrating the fulfillment of a lifelong dream—only to see it shattered.

· PART 2 ·

THE DEALS

I hailed a cab in Manhattan the other day. It was across the street and the driver nodded and blinked his lights in acknowledgment. Then another cab came along on my side of the street and stopped in front of me. I waved him on, having made my deal with the other cabbie.

Then a woman in a mink coat opened the door of the cab across the street. I could see the taxi driver pointing to me, explaining that I had already hailed him, but the lady was adamant and the rule is the first one in the cab gets it. The cabbie looked at me and shrugged, leaving me shivering in the wind and thinking about the sanctity of contracts.

—William Safire

My son, if sinners entice thee, consent thou not.
—Proverbs 1:10

· 10 ·

A contract can be two things. A sheaf of papers stapled together with signatures at the bottom and the word "contract" typed on top is a contract. But in the eyes of the law, a contract can also be a state of mind, something that may come into existence without paperwork and without lawyers. The main requirement, put simply, is that the parties achieve a *meeting of the minds* on the essential terms of the deal. In addition, they must demonstrate their *intent to be bound*, perhaps by shaking hands, by putting up earnest money, by hoisting a glass of champagne or—as in the case of the Hassidic diamond merchants of Forty-seventh Street—by uttering *mazel und broche*, Yiddish for "luck (for the seller), blessing (for the buyer)."

Bingo: contract.

For nineteen weeks through the summer and fall of 1985, the Houston jury would hear the first three days of 1983 as spent in midtown Manhattan reconstructed from twenty-three points of view. And although the jury would consider several matters besides whether a multibillion-dollar contract had come to life in those seventy-two hours, it would be that issue they would have the least difficulty deciding.

The forty-two-story Hotel Pierre, overlooking the southeast corner of Central Park at Fifth Avenue and Sixty-first Street, has been the international jet set's home in New York ever since J. Paul Getty bought it in 1939. The mirrorlike lobby floor, made of black marble, is draped with Persian rugs, and the perimeter is defined with Doric columns and Moorish arches, the touch of antiquity that J. Paul appreciated in everything he bought. The hotel management of modern years maintains a mile-high arrangement of fresh-cut flowers in the center of the lobby. You can drop in on Le Chocolatier shop, the Gio Skin Care Salon or Bulgari Jewellers without ever leaving the lobby.

In Gordon Getty's suite on New Year's Day, the college bowl games were glowing on the television set and there was a little friendly betting under way. The regulars in Gordon's group were all present: Cohler and Woodhouse, the Amherst and Harvard Law classmates, along with Marty Siegel. And Gordon's circle of advisors had grown to include a family friend whom Ann particularly admired: Mark Leland, a lawyer who was now serving as an assistant secretary of the U.S. Treasury.

In his meeting two days earlier with Liedtke's lieutenants, Siegel had warned Pennzoil that Gordon would never settle on remaining a minority owner of Getty Oil. If the trust went into partnership with Pennzoil, the trust somehow had to emerge with majority control. To test Pennzoil's willingness, Siegel had tossed out a vague restructuring proposal. The trust could continue to own the same 32 million shares it currently held out of the 79 million total shares outstanding. Pennzoil could acquire, say, 24 million shares, and Getty Oil, then controlled by the two of them, could itself buy in the museum's shares and everything else remaining in public hands—in effect canceling out the entire ownership except what Gordon and Pennzoil had. In the end, Gordon and Pennzoil would own Getty Oil in a perfect 4-to-3 ratio, although Siegel would recall picking the numbers "out of the air."

Remarkably, what Siegel conceived of arbitrarily, Gordon independently had thought up rationally: Pennzoil buys exactly half of the 48 million shares that Gordon didn't own and Getty Oil buys the remaining half. *Voilà!* Gordon ends up with four-sevenths of Getty Oil, Pennzoil with three-sevenths. "I'm a numbers fan," says Gordon. "I'm comfortable with calculus."

But Gordon and his advisors knew that this was no mere numbers game. This would be a "going private" transaction, in which not only Pennzoil but *the company* would be buying shares. The whole package deal would require *up-front* approval—by Getty Oil. Gordon would have to appeal for the blessing of the same "bunch of snakes" that only three months earlier he had tried to throw out, that only two months earlier had tried to throw *him* out as sole trustee. This board certainly wouldn't approve the most extraordinary transaction in the company's history purely out of fondness for good old Gordo. Moreover, the directors had a legal obligation—a "fiduciary duty"—to insist on the best possible price for the public shareholders. Making matters worse, the board had a meeting scheduled for the following afternoon in New York. Gordon's group and Liedtke's lieutenants would have exactly one day to detail the restructuring plan and to conceive a strategy for getting it by a board that was feeling more headstrong than ever.

Liedtke's lieutenants were late in caucusing prior to the coveted meeting with Gordon. A Pennzoil plane had left Houston that morning transporting more of Liedtke's lieutenants to New York, but the landing gear malfunctioned, causing one set of wheels to ascend whenever the other descended. The additional troops finally landed in New York after an explosive device was used to blow down all the wheels at once.

Hugh Liedtke had been thrilled at the preliminary proposal that Pennzoil would own three-sevenths of Getty Oil's stock—and, more specifically, three-sevenths of its oil-in-the-ground, or exactly one billion barrels. In one fell swoop Pennzoil would *quadruple* in size, fulfilling Hugh Liedtke's lifelong ambition to reach the major leagues of the oil patch. Sure, the deal would be costly; Pennzoil would have to take down its entire $2.5 billion line of credit to buy the number of shares contemplated under the proposal. Liedtke couldn't pay much more than $100 a share without going into hock even deeper. But oh, such a sweet arrangement this was! Access to a billion barrels for a cash outlay of $2.5 billion, give or take—that would be only $2.50 a barrel! By contrast, it cost Pennzoil roughly $10 a barrel, on average, to actually *drill* for oil-in-the-ground. Getting $10 billion worth of oil for something like $2.5 billion would in effect put $7.5 billion in Pennzoil's pocketbook. J. Paul Getty had boasted of buying oil on Wall Street with 50-cent dollars; J. Hugh Liedtke was now in line to do the same, only with *25-cent* dollars.

As Liedtke thrilled over the economics, he contemplated some of the risks. "We had heard various rumors that Mr. Getty was bright, but kind of flaky," Liedtke would later say. There was every chance that the alliance would work out beautifully—but Liedtke recognized that there was also a chance that Gordon might want to operate the partnership with "an iron claw—and get into matters that would run afoul of me pretty fast." Thus, Liedtke would insist on a divorce clause, something that permitted Pennzoil and the trust to divide the assets—to "split the blanket," in Liedtke's words—on a 4-to-3 basis and part their ways if things didn't work out after a year. That way Liedtke could always walk away with his billion barrels of oil-in-the-ground—free and clear. Liedtke did not realize that one day his insistence on a split-the-blanket provision would be worth more than $7 billion to him.

But he did know that the deal would require Gordon's unwavering commitment, particularly to see it past the company's strongwilled board of directors. Thus, he decided that he would touch the trustee's hot button by offering one additional proposal: making Gordon Getty the chairman of the board, a position that Liedtke was sure the heir had wanted for years. "You know," Liedtke would explain, "if you give someone what they want, why, they're apt to push for it."

With Gordon in a chair to his left, "Chairman Mao" sank into the sofa in the Hotel Pierre suite and spoke first, switching on that masterful Oklahoma, aw-shucks charm, warming his quarry with the story of the Old Man's attendance at the Liedtkes' engagement party in Tulsa, detailing his longtime acquaintance with the Old Man's aides, regaling Gordon with the story of how he doubled his father's investment in the old South Penn Oil Company as if to suggest that the same could happen at Getty Oil. Siegel, who was seated to Liedtke's right on the couch, marveled at Liedtke's touch, at how the bearlike businessman succeeded in visibly dissolving the tension in Gordon. *This guy is good,* Siegel thought.

Liedtke then changed tone, arousing Gordon's passion. "That was a horrible thing they did to you in Houston," he said of the "back-door" board meeting. He expressed incredulity that Petersen's people—the hired hands—could treat a 40% shareholder with such shabbiness. Liedtke said that he, on the other hand, appreciated the rights of ownership; after all, he himself was a major shareholder in Pennzoil. "I think it's high time you asserted yourself," he said to Gordon. The heir listened quietly, finally expressing his gratitude for the friendliness and constructiveness of Pennzoil's overture. Siegel, however, did most of the talking for Gordon. His client was a *buyer* of Getty Oil stock, Siegel said, not a seller. "Well, if the stock gets to one thousand dollars a share, I might be a seller!" Gordon piped up, drawing laughter all around.

But the matter of price was serious business. Liedtke felt he had been generous to offer $100 a share for a stock lately selling for $80. Raising the price paid for the public shares not only made the deal riskier for Pennzoil, it put Liedtke in the uncomfortable position of bidding against himself; to him, raising the price would be like offering $100,000 for a home and then increasing the bid by 10% before ever receiving a counterproposal. Gordon's group, on the other hand, recognized that besides being the head of the family trust, Gordon was a director of Getty Oil—and thus had a legal duty to treat the public shareholders fairly. Moreover, concern persisted in Gordon's group that the Getty Oil board would look dimly on *anything* proposed by Gordon.

"The minimum price that the board will find acceptable is one hundred and ten dollars," Siegel said.

"How about one hundred and five?" Liedtke countered.

Siegel was stunned. *We're trying to get on the same side and this guy is negotiating with me,* he thought.

"The museum also will want one hundred and ten," Siegel added. After all, Marty Siegel was sure that he knew what his friend Marty Lipton, the museum attorney, would want out of the deal.

Liedtke was hardly in a position to ignore the museum's interests. He recognized that the proposal had a much better chance of clearing its biggest hurdle—the Getty Oil board—with both the museum and the trust committed to it; ever since early December, their combined 52% interest carried the implicit threat that at any time the directors could be pitched out on their ears. As the discussions wore on, Liedtke finally relented: *A hundred and ten bucks. And that's it.*

On the subject of titles, however, there was little disagreement. *Your name is on the door of this company, and I can certainly understand why you'd want to keep it that way,* Liedtke said. If Gordon became chairman, Liedtke would serve as chief executive officer—the number-two position on any corporate organization chart, to be sure, but the one that carried the real authority to operate a company on a day-to-day basis. Liedtke would happily spare Gordon the need to come into the office every day—but assured him that he would be regularly and closely consulted on policy matters. Gordon, Liedtke recalls, was thoroughly delighted with this arrangement.

It had taken barely two hours to sketch out one of history's largest corporate reorganizations. The Getty Oil board would be meeting in exactly twenty-four hours. Liedtke's lieutenants and Gordon's group undertook to recruit the indispensable support of the museum to the deal.

Nine blocks south of the Pierre, Petersen's people spent the day in a law office at Madison and Fifty-first Street. They were awaiting Gordon's signed copy of the agreement they had spent the prior four days negotiating, the three-way treaty calling for the controlled sale of Getty Oil to the highest bidder in ninety days. It never showed up.

Liedtke's willingness to make Gordon the chairman would quickly strike some people as the kind of folly of which they thought Liedtke was incapable. "Hugh, how *well* do you know Gordon Getty?" asked Liedtke's old friend Harold Berg, the retired chairman of Getty Oil, when Liedtke briefed him on the arrangement in advance of the Getty Oil board meeting.

"I've never met him until just now," Liedtke calmly answered. "But we've always been able to get along with difficult people, and we've got a one-year liquidation provision in this deal. If Gordon Getty wants to be chairman, that's fine with me. *I'm* going to be the chief executive." As Berg recalls it, Liedtke continued: "I don't care if my title is janitor. This may be the quickest marriage and the quickest divorce in history." Liedtke doesn't believe he ever made that last comment. But in any case, making Gordon chairman was a small price to pay for the chance to walk away with a billion barrels of oil-in-the-ground in a year.

Liedtke's motivation was plain enough; Gordon's, however, would become the source of endless speculation.

It would become a foregone conclusion among many of Petersen's people that Gordon struck his partnership with Hugh Liedtke in order to get himself the title of chairman.

But to what end?

There was the theory that *Ann wears the pants in that marriage*. Harold Stuart, the longtime Getty Oil director, was convinced that Ann Getty accounted for "80 to 90%" of Gordon's motivation. Many others were convinced that Ann longed for the distinction of a marriage to a corporate chairman rather than a mere corporate director. Certainly, there was no question that Ann had assumed an unusually active role in her husband's affairs. But she could never have accounted for Gordon's entire motivation, or even most of it. Ann may have wished, for instance, that her husband discontinue taking leave of her parties and inviting her guests into his music room, but her wish did not prevent him from doing so.

There was the spite theory: that by becoming chairman, Gordon would have the devilish pleasure of dismantling Sid Petersen's "Imperial Chairmanship." To be sure, Gordon's actions made it plain he harbored resentment toward Petersen, particularly after being "snookered" at the board meeting in Houston two months earlier. "It's true I found Sid to be cold and rude," Gordon says today. But he adds: "Sid is not a bad man. Sid is a nervous man, a tense man. Sid by nature simmered and festered. But he is a bright man, and I'm convinced a very decent man." Indeed, at the height of the battle, Gordon refused, even to his closest friend, Bill Newsom, to bad-mouth Petersen.

Finally, there was the *dead Daddy* theory, that Gordon had a burning desire to prove himself worthy of his lineage, or even to get the last laugh on the overbearing figure in whose eyes Gordon seemed so imperfect. This theory was especially appealing to the Getty Oil officials who had put on Freudian eyeglasses while reading copies of the awful letters Gordon had received from his father. Indeed, Gordon often did measure his actions against those of his father. And there was no doubt that the executive-suite arrangements proposed by Liedtke would codify a personal role for Gordon identical to the one his father had assumed in his final years, when he served as chieftain of Getty Oil from a castle across the Atlantic. "I don't make the day-to-day decisions in my companies," the Old Man said in 1974, two years before his death, "but I like to think I am well posted on what they are doing, and express my views from time to time." Ashes to ashes, Getty to Getty.

But bargaining his way into a title solely to prove himself to his father? This theory was inconsistent with other actions by Gordon. For instance, while the Old Man had failed entirely in his lifelong ambition to become a diplomat, Gordon had once been offered an ambassadorship in the administration of Richard Nixon, a politician his father greatly admired. Gordon, however, had declined to make the requisite campaign contribution. Getty Oil itself had offered Gordon the title of chairman, yet he had flatly refused.

The speculation about titles clouded the real issue. Even Hugh Liedtke probably overestimated Gordon's interest in the title. The *real* appeal of the Pennzoil deal was that it gave Gordon something no one else had ever offered him: control.

The day before Gordon's meeting with Hugh Liedtke, the Sarah

C. Getty Trust had formally entered its fiftieth year. As quick as Gordon's adversaries were to recognize the clumsiness that he brought to the job as sole trustee, they failed to appreciate the passionate sense of duty that he felt. The mission had been drummed into him in one family conflict after another. From his father: *It is very important for the trust to have control of the corporation. . . . The trustee should be buying Getty Oil shares rather than selling them.* From his brother George: *The trust's strength is its control over the destiny of a very large corporation.* From his niece Claire: *The trust clearly prohibits the sale of Getty stock.* Petersen's people had spent a year trying to block Gordon from fulfilling what he saw as the one true destiny of his grandmother's trust—and his own destiny as the ruler of that trust.

To a degree, Petersen himself would come to realize this. "In fact," he would say, "Gordon had an agenda all along. Gordon wanted power, whether it was in a royalty trust, through a self-tender or a deal with Liedtke. It wasn't money. It was power."

Alas for Gordon, his attempt to do what he thought was best—for the trust, for himself as trustee—would only heap family contempt and legal ridicule upon him. But Gordon had long ago come to realize that doing what seems the proper thing did not always bring the best result. In his college years he had written an epitaph to an imaginary seafarer, William Tynes, which read in part:

> With Drake and Robert Poole I sailed,
>   A pirate bold and free,
> And many a dreadful deed performed
>   Upon the sightless sea,
> But Christ forgives, if any sins,
>   The sins of bravery.
>
> I sang to harlot or to maid
>   As gently as a dove,
> And many a promise false I swore
>   By the false moon above,
> But Christ forgives, if any sins,
>   The sins of too much love.
>
> When Drake returned they made him knight
>   And every churchbell rang;
> And Robert Poole went free, but I,
>   For doing the same, did hang.

Thus ends the song of William Tynes,
  Remember well my name;
My flag was e'er my kerchief green,
  My honor was my shame.

Gordon's group and Liedtke's lieutenants had exactly twenty-four hours to rope the museum into their plan and commit the proposal to writing for consideration by the Getty Oil board. Doing so would arouse all the suspicion, misdirection, bluff, bluster, innuendo and playacting that these big-time negotiators were capable of. The emerging document would become the full expression of Hugh Liedtke's billion-barrel dream and Gordon Getty's assurance of majority ownership. The document, plainly entitled Memorandum of Agreement, would before long assume the additional moniker of Plaintiff's Exhibit #2.

6:10 P.M. Sunday, January 1: (note: times are approximate). Marty Lipton's New Year's Day party is in full swing, with a veritable rogues' gallery of takeover professionals in attendance. The host assumes his persona as museum lawyer, excusing himself to a quiet hallway near the kitchen. He huddles beneath his Frank Stella lithographs with Marty Siegel of Gordon's group, who has just arrived from the Pierre. Arthur Liman, one of Liedtke's lawyers, has also made the scene. None of Petersen's people happen to be present. The joint offer of Pennzoil and the trust is conveyed: $110 for each of the museum's shares—a total of $1,025,237,400, as in billion. Lipton is noncommittal, but leaves the Pennzoil and Trust emissaries with the impression that the deal might be doable. They return to the Waldorf to relay the good news to Hugh Liedtke.

Only that day, one of Lipton's guests, Jay F. Higgins of the venerable Salomon Brothers Incorporated partnership, had been hired to function as the museum's investment banker. Higgins will issue an opinion of whether the Pennzoil deal is fair from the museum's point of view.

8 P.M.: A dozen lawyers and investment bankers representing Gordon and Pennzoil have gathered in Liedtke's suite at the Waldorf. Moulton Goodrum, the longtime Baker & Botts lawyer on the Pennzoil account, is reading aloud the main points proposed for the memorandum of agreement. They call Lipton, once again seeking to extract the museum's agreement to sell at $110 a share. Lipton remains unaware that some of his party guests have already spread word through the Pennzoil suite that Lipton reacted warmly to the proposal. Like an emperor without clothes, Lipton insists that he will have to call Los Angeles to confer with his client, Harold Williams. However, Williams will never recall hearing from Lipton.

8:05 P.M.: Lipton calls back to the Waldorf and leaves everyone from Pennzoil with the distinct impression that the museum has committed itself to the deal. Everyone is convinced that Lipton had put on a little charade to avoid appearing as an overenthusiastic seller. Everyone gets a big laugh—everybody except Liedtke. *What the hell is the point of going through this drill?* he wonders.

9 P.M.: Aided by strong tailwinds, Sid Petersen arrives in New York from Los Angeles after only five hours on a commercial flight. He checks into the Helmsley Palace Hotel at Madison and Fiftieth. Unaware of the discussions under way in the Waldorf a block to the east, Petersen is hopeful that the hard-fought agreement for the controlled sale of the company will be reached in time for tomorrow's board meeting.

10 P.M.: Anxiety grows in the Pennzoil suite as Liedtke's lieutenants and Gordon's group ponder the reaction of the directors to the joint "going private" proposal. The executives, lawyers and investment bankers recite their worst fears in turn. *What if it tries to torpedo this transaction? . . . They might not want* ever *to let Gordon get control. . . . The blood is so bad. . . . The board might react emotionally rather than rationally.* There was only one solution: get the museum to commit to exercising a new majority-shareholder consent with Gordon to fire the entire board if it tries to stop the deal.

Another call goes out to Lipton. It's not part of the program, Lipton replies. "I'm not authorized to give that and I won't give that." Harold Williams, the museum president, has little stomach for such a brazen act of aggression. *We're not going to let the board stand in the way of a transaction,* Lipton assures them. But there was no way the museum would commit in advance to firing the entire lot of directors.

Hugh Liedtke is consumed with suspicion. "If Lipton had the authority to accept our price, why in hell doesn't he have the authority to sign a consent?" Liedtke demands of his assembled advisors. "What if this is a setup? What if they're just trying to get me to raise the price? Hell, I've already increased this god-damn offer ten percent!" Liedtke bears down on Glanville and Liman—his two advisors who were close friends of Lipton. "You know Lipton? Do you trust him?" Both men assure Chairman

Mao that Marty Lipton is an honest and honorable man; if Lipton is committed to supporting the deal, he'll support it.

Midnight: Everyone clears out of the Pennzoil suite. Moulton Goodrum, Pennzoil's man at Baker & Botts, goes to his own room at the Waldorf, scatters his notes and files over a table and bedspread and begins writing out the memorandum of agreement by hand.

9 A.M. Monday, January 2: The wastepaper basket brimming with crumpled paper, the sleepless Goodrum completes the memorandum of agreement, leaves the Waldorf and walks seven blocks uptown to Third Avenue and Fifty-fifth, where he leaves a sheaf of scrawl in the thirty-second floor word-processing pool at Paul, Weiss, Rifkind, Wharton & Garrison—Pennzoil's New York law firm, where Liman works.

9:30 A.M.: The morning after his party, Marty Lipton makes a note to demand that Pennzoil and the trust provide "favored nation" status for the museum—an assurance that if the museum sells to Pennzoil at $110, it will receive more money if eventually some interloping bidder offers more money for the public shares. You could never tell how these things will turn out.

10 A.M.: Liedtke and his lieutenants are more worried than ever about the board meeting, now only eight hours away. Petersen could be a problem; he, after all, would be chairing the meeting. There was no telling if Gordon would knuckle under the pressure and back out of the deal. Liman goes off by himself to begin drafting a private "side agreement" to present for Gordon's signature. Gordon's group arrives at the Paul, Weiss offices on Third Avenue to begin reviewing Pennzoil's draft of the memorandum of agreement.

Noon: A Pennzoil airplane takes off from Acapulco after picking up Stephen Booke, a vacationing public-relations consultant on Pennzoil's retainer. The plane drops in Houston to pick up three of Pennzoil's in-house PR people. They would make sure that the world knew about Pennzoil's deal as soon as it was done. Meanwhile, Liedtke and his top in-house advisors eat lunch at Peacock Alley in the Waldorf. Multibillion-dollar deals cause big appetites. "They must've eaten a whole side of beef that week," says Liedtke, who is picking up the tab.

2 P.M.: Next door to the Waldorf, in Marty Lipton's law firm, Petersen and his people are finally informed by Marty Siegel, Gordon's investment banker, that Gordon has his own agenda for the board meeting. (Siegel will claim that he had informed them much earlier in the day.) The four days of negotiations for the controlled sale of the company—an agreement that would have preserved Petersen as chairman for at least ninety days, until the highest-paying buyer could be found—have been rendered meaningless. Petersen's people are stunned that Gordon's group is consenting to a deal in which the public will be bought out at a paltry $110 a share.

"Well, Mr. Liedtke is a tough negotiator," Siegel replies, intending a tongue-in-cheek remark.

"Well, he may be," counters Bart Winokur, Petersen's outside lawyer. "But in this situation *he's* a buyer and *you're* a buyer. And with all due respect to how good a negotiator you are, it's hard to expect two buyers to agree on a fair price that the seller wants to sell his stock at."

3 P.M.: Three of Getty Oil's representatives—Winokur, Boisi and special counsel Herb Galant—show up at the Waldorf, informing Pennzoil that the $110 price is unfairly low. *Just a little bit more,* they implore Liedtke's lieutenants, *to get the deal past this board.* Galant thinks that only $5 is all it might take. They also propose an alternate arrangement with Pennzoil: If Pennzoil wants oil-in-the-ground, then perhaps Getty Oil can sell it some outright, but at a fair price. Liedtke's lieutenants reply that they are satisfied with their arrangement with Gordon. *In that case let us tell you a little about what it's like to work with Gordon,* Petersen's people say. *This guy changes his mind from one day to the next. He doesn't understand the implications of his actions.* Perry Barber, Pennzoil's general counsel and number-three executive, makes it clear that Petersen's people have no standing to bargain. "Barber left us with the impression that 'We've got you against the wall. Not another nickel,'" Galant will recall.

3:30 P.M.: Liedtke's lieutenants have completed drafting a letter that they will ask Gordon to sign, binding him to "the Plan" despite whatever hell or high water is inflicted on him by the board.

Dear Hugh,
    . . . I agree that I will support that Plan before the board and will oppose any alternate proposals. . . . If the board does not approve the Plan today, I will execute a consent to remove that board and to replace the directors with directors who will support the best interests of the sharholders, as reflected in the Plan. I will also use my best efforts to urge the J. Paul Getty Museum to execute a consent to the same effect.

It is implicit in the arrangement that if the museum can't be importuned to join with Gordon in the consent action, Pennzoil itself can do so. The original $100-a-share tender offer has not yet been canceled—it remains as a backup measure that could be used to buy enough shares to combine with Gordon and forge the necessary 51% block.

In exchange for Gordon's promise to see the plan through to the point of ousting his father's old friends and the other members of the board, Liedtke adds a paragraph swearing, in effect, that Pennzoil will not leave Gordon high and dry as a locked-out minority shareholder—a disturbing, if remote, possibility in the back of Gordon's mind. The secret "Dear Hugh" letter will one day become Plaintiff's Exhibit #3, characterized on the one hand as evidence of a contract between Pennzoil and the trust, on the other hand as a shameful tactic to neutralize a board of directors elected to represent the investing public.

4:30 P.M.: Harold Williams—executive, academic, federal regulator, museum president—reaches his lawyer Marty Lipton's Park Avenue office after a late arrival on the flight from Los Angeles. He will be attending his first Getty Oil board meeting ever, having become a director as part of the standstill agreement reached among the warring Getty parties three months earlier. As Lipton will recall it, Williams has been fully briefed on the Pennzoil deal before his arrival, but Williams will recall arriving in New York and learning for the first time that his 9.3 million shares are already spoken for. Lipton has scratched out the language in Pennzoil's proposed memorandum of agreement saying that the museum has "developed and approved" the plan, making it read that the museum "recommends" its adoption by the board. The former SEC chief considers it a trifle heavy-handed that he would commit to a plan to overhaul the company five minutes into his first board meeting.

Lipton has added some language on the signature page, noting that Williams would be signing "upon condition that if not approved at the Jan. 2 board meeting . . . the museum *will not be bound in any way* by this plan and will have *no liability or obligation* to anyone hereunder." One of Lipton's associates phones in the changes to Liedtke's lieutenants at the Paul, Weiss firm, forcing the word-processing pool to produce yet another draft. A copy of the version showing Lipton's handwriting will eventually evolve into *Defendant*'s Exhibit #52.

5:30 P.M.: The directors are gathering at the Intercontinental Hotel, just across Forty-ninth Street from Pennzoil's command post at the Waldorf. A few blocks north, at the Pierre, a dozen lawyers, investment bankers and other advisors mill around Gordon's suite figuring out who will ride with whom to the board meeting. Gordon is handed a copy of the Dear Hugh letter, in which he promises to oust the board if necessary. In all the tumult, he repairs to the privacy of the bedroom with his lawyer Cohler and investment banker Siegel. The board will be furious if it ever learns about the side agreement, Siegel warns. But the letter also prevents Liedtke from taking any action to hurt Gordon. Moreover, by this time Gordon has more severe misgivings than ever over whether he can count on the board to treat him fairly. "I'm not suggesting that there is anything wrong with their characters," he will explain, "but emotions had built up pretty high."

Gordon signs the Dear Hugh letter—but only after instructing his lawyer to make a few last-minute additions by hand. The letter now declares that Gordon will muscle the board if necessary, but that he will do so "subject only to my fiduciary obligations." As Gordon will explain: "The world changes." If something unexpected were to happen with the Pennzoil plan before it was signed, sealed and delivered, Gordon will say, "then I was *darned* if I was going to be obligated to continue to support it."

6 P.M.: Arthur Liman of Pennzoil rushes to the Pierre in a cab, locates the signed documents and leaps into a waiting car along with Gordon and three of his advisors. It races down Park Avenue to the Intercontinental Hotel, where everyone else is waiting for the meeting to begin.

There is some other last-minute paperwork undertaken just before the board meeting, only in this case Petersen is involved. After working nearly a year as Getty Oil's investment banking firm, Goldman, Sachs presents a fee letter, which Petersen readily signs. Negotiated by Geoff Boisi of Goldman, it is a fairly standard fee arrangement, with payment based partly on the number of shares changing hands in certain restructuring proposals, and on a percentage of the transaction in other cases. Goldman now stands to make a fee of about $9 million if the unwanted Pennzoil proposal goes through.

But if the entire company is sold—if some other bidder should overtake the deal between Pennzoil and Gordon—Goldman stands to earn a fee of nearly twice that amount. Maybe more.

· 11 ·

The Hotel Intercontinental is a study in old-world masculinity, replete with brass rails, mahogany paneling, Chippendale chairs, wall sconces and leather button-back sofas. On the third floor, past the coat-check room and the pastoral horse-and-trainer paintings, the hallway leads to a labyrinth of passageways and conference halls, each closed off from the next by carved double doors. This mazelike setting was the perfect venue for a meeting of the frustrated, frightened and put-upon directors, who would spend the next twenty-five hours recessing, caucusing and commiserating more than they would spend actually meeting.

Several special factors guaranteed tumult. For one thing, most of the directors arrived without the vaguest idea that their nemesis, Gordon Getty, had combined with Pennzoil, the promulgator of a hostile tender offer against the company. Instead of meeting to respond to a tender offer they had had a week to ponder, they would be forced to act on an extraordinary and complex proposal that had emerged from a word processor only minutes before the meeting began. For another thing, this wasn't merely a directors' meeting. A phalanx of outside advisors flocked to the meeting, assembling like cornermen behind boxers: Gordon had two lawyers and an investment banker; Petersen had seven lawyers and three investment bankers; and Harold Williams of the museum had two of each. For the embattled Petersen, maintaining order and purpose would be like conducting an orchestra with the strings playing Stravinsky, the woodwinds playing Wagner and the brass playing Brahms. Not counting the hangers-on whose brief appearances were never recorded, a total of thirty-one men and one woman—Patricia Vlahakis, of Marty Lipton's firm—would spend all night and then some clamoring in a room just seventeen feet wide and forty feet long.

There were some additional strangers in the directors' midst, and unlike the advisors these people could vote. This was the first board meeting not just for Harold Williams of the museum, but also for the three additional directors appointed by Gordon under the standstill agreement signed three months earlier—and the new directors were hardly wallflower types.

Ann Getty, taking a more active role than ever in her husband's affairs, had personally recruited her husband's new designees to the board. Although Petersen counted many big-name business figures among his loyalists on the board, those who wound up serving at Gordon's pleasure matched them easily for credentials.

One was A. Alfred Taubman, one of the nation's largest real estate developers and the owner of A&W Root Beer, a U.S. Football League franchise and, as of only a few months earlier, Sotheby Parke Bernet, the art-auction house—on whose board he had invited Ann Getty to sit. He was one of the nation's wealthiest men, with a fortune estimated by *Forbes* at $650 million. Although Taubman was becoming known to the public through the social columns, his business reputation was sufficiently large that he would once succeed in obtaining upwards of $2 *billion* of financing from the pension funds of General Motors and AT&T.

Gordon's second new man on the board, intended to fulfill his wish of appointing an academic, was Graham Allison, dean of the John F. Kennedy School of Government at Harvard, whom Ann had met and befriended as a result of her son's enrollment at Harvard College.

The third of Gordon's intended directors had been chosen from a list prepared by the two Martys—Siegel and Lipton—who had suggested candidates who owned major shares in business enterprises and who therefore appreciated the "rights of ownership." He was Warren Buffett, who had become a titan among Wall Street players by operating a hugely successful investment partnership from Omaha. Ann had flown to Omaha to recruit him, but a conflict with another business affiliation prevented him from accepting the invitation. He never did take his seat on the Getty board.

The last of Gordon's designees had the largest reputation of all: Laurence Tisch, the chairman of Loews Corporation, who would quickly go on to become the controlling shareholder and chief executive officer of CBS. With one of the largest investment portfolios on Wall Street and a cue-ball pate that gave him the likeness of Telly Savalas, Tisch was a larger-than-life figure, a dropout from Harvard Law School who'd built with his brother Robert a fortune that *Forbes* would tote up to $1.7 billion—on tactics that wavered between gentility and ferocity. Tisch invested money with actors Michael Landon and Sidney Poitier and producer Norman Lear, among other celebrities. He and Robert Tisch were the acknowledged inventors of the "power breakfast," a phenomenon they cultivated at their 540 Park Restaurant, where investment bankers and corporate chieftains cut more deals over $2.50 cups of coffee than the ward bosses of Chicago conduct over 75-cent beers. Through Loews, his interests included Bulova Watch, CNA Insurance, Kent and Newport cigarettes and a chain of luxury hotels. Tisch was the chairman of the board of trustees at New York University, the nation's largest private school. And like nearly every other person of the greatest means and reputation on Wall Street, Tisch was a friend of Marty Lipton, the museum lawyer.

The presence of four new directors—three of them apparently beholden to Gordon—made this odd and uncomfortable situation seem all the more surreal. Here were some of the wealthiest people in America—Gordon, Taubman and Tisch—sitting among the retired chairman of the board (Berg), his hand-picked successor (Petersen) and the man passed over by Petersen (Bob Miller). There was a former SEC chairman (Williams) and a former assistant Air Force Secretary (Stuart). A dean from Harvard (Allison), a dean from UCLA (Clayburn LaForce) and a former UCLA dean (Williams again). All the directors had other business problems on their mind; Greyhound's John Teets, for his part, was in the throes of a major labor dispute that could make or break the bus-line operations of Greyhound.

There would be many different opinions about what actually transpired that day, but in the end the only one that really mattered was the jury's.

*   *   *

Initially, Petersen stuck with his script and called on his invest
ment banker, Geoff Boisi of Goldman, Sachs, to give a report o
Liedtke's week-old tender offer. Boisi conceded that Pennzoil ha
an exemplary reputation as an oil patch operator, but said tha
Liedtke had put Getty Oil behind the eight-ball. The $100 price o
Pennzoil's pending tender offer was "grossly inadequate," ye
panicky shareholders would probably drown Pennzoil with stock
enabling it to snatch up 30% or more. If that happened, no othe
oil company in its right mind would take an interest in buying
Getty Oil at a higher price. Bart Winokur, Petersen's outside
lawyer, then told the directors that the company had spent fou
days with Gordon's group and the museum working out a plan t
buy time against Pennzoil—but that Gordon had dashed the plar
at the last minute by defecting to the enemy.

"There's a better proposal available," Gordon said. "And it'
available to all shareholders." Marty Lipton then announced tha
the museum had agreed to sell its shares to Pennzoil and the
trust—a disclosure that startled and dismayed those in the room
who were hearing about the deal for the first time. Once again
52% of the company's stock—the same 52% that had once forced
new bylaws on this same board—was acting in concert.

But in a speech that had been carefully rehearsed before the
meeting ever began, Lipton's client, Harold Williams, was quick
to add that "the museum would not be bound by its signature"
unless the board approved the Pennzoil plan. The museum was
walking a tightrope, putting its imprimatur on a plan that the
company's management considered unfair, while at the same time
appearing to apologize for doing so.

After a recess some extra copies of the memorandum of agreement had finally made it to the Intercontinental. With too few
available for everyone in the room, the four-page document was
read aloud. The directors were stunned at the details: Besides
agreeing to pay only $110 a share to the public, Pennzoil and
Gordon had put a time limit on the board. *The proposal was good
only as long as the board meeting was under way!* If the directors
failed to act, Pennzoil could fall back on its original $100 offer.
This is blackmail! many of the directors thought.

Henry Wendt was convinced that Gordon was selling the public
shareholders down the river simply to achieve a long-coveted
ambition to become chairman of the board. "Do you know what
this company is worth?" he demanded. "Have you tried to shop
the company?"

"I've considered and rejected many things," Gordon calmly answered. "This approach is best." Another recess, during which the finger sandwiches and little weenies came in.

Dr. LaForce of UCLA resented the whole thing. "We're being placed in a position of having to make a decision in three or four hours on a take-it-or-leave-it basis," he said. "This involves enormous sums of money and some complex issues. Why was this proposal structured to be withdrawn if it wasn't accepted right now?"

"In the terms of the trade," explained Boisi of Goldman, Sachs, "Pennzoil is using a 'bear hug.' They're using speed and pressure to get a good deal for themselves. *That's* the tactical reason for putting a deadline on the deal."

Marty Siegel rose to Gordon's defense. "Look, the worst thing that can happen is if there's no contract. It's critical to reach an agreement to avoid litigation and turmoil." This whole concept came up only last night, Siegel insisted—and it was a bird in the hand worth $110 a share. "If you want to take a chance on getting more and you're wrong, it's back to one hundred dollars," under the pending Pennzoil tender offer.

Then arose the question of "fairness." Most forms of takeover—particularly "going private" transactions, in which insiders such as Gordon propose to buy in shares—call for an investment banker to issue an opinion stating that the price is fair to shareholders. Fairness opinions are CYA documents, as in "cover your ass," because without them directors are liable to claims that they sold out on the cheap. In recent years the flood of takeovers and management buyouts has made the rendering of fairness opinions a profitable little niche in investment banking—and has touched off concern that opinions are written primarily to justify whatever action the client intends to take. One economic commentator would publicly call them "one of the most ethically dubious corners of the financial world."

As it turned out in this case, the willing seller—the museum—had just obtained a letter from Salomon Brothers calling $110 a fair price. The company, an unwilling seller, was represented by Goldman, which refused to affirm the fairness at that price. If the board approved the Pennzoil deal, it would be in the absence of a fairness opinion—exposing the directors to liability in suits by shareholders.

The debate raged for hours, interrupted by recesses in which the various factions huddled in the nearby meeting rooms. Finally, Gordon's lawyer Tim Cohler decided to force the issue: "You

must now consider what is before you in reality.'' Gordon moved
adoption of the Pennzoil proposal. His wife's friend, Taubman,
seconded. Tisch and Allison, Gordon's other designees, voted
yes. So did Williams. So did Harold Berg, the retired chairman.
Everyone else, however, voted no, and the measure was defeated
ten to five. (One director, David Mitchell, chairman of Avon
Products, had not yet made the meeting.)

There was, perhaps, a weapon the board could wield against
Pennzoil in an effort to obtain more money.

Petersen's people resurrected the concept of a ''self-tender''
offer at a higher price than the $100-a-share offer that Pennzoil
continued to hold open. Under this plan, however, the company
would never actually purchase any of its shares. The board would
simply authorize the self-tender, giving its agents a club to go to
Pennzoil with the threat to knock Liedtke out of the game unless
he agreed to cough up more money.

''Does everyone understand the proposal?'' Petersen asked.

''No,'' Gordon said.

But the proposal was a dead issue without Gordon's endorsement, for the Getty Oil board was functioning for the first time
under new rules. A month earlier, when Gordon and the museum
exercised shareholder consents in their outrage over the company's involvement in the Tara suit against Gordon, the company's bylaws were forcibly changed. Any significant board
action now required fourteen of sixteen votes. Any three directors
could block any action. And Gordon now had four votes.

Among some directors, there was a certain justice in the situation. The museum had helped impose the ''supermajority'' voting
requirement—the very rule change that now prevented the company from getting more money from Pennzoil, for the museum
and all other shareholders. *The museum had shot itself in the foot,*
they thought.

The board was utterly stalemated. The Pennzoil proposal had
been defeated. The plan to buy time through a self-tender did not
have the votes to pass. A recess was called at 1:35 A.M.

None of the directors realized that earlier in the day, Gordon had
signed a letter to Liedtke swearing that he would see the Pennzoil
proposal through—even to the point of trying to oust the board.
Without knowing this, the directors mistook Gordon's unwillingness to consider any other proposal as pure stubbornness,
as an unwillingness to take account of the interests of the public
shareholders.

As a result, even some of Gordon's support on the board was beginning to waver.

Harold Williams of the museum had never witnessed a power play so intense—neither in his experience as a corporate chieftain nor in his years as Jimmy Carter's top business regulator. "Despite my familiarity with tender offers and the takeover rules," Williams would later note, "they feel different when you are in the middle of it than when you are sitting in Washington." Williams resented the implicit threat that Pennzoil could go forward with its tender offer, buy shares in the open market, and combine with Gordon to squeeze out the museum and all remaining shareholders.

Even Larry Tisch, Gordon's most high-powered appointee to the board, was beginning to have doubts about the man at whose pleasure he had been nominated to serve. Among Wall Street's movers and shakers, few enjoyed a reputation for integrity and independence as massive as Larry Tisch's. Certainly, Tisch had been involved in his share of hostile-takeover scrapes over the years, and in 1978 he was caught up in a lawsuit alleging a questionable transaction with Michele Sindona, who was eventually convicted of fraud in connection with the collapse of Franklin National Bank. Yet in dozens of other deals Tisch emerged as someone always willing to deal squarely, and even the targets of his takeover attempts were usually at a loss to criticize him. Thomas Wyman, the chairman of CBS, would one day note: "When Larry Tisch looks you in the eye and says, 'I'm your friend,' you either believe that or you don't. I do." (Wyman, alas, would one day be ousted by Tisch.)

Gordon, according to his friend Bill Newsom, had been filled with pride that a businessman of such a standing would accept his invitation (actually, Ann's invitation) to join the board, and in only a brief acquaintance over the past few weeks had come to count on Tisch as a friend. But Gordon was about to learn that mere friendship did not make Larry Tisch a yes-man.

At 1:45 A.M. the meeting resumed in executive session, forcing the throng of advisors to hang out in the hallways and smaller meeting rooms.

"We're in a very difficult position," Williams said. "There's no way we can do *nothing*. The trust will be in control with Pennzoil. And at the same time, we've got a problem because Goldman won't give a fairness opinion."

Tisch saw only one solution to the board's dilemma: sending

Gordon back to Pennzoil with a demand to cough up more money—$120 a share, perhaps, enough that the company's investment banker could finally render a fairness opinion. That way everyone would be satisfied. Another ten bucks a share, another $400 million for the public investors. That kind of money shouldn't break a $4 billion deal.

Tisch called on the directors to put reason ahead of emotion. "If we're voting for a self-tender just because we're upset at Pennzoil and Mr. Getty, that's not a valid reason," Tisch said. He then turned to Gordon and appealed to his sense of fair play; Gordon, after all, was a director himself who owed a duty of fairness to the public shareholders. "You may have suits if you do this by threat, and you should discuss this with your attorneys," Tisch told Gordon. The board *had* to have a fairness opinion from Goldman. "If someone challenges this transaction, we will say you forced us, Mr. Getty."

"I have done nothing unethical!" Gordon protested.

"This is not ethics. You have not given the board the opportunity to seek a fair price. . . . A small ten-dollar sweetener. *Something* to satisfy this board."

After months of acrimony with the old directors, Gordon was now engaged in the sharpest exchange of the meeting with a brand-new director he counted as a friend.

Gordon lashed back. "When I became sole trustee the stock was selling at fifty, and if I weren't the sole trustee it would *still* be selling at fifty." And now Gordon had delivered an opportunity for the public shareholders to receive $110!

"Go back and bargain for more," Tisch replied.

Gordon asserted that despite what the company thought, $110 a share was fair; after all, the museum's investment banker, Salomon Brothers, said the price was "okay."

"Yes, well they studied it for only *two days*," Harold Stuart retorted.

"Is it reasonable to ask Pennzoil to go over one hundred and ten dollars a share when a major investment banker says one hundred and ten is OK?"

"*Our* investment bankers haven't said that," Teets replied.

As the discussion wore on, Harold Stuart—seventy years old, the board's most senior outside director and a horse trader from way back—became curious about something. Gordon appeared to be behaving unconventionally even by *Gordon's* standards. There was no logical explanation for his unwillingness to extract more money from Pennzoil. *There's something going on here that Gordon isn't telling us about.*

"Do you or any of your bankers, lawyers or agents have any secret agreement with Pennzoil?" Stuart harshly demanded, pointing his finger at Gordon from across the table.

"I've got to see my lawyers," Gordon said.

Gordon's group had reserved its own suite on the third floor of the Intercontinental at the opposite end of the floor, past the elevators and beauty parlor. While the board was in session it functioned as a holding area for two of Liedtke's lieutenants on the scene, Arthur Liman and Seymour Hertz of Paul, Weiss, as well as for Mark Leland, Gordon's friend from the Treasury Department. Whenever the board meeting broke up, Gordon's type-A advisors would hustle back to the suite with their exhausted client in tow. Liman, in turn, would relay the latest news across East Forty-ninth Street to the Pennzoil Suite in Waldorf Towers.

The people in Gordon's holding area were immediately disturbed when they heard about the probing questions that Harold Stuart was asking. The Dear Hugh letter, in which Gordon promised to throw out the board if necessary to get the Pennzoil proposal through, now threatened to cause all the trouble that Gordon's group had initially feared.

"You know," said Gordon's friend and advisor Mark Leland, "this agreement looks terrible. . . . Instead of everyone focusing on the merits of the Pennzoil proposal, they're all going to be so outraged at Gordon Getty that the letter is going to become the focus of the dispute. This agreement is unnecessary. Why don't we cancel it?" Liman, even though he was the author of the Dear Hugh letter, immediately grasped Leland's point. "I thought that the letter would be an embarrassment to Mr. Getty," Liman would recall, "and would create animosity toward him by the board and prevent the board from considering this proposal on its merits." The board was about to resume the meeting, and Stuart would again demand to know whether Gordon had any secret deals with Pennzoil. Liman immediately phoned the Pennzoil suite.

He reached Moulton Goodrum, Pennzoil's lead outside lawyer from Baker & Botts. That letter was Pennzoil's assurance that Gordon would not shrink from the battle, Goodrum explained. *We won't cancel it.*

*   *   *

When the board meeting resumed, Harold Stuart wanted to settle unfinished business. *OK, Gordon, tell us whether you've got a side deal with Pennzoil.*

Gordon's lawyer Tim Cohler responded by reading the Dear Hugh letter. *I will support the Pennzoil plan . . . I will oppose any alternative proposals. . . . I will execute a consent to remove the board if necessary. . . . If Pennzoil doesn't yet own enough shares to create a majority block with me, I will try to get the museum to throw out the board with me. . . .* There was a stunned silence, followed by mumbling and grumbling, followed by open outrage. Henry Wendt shook his head. Medberry was livid at the disclosure of the ''conspiracy,'' and sensed that Gordon's own lawyer was embarrassed at having read the letter. ''The Old Man must be turning over in his grave,'' said Larkin. Williams was startled at the level to which the power-playing had reached.

Harold Stuart wasn't as shocked at Gordon's entry into so ''repugnant'' an arrangement as he was at Liedtke's. That wonderful young man from Tulsa, who had grown up next door to Stuart's in-laws, the Skellys? The son of a judge? Stuart knew that even as a young man Liedtke had yearned to be a big-time oilman, but trying to attain such an admirable goal through such *dishonorable* means—to Stuart, it was like shooting game birds on the ground. ''Pennzoil has violated the code of decency,'' Stuart would say, ''the code of the oil patch.''

Teets, the Greyhound chairman with the seventeen-inch biceps, could not contain himself. ''You mean we've been going through a *dumb show* here, that this discussion we've supposedly been having hasn't really been a discussion at all, that all along when you were pretending you were having discussions . . . you had no intention of listening to *anything* we said?''

Gordon sat silently. Now maybe the directors could begin to understand how he felt when he had been led out a back door. ''Holding back on the Dear Hugh letter may have been bad form,'' Gordon acknowledges today. ''But maneuvering me out of the boardroom in Houston was *bad faith*.''

Lipton immediately sensed that everything would fall apart— and with it the chance for the museum to sell its shares—unless the board's attention was quickly refocused on the deal. Lipton suggested the following counterproposal: Pennzoil withdraws its tender offer, removing the gun it has pointed at the board. Pennzoil then pays not only $110 in cash, but a debenture—a form of a bond—with a value of $10, making the total deal worth $120 a share. Goldman, Sachs would consent to issuing a fairness opinion at $120.

Stuart asked Lipton whether the other provisions of the Pennzoil deal would remain in effect. According to the board minutes, the exhausted Lipton tersely answered, "Who cares?" Two years later, Lipton would insist that he really said, "Who knows?" Exactly *what* Marty Lipton said would become a highly relevant issue when a jury would have to figure out whether *anyone* at the marathon board meeting knew what was really happening.

When Petersen asked Gordon for his comment on the counteroffer, it was 2 A.M. "I'll sleep on it," Gordon answered. Petersen and Wendt couldn't believe their ears. The board was about to begin its sixth recess of the night, but nobody was going to bed yet.

Amid all the traffic of directors and advisors outside the boardroom, Larry Tisch ambled past the hair salon and the elevator bank to the private suite where Gordon's group and Liedtke's lieutenants had set up their headquarters for the meeting. There he had a talk with Pennzoil's main person on the scene, his old friend Arthur Liman.

"Arthur," Liman recalls hearing from Tisch, "get your client to go up the extra ten dollars and we can close the deal tonight. It doesn't have to be ten dollars in cash." The sweetener could involve "optics," something that looks like real money but isn't—a piece of paper entitling shareholders eventually to receive another ten bucks. "You go and persuade them that they should go up the ten dollars and we can all go home tonight and have the deal done."

When Liman relayed the demand across the street to the Waldorf, Liedtke was furious. He had already increased his price from $100 to $110—and now, without any competing offers on the table, he was being shaken down for another $10. "Tell them we'll get back to them in the morning," Liedtke said. "Meet me for breakfast at eight o'clock."

At 2:30 A.M., fourteen directors, exactly the number required under the new bylaws, voted to authorize a formal $120-a-share counteroffer to Pennzoil. When the meeting broke up, the directors returned to their rooms at the Helmsley Palace to hit the sack for a few hours until the meeting was reconvened. But Harold Stuart, the longtime outside director, and Harold Berg, the retired chairman, were so wound up over the Dear Hugh letter that they couldn't sleep. They ordered bacon and eggs from the twenty-four-hour room service at the Helmsley; even though the hotel bars were closed, they persuaded room service also to deliver them a few drinks. "We were just about out of our minds," Berg would recall.

Two of Petersen's people—Geoff Boisi and Herb Galant—stayed behind at the boardroom to set out the $120-a-share counterproposal in a handwritten letter to Jim Glanville, Pennzoil's investment banker.

> Jim,
>   Attached is the transaction the Getty board approved. After you have had a chance to review this with your client, or if you have questions, please contact me at the Helmsley Palace or at Dechert-Price.
>
> Geoff Boisi

Petersen's people walked across Forty-ninth Street, rode to the forty-first floor of the Waldorf and slipped the proposal under the door of Pennzoil's slumbering investment banker. It was 4:30 A.M.

· 12 ·

"Goddamn sharks!"

The occasion was breakfast at the Waldorf, the speaker was Hugh Liedtke and the subject was takeover lawyers and investment bankers. Marty Lipton had chiseled him into increasing his initial price by $10. Now, Getty Oil's investment banker apparently figured that *he* has to do something to justify his fat fee. Liedtke had gone to bed believing that Getty Oil would settle for some "optics"—but then Boisi shoves a letter under a hotel-room door at 4:30 A.M. demanding yet *another* $10! Hugh Liedtke was bidding against himself—and experienced horse traders like Liedtke absolutely do not lower their guard to such seduction.

"Goddamn sharks!"

In investment banking parlance, optics signifies part of a deal, often a sweetener, that isn't everything that it appears. It may be a promissory note, a bond, a debenture, a "stub," a "boot," a plain old piece of paper—anything entitling the bearer to receive some money ($10, say) in a certain amount of time (twenty years, for instance). However, this piece of paper earns a subpar rate of interest, which means it really isn't worth $10 cash at all. Liedtke might have agreed to sweeten his offer with such a stub, but the pre-dawn counterproposal received by his investment banker wasn't optics. This stub was supposed to pay interest at the *market* rate. *This was real money they were after!*

Liedtke wouldn't have it.

It was a credit to Liedtke's genius as a negotiator that before the day was out, he would finally celebrate a deal giving him three-sevenths ownership of Getty Oil. Yet in this case, even Liedtke would prove too clever for his own good.

Ever since Gordon's abortive takeover attempt in London, Geoff Boisi had been pawing at the ground, ready to gallop across the oil patch in search of a white knight to rescue his client Getty Oil. In fact, he had already drawn up a list of major oil companies that could qualify to serve as Getty Oil's Lancelot against the dragonlike Gordon.

Boisi's incentive was now greater than ever. He passionately believed that the $110 a share being offered by Pennzoil was an unfairly low price. Indeed, months earlier, when Gordon was proposing a "leveraged buyout" that would have left him in control of Getty Oil, Boisi had determined that only a buyout price of $120 or so could be declared fair to the public shareholders. How could Boisi now cave in? How could he possibly certify Pennzoil's price of $110 to be fair?

"Sometimes," Boisi would say, "we have to be the bastards in the background who hold the line."

But it went without saying that Boisi had an additional incentive to seek a white knight. If the entire company could be purchased—Gordon's shares, the museum's, the public's—at a higher price than Pennzoil was offering, Boisi's multimillion-dollar fee would escalate into a multi-multi-million-dollar fee. Selling Getty Oil in its entirety would be far and away the biggest transaction in history—one that could earn the biggest investment banking fee in history.

But under what authority could Boisi shop Getty Oil?

The company had spent four days trying to gain Gordon's concurrence in a plan to stave off Pennzoil and to arrange the "controlled sale" of the company—and had failed. The previous night, Boisi had again argued for a "self-tender offer" to stall Pennzoil so the company could find another buyer, but that proposal too had collapsed. Obviously, under the company's new bylaws, Getty Oil could never muster the fourteen votes necessary to authorize Boisi to break Gordon's deal with Pennzoil through the sale of the whole company.

Yet, Boisi thought, there had been a *consensus* of the board to search for a white knight. Moreover, Sid Petersen had expressed enthusiasm about looking for another buyer. And finally, Boisi knew that under certain conditions, Gordon could easily be switched from being a *buyer* to a *seller* of his Getty Oil shares. Gordon would never tolerate remaining a minority owner if another company bought up all the other shares.

At 7 A.M., after barely two hours' sleep, Boisi decided there was no point in waiting any longer. He had been informed that Pennzoil was "irked" by the $120 counterproposal he had slipped under the door barely two hours earlier. Boisi believed that the time had come to test the market, to play the field a little bit, to see whether this company, as he believed, was worth more than Pennzoil claimed. The calls went out to Exxon, Mobil, Chevron, Shell—even General Electric and the government of Saudi Arabia. And one to Texaco.

John McKinley was still vacationing in Alabama when Boisi tried to reach him at Texaco headquarters in White Plains, so the call was switched to Alfred DeCrane, the president. Boisi was delighted to reach DeCrane, whom he'd come to know and respect when Goldman, Sachs helped arrange a merger between INA and Connecticut General, an insurance company on whose board DeCrane served. DeCrane made copious notes of the call: "recessed board meeting: the sense of the board is, if there were an attractive offer for the entire company [they] would sell. The museum is a definite seller. Gordon is a seller *or* a buyer . . . he doesn't want to be a *minority party*. . . . Is Texaco interested in principle?"

Five days after Texaco's final Mukluk misadventure, DeCrane knew as well as any oilman the value of Getty's reserves. He told Boisi that Texaco couldn't get excited about Getty's cable TV network or insurance company. Nevertheless, he said, as long as Goldman was putting together a list of companies to keep posted, Texaco would be glad to be included.

It was remarkable how many things were coming full circle. One of the first, fateful signs of trouble at Getty Oil had occurred four years earlier, when Gordon unexpectedly voted against Sid Petersen's brilliant move to acquire Employers Reinsurance Corporation. Now ERC would become the basis for settling the price dispute with Pennzoil—and ultimately enable Liedtke's deal with Gordon to go through on Pennzoil's terms.

There was no denying the extraordinary success of Petersen's ERC acquisition. Getty Oil had paid $570 million for the company, whose value had immediately begun swelling on a tide of

ever-higher underwriting profits. Now, in 1984, Boisi of Goldman, Sachs reckoned that ERC was worth as much as $1.4 billion.
In fact, one of the chief objections expressed against Pennzoil's
price during the previous night's board session was that $110 a
share didn't begin to take account of ERC's true value. Liedtke
and Gordon obviously intended to sell ERC; why should they reap
the profits of an adroit investment made by someone else?

Marty Siegel of Gordon's group also knew something about
ERC; he had been ERC's investment banker four years earlier
when Getty Oil bought the company. The board was dreaming if it
considered ERC worth $1.4 billion, Siegel told Liedtke's lieuten-
ants.

Suddenly, the Pennzoil group had a brilliant idea. Hugh Liedtke
had spun off more companies over the years than most chief
executives even think of buying. Why not spin off stock of ERC to
the public shareholders and to the museum? That ought to settle
this valuation dispute. *Let them eat ERC.*

In the hours before the Getty Oil board was to reconvene at 3
P.M., Liedtke's lieutenants drew up a new proposal giving shareholders a choice: $110 a share, as proposed by Pennzoil the day
before, or $90 a share plus a piece of ERC. Liedtke's lieutenant
Liman would carry the proposal to Lipton, who had clearly assumed a role as the chief go-between.

"It won't sell," Lipton told Liman. "It's too cute."

The scene was tumultuous, crazy. The most important board
meeting in the history of Getty Oil was minutes from resuming
after a long recess. A crowd of nearly a dozen has congregated in
the suite used by Gordon's group and Liedtke's lieutenants as a
holding area. Hugh Liedtke, one of the wiliest dealmakers in the
oil patch, is convinced he is being looted; Getty Oil is convinced *it*
is being looted. Marty Lipton, the nation's premier takeover lawyer, is bantering over a multibillion-dollar deal with his friend
Arthur Liman, a trial lawyer of nearly equal distinction in his
field. Lipton's client, the museum, has been assured the Pennzoil
counterproposal is fair, yet Lipton himself is saying that Getty Oil
will *not* consider the same proposal to be fair. Liedtke . . . Liman
. . . Lipton. . . . It would be no wonder that a jury would spend
four and one-half months making sense of it.

Liman did not relish calling across the street to the Waldorf and
telling his client that he couldn't even get the new proposal inside
the four walls of the boardroom. Even to a high-powered lawyer

like Liman, "Chairman Mao" was an intimidating figure. But Liman dialed the phone anyway and got Liedtke up from the lunch table. Liedtke let loose a chain of expletives.

In the face of the tirade, Liman drew his breath. "Look," Liman told his client, "there is no way we are going to get acceptance of the proposal today unless you come up with a sweetener. You may not like it, but that's what they're doing to us."

Liedtke thought for a moment. *There comes a point in trading where some people feel they have to chisel some more. That's what this was—pure chiseling.* But Liedtke was so close he could practically taste that lovely, heavy, sour West Coast crude.

"Okay," he finally told his lawyer. Here was the deal: Pennzoil would pay the same $110 in cash it had already offered. ERC would be sold. All the proceeds over $1 billion would be paid to the former shareholders of Getty Oil. Pennzoil and Gordon would promise that at the very least, the shareholders would get an extra $3 a share within five years.

Liedtke hung up the phone. Liman, the Pennzoil lawyer, turned to Lipton, the museum lawyer. Board members were already filing into the meeting room. Amid the commotion and banter in Gordon's little suite, Lipton did some fast mental arithmetic. *Three dollars within five years. . . . In today's dollars, adjusted for interest over five years, that's only about $1.50. . . . This still will not get past the board. . . . If Pennzoil guaranteed to pay an extra $5 in five years, that would be worth $2.50 in today's dollars. Maybe, just maybe, the board would sign off on that.*

"Get them to make it a guarantee of five dollars," Lipton said. That would give the deal a total present value of $112.50.

Lipton did not realize what he was putting his friend Liman through. *I value my life now,* Liman thought.

"I'm not going to go back to Hugh Liedtke asking him to bump this again," he told Lipton.

So the Pennzoil lawyer took matters into his own hands—always a grave risk where "Chairman Mao" was concerned. If the board completely signed off on the $112.50 deal, *then* Liman would present it to Liedtke. "But I'm not going back asking him to keep raising it without having a firm deal," Liman repeated. Lipton agreed to seek the board's approval.

Lipton took his seat in the boardroom and slipped a hurriedly written note across the table to Geoff Boisi, the company's invest-

ment banker. Lipton's note outlined the "stub" portion of the
$112.50 deal as follows:

> Believe if we go back firm at $5 instead of
> $3, will accept.
> Present value $2.50-$3.
> Museum accepts.
> Salomon advises fair.

This is just typical, Boisi thought. *Once again Pennzoil has
gone directly to the museum to try to conjure up a deal.*

The directors weren't quite sure what to expect when the meeting finally resumed. And in fact it would be impossible to say
exactly what did happen when it resumed. Boisi would recall
informing the directors that he had spent the morning making
contact with potential white knights; the sixty-eight-page meeting
notes, however, reflect no such report. Lipton took the floor to
outline the $112.50 deal he had worked out minutes earlier with
Liedtke's lawyer. *Lipton seems vague, inconclusive,* Harold
Stuart thought. "Mister New York Lawyer," as director Chauncey Medberry would call him, "was delivering a speech in the
presence of three conflicting conversations," Medberry would
recall. Even the chairman of the meeting had trouble following the
proposal, but that was hardly unusual under the circumstances.
"Mr. Lipton said a lot of things I didn't understand," Petersen
would say.

There was one thing that everyone in the room understood,
however. Boisi still was refusing to give the board the comfort of
a fairness opinion.

As the board went into its final recess, the third-floor hallway of
the Hotel Intercontinental was like a maze teeming with frightened
lab rats. Lawyers and board members worked their way to Gordon's suite to demand more money from Pennzoil's lawyer;
Liman would not budge. Boisi's associates from Goldman, Sachs
took up positions at the public pay phones, calling their office to
see whether any expressions of interest in taking over Getty Oil
had occurred. No competing offers appeared imminent.

Boisi himself, still refusing to give the board a fairness opinion,
was cornered in the nearby Vanderbilt Room by Lipton and
Harold Williams of the museum and Larry Tisch, Gordon's designee on the board. The museum officials implored Boisi to

declare the $112.50 to be fair. Salomon Brothers had determined the deal to be fair to the museum, so why couldn't Goldman do the same for this board? *No way,* Boisi responded. Tisch blew up and read Boisi the riot act, pointing out that oil prices were beginning to edge lower, demanding a fairness opinion to give the board some peace of mind. *No way,* Boisi said. Here was a gawkish investment banker in his thirties standing up to one of Wall Street's giant figures. *No way.* At the end of the confrontation, Lipton, despite failing to convince Boisi, congratulated him on his resolve.

Before filing back into the boardroom, the "old" directors—everyone but Gordon and his new board appointees—held a rump meeting with the company's advisors to figure out just what to do. Boisi told the directors that it could be a matter of time before another company swept in with a fair offer. Chevron was a possibility. Texaco was a possibility. But these companies needed just a little more time to study the situation. Bart Winokur, Petersen's Philadelphia lawyer, explained to the directors the risk of approving the Pennzoil proposal in the absence of a fairness opinion. Lawsuits by shareholders were probably inevitable, he said.

But the battle-weary, sleep-deprived directors had had enough. Their resolve was dissolving. Some were growing annoyed with Boisi's recalcitrance; although he was on their side, his refusal to sign off on the deal was depriving the directors of any peace of mind. There was no assurance that a competing bid—a fair bid—would emerge in enough time. Henry Wendt remained convinced that Gordon would fire the board—either by combining with the museum or with Pennzoil, if Liedtke went forward with his tender offer, which was still hanging like Damocles' sword over the board. Harold Berg, the retired chairman, said that if the board walked away from the Pennzoil deal, the stock price might go back into the tank. *We'll have lawsuits all over the place,* Berg thought.

"It looks to me like there's not much else we can do," said Berg, who carried the credibility of having been J. Paul Getty's handpicked successor.

The directors ended the last recess at 6:15 P.M. and went back into the boardroom so that the $112.50 deal could be finally put to a vote.

Liedtke's man Liman watched the directors filing through the hallway. "Well?" he demanded. "Do we have a deal? Can I go home? Can we get it over?'' The last director pulled shut the door behind him.

"I felt like a hostage," Liman would later recall.

Across Forty-ninth Street, Pennzoil's PR people had set up a little pressroom and had begun roughing out a press release, just in case the Getty Oil board relented. Every good PR man knows that newspapers like quotes, so the Pennzoil publicists began fabricating a few of them to issue in the name of Sid Petersen; obtaining Petersen's approval would be a mere formality once the Getty Oil board had agreed to put Pennzoil and Gordon in command. "We are pleased that through this transaction, our shareholders will receive an excellent price for their holdings," the PR people had Petersen saying. "The differences which have caused so many distractions for the management and board of this company have been concluded."

Petersen would never get the chance to approve the remarks. After going to all that trouble, the Pennzoil PR people were told that their release would not be needed after all. One of the writers, Terry Hemeyer, ripped it to shreds and threw the pieces in his briefcase. They would take another crack at it later.

All day, the shares of Getty Oil had been suspended from trading on the New York Stock Exchange at the company's request—but without explanation. The outside world certainly knew something was happening, but remained oblivious to what it was. There was widespread speculation that Liedtke might be proposing creation of a royalty trust at Getty Oil. Many believed that he was working up a leveraged buyout with Gordon as his partner. Others thought that perhaps Getty Oil was trying to heal the rift within itself. Some were sure the company was planning an offer for its own shares to defeat the $100-a-share Pennzoil tender offer. Elements of truth existed in all the speculation.

A *Los Angeles Times* reporter sat down to the unhappy task of writing an article about the board meeting for the next day's editions:

> Directors of Getty Oil Co., embroiled for months in a bitter power struggle, met Tuesday in New York for the second consecutive day of closed-door meetings. . . .
>
> After issuing a terse statement in the morning . . . a company spokesman said late Tuesday, "There will be no announcement tonight."
>
> The official refused further comment on any facet of the board meeting, including whether the meeting had concluded Tuesday,

whether it will resume today and where it is being held in New York.

The nation's former number-one watchdog of shareholder rights was uncomfortable moving the adoption of a proposal that had not been blessed as "fair" to the public. But Williams nevertheless did take it upon himself to move acceptance of the $112.50 Pennzoil deal. Director David Mitchell of Avon Products, who had been absent for the entire meeting until now, seconded the motion.

In the formal discussion, Gordon's designee Al Taubman tried to placate the embittered directors. The public invested in Getty Oil knowing full well that control lay in the hands of a 40% shareholder, Taubman said. It's like the boy who dates a girl with pimples, he added; after marrying her, the boy has no right to complain about her face.

"In this case," said director Harold Stuart, taking his parting shot at Gordon, "she didn't used to have a pimple."

The vote was fifteen in favor of the Pennzoil offer, with one—Chauncey Medberry of Bank of America—against. "I wanted to save that for Medberry," he would explain. "I wanted to get it in the record. I thought that there would be lawsuits against the individual directors very easily in today's society." Aptly, Medberry would be called upon to make this explanation before a jury.

The board unanimously commended Petersen. It undertook the typical step of awarding him and his top executives improvements to their golden parachutes, the employment contracts that would stick the new owners of Getty Oil with millions of dollars in extra severance payments. There was, however, a piece of paperwork that was not attended to.

The five-page, typewritten memorandum of agreement containing the original $110 price—a document signed by Gordon and Liedtke, and by the museum's Williams "subject to" board approval—contained very little on the final page. It read simply:

Joinder by the Company
The foregoing Plan has been approved by the Board of Directors.

Getty Oil Company

January 2, 1984   By _____

As the directors and advisors began gathering up their overcoats to leave the boardroom they had first entered twenty-five hours earlier, nobody from Getty Oil stopped to fill the blank space above the signature line.

Was it a done deal? Was the ballgame over? Had a meeting of the minds occurred? Had Gordon, the museum and the board of Getty Oil expressed their intent to be bound? If the account of Pennzoil's Arthur Liman could be trusted, there wasn't much doubt by the way that everyone acted at the conclusion of the marathon board meeting.

As Liman would later reconstruct the event, Lipton of the museum and Siegel of Gordon's group raced from the boardroom, informed Liman that the board had agreed to a firm deal at $112.50 along with all the other terms in Pennzoil's memorandum of agreement. Liman called Liedtke, who agreed to the price change. Liedtke also consented to picking up the tab on Sid Petersen's golden parachute. By Liman's recollection, everyone then returned to the boardroom, leaving Liedtke's lawyer waiting outside.

A moment later, the doors again flew open.

"Congratulations, Arthur. You've got yourself a deal!"

Liman would distinctly recall hearing the words from one of the Martys, although he would not remember from which. Lipton would never recall making this remark. Siegel wouldn't remember for sure one way or the other.

Liman would recall asking for permission to enter the boardroom, where the directors were finally getting ready to leave, and shaking hands with everybody in sight.

"Congratulations," Liedtke's lieutenant said to the directors in turn.

"Congratulations to you," they answered.

"Thank you for putting in all these hours," he told them.

Never, however, would so many handshakes be remembered only by one-half of the clasp. Lipton would not remember handshaking. Neither would his associate Pat Vlahakis. Neither would Boisi, Medberry, Tisch nor others when asked under oath. Marty Siegel, however, would eventually tell a reporter: "Liman may have shaken some hands."

What *really* happened? It would be the jury's job to decide.

*   *   *

At the conclusion of the board meeting, Gordon approached Petersen to offer a friendly good riddance. "Good luck in your career," he said. Petersen's thoughts at that moment were, he says now, "unprintable." Marty Siegel also approached Petersen with an outstretched hand; Petersen refused to take it.

That evening Petersen left for Los Angeles aboard the company's Falcon jet with Chauncey Medberry, Fritz Larkin, Bob Miller and several others. The following day a middle-level manager who revered Petersen walked into his office and was startled at the weary and haggard sight of the Imperial Chairman. A blemish had appeared on Petersen's face.

"Sid, we're a hundred percent behind you," the manager said.

Tears welled up in Petersen's eyes.

"Thanks," he said.

Hugh Liedtke was euphoric. Those billion barrels of oil were now practically his. In fact, as far as he was concerned, they *already were* his, except for the need to clear a few run-of-the-mill hurdles. Obtaining the necessary federal antitrust blessing was a foregone conclusion; by strengthening Pennzoil and Getty Oil, this merger would increase competition rather than lessen it. Delaware law required that a majority of shareholders approve the transaction, but there wouldn't be much suspense over the shareholders' vote—not with Gordon and the museum voting 52% of the stock in favor of the deal. There was, of course, always a danger that some interloper could make a grab for the publicly owned stock; there could be no "contract" for those shares. But Pennzoil's memorandum of agreement contained a provision giving Pennzoil the option to buy eight million Getty Oil shares that would be issued if it became necessary to strengthen Pennzoil's control. These options are common in a takeover deal. "Its purpose is to lock up the deal and keep predators from sniffing around," Liedtke would explain.

Finally, there *was* a host of final implementing documents that had to be prepared. But Liedtke considered that mere lawyers' work, and the lawyers had begun the job immediately following the board meeting.

For Liedtke, tonight—January 3, 1984—was the night to celebrate.

Liedtke called his new partner, who technically was his new boss.

"He's not here just now," said Ann, the wife of the chairman to be. Ann was more effervescent than ever. "Would you care to come up and join us in some champagne to celebrate the occasion?" Larry Tisch, in fact, was also coming over to toast the deal; neither he nor Gordon bore any grudges from their sharp exchanges of the night before. They'd both done only what they had to.

"I very much appreciate that," Liedtke told Ann. "But I'm running very late for a previous commitment to have dinner with some other people." Liedtke had promised to head downtown to the restaurant "21" for a night out with the boys, with his innermost circle: Baine Kerr, Pennzoil's president, and Perry Barber, the company's general counsel, along with Glanville and one of his partners from Lazard Frères.

"Maybe I can take a raincheck," Liedtke told Ann.

In New York, as in many other jurisdictions, the formation of a contract generally requires something else besides a meeting of the minds and an intent to be bound. In case after case, for more than a century, the courts have found that a contract does not exist until parties have come to terms on each and every *essential* term of the deal. Take the sale of a house. If a higher offer comes along after the sales agreement has been signed, the seller cannot break the original deal just because of a misunderstanding over whether the door knocker stays; the closing must go forward on the original terms. But if the sales agreement does not address who pays the last month's property taxes of $1,000, the seller may very well remain free to walk away from the deal. It is with such good reason that sales agreements for real estate are fill-in-the-blank affairs, with all the essential terms except price written out in standard form.

Alas, no ready-made forms exist for multibillion-dollar deals. As soon as the Getty Oil board meeting had broken up at 7 P.M., the word was passed: *Everyone involved in drafting documents, please meet right away.* Before the night was out, more than thirty lawyers, investment bankers, PR people and others would be in on the act.

The main office of Paul, Weiss, Rifkind, Wharton & Garrison— Pennzoil's New York law firm, where Arthur Liman was a senior partner—lies just outside the high-tone corporate district running west to east from Seventh Avenue to Park. The building occupied by the firm, at East Fifty-fifth Street, just across Third Avenue from Ray Bari Pizza and Clancy's Bar and Grille, is inhabited more by carpet companies than any other variety of enterprise. The building also houses the East Side Cinema, where the early showing of Walt Disney's *The Rescuers* was getting under way when the lawyers began arriving.

Paul, Weiss also has offices on Park Avenue, which is where Getty Oil's lawyers initially went by mistake. When they finally showed up at the right place, Pennzoil already had six lawyers at work: the host, Arthur Liman, who played traffic cop to the arriving hordes; two of his Paul, Weiss partners, and three lawyers from Houston's Baker & Botts led by fifty-year-old Moulton Goodrum, a softly spoken and painfully careful M&A lawyer. When Bart Winokur of Getty Oil showed up (with at least three lawyers in tow), he recognized another of Pennzoil's Houston lawyers, Bill Griffith, as the flat-topped tax man who, months earlier, had showed up at the Century Plaza Hotel in Los Angeles with Boone Pickens to present the ''Blue-Gray'' merger proposal. Gordon's lawyers were also out in force.

While the lawyers gathered in a nondescript conference room to divide the drafting duties, a bevy of PR people gathered elsewhere in the offices to bang out a news release. Hank Londean, Getty Oil's chief PR man, introduced himself to his counterpart from Pennzoil, Richard Howe. ''Congratulations,'' Londean said. ''*You* are stuck with Gordon Getty now.'' Two of Wall Street's premier PR consultants were also in attendance: Gershon Kekst of Kekst & Company, representing Gordon, and Richard Cheney of Hill & Knowlton, the world's largest PR firm, representing Getty Oil.

If the Keystone Kops had ever drafted a multibillion-dollar press release, it might have been like this. Terry Hemeyer of Pennzoil recalled the press release that he had drafted and ripped into little pieces earlier in the day. He pulled the scraps from his briefcase and undertook to reassemble them with tape, trying to align *$5* with *billion* and *Sidn* with *ey Petersen*. This quickly proved more trouble than actually writing, so the PR people began roughing something out by longhand.

Gershon Kekst was disgusted at the whole affair. He would later swear that he detected a strong streak of regional bigotry in Pennzoil's Houston-based representatives, with their derisive-sounding references to the *New York* lawyers and the *New York* PR people. Indeed even a little well-intentioned Texas pride can come off as outright chauvinism in a place like New York. But Kekst would not be the last New Yorker to raise the issue where Pennzoil was concerned.

After settling such burning issues as whether the announcement should carry a New York or Los Angeles dateline, the PR people discovered that Paul, Weiss had scheduled but one overworked secretary on the night shift. Moreover, after waiting what seemed an eternity for the typedraft of their announcement, they discovered that the typist had a trigger-happy right middle finger that insisted on identifying the museum as the *musieum* and the proposal as a *proposial*. The intrepid publicists then tried to make some copies to pass around, but only after they had completed a frantic search for the one man in the building who at that hour had a key to the photocopying room.

When finally, at about nine o'clock, the press release was circulated, it was universally pronounced to be unacceptable by the lawyers for Getty Oil. It said the Getty Oil board had "voted to accept a plan" when in fact the details of the plan had yet to be worked out, at least as far as the Getty Oil lawyers were concerned. It suggested that shareholders would receive a stub worth only $5 within five years when in fact $5 was the *minimum* amount; depending on how the sale of ERC came out, the stub might be worth a great deal more than that. The press release also suggested that the museum would be voting its shares in favor of the merger *after* those shares had already been sold back to Getty Oil as part of the going-private proposal.

The Pennzoil PR people left the Paul, Weiss offices and went to dinner.

What happened next at the law firm would eventually become the subject of more contradictory testimony than a murder committed in a hall of mirrors. This much, however, would be clear: Instead of laying to rest the fear and loathing that had gripped Getty Oil for a year, the Pennzoil deal had assured its perpetuation.

The sleepless Marty Lipton had gone home to bed after the marathon board meeting, dispatching his young associate Pat Vlahakis—a Bryn Mawr math and English graduate barely two years out of Columbia Law—in his place. "I was very tired and asked her to sort of hold the fort," Lipton would explain, "until it was necessary for me to participate." Vlahakis, of course, was plenty exhausted herself, besides recovering from a severe case of laryngitis. The only women who had been present among the three

dozen participants in the board meeting, the newly minted lawyer
was now also the only woman present in nearly as large a group—
and the only representative of the museum in a sea of killer
lawyers.

Vlahakis wasn't about to let down her guard while holding the
fort for her boss.

While the PR people were drafting their release, the retinue of
lawyers representing all sides agreed that Pennzoil, as the acquiring company, would produce the first draft of the final merger
documents. Having settled that, the lawyers divided into their own
little groups, leaving Vlahakis by herself to wander amid the
overstuffed chairs in the Paul, Weiss reception area. At one point,
she wandered by a room to discover that Gordon's investment
banker, Marty Siegel, was on the telephone discussing the deal
with a *Wall Street Journal* reporter. "Hang up that phone or I'll
hang it up for you!" she yelled.

Finally, Vlahakis was collared in the hallway by Bart Winokur,
the Getty Oil lawyer, who had a big problem on his mind. Under
Pennzoil's memorandum of agreement, Getty Oil was going to
buy the museum's nine million shares before a single share had
been purchased from public hands—the museum had insisted on
being "taken out" immediately. But once the museum was out of
the picture, Gordon would be dangerously close to majority control, and there was no telling what kind of unilateral act Gordon
might do with that kind of power; merely by picking up a few
shares on the stock exchange, he and Liedtke could easily grab
majority control and do as they pleased, changing the terms of the
deal, perhaps, and ripping off every other shareholder in sight. In
its paranoia, Getty Oil did not want the museum's "at large"
shares to fall under Gordon's control until the public shareholders,
too, had their money safely in their hands.

As Winokur described the company's anxiety, Vlahakis saw
her career flashing before her eyes. Suddenly people were raising
questions about how and when the museum shares would be purchased. But the museum by God wasn't *about* to wait around for
its money. *Somebody* was going to have to take out the museum
immediately, no matter how the deal were structured; that's what
Marty Lipton and Harold Williams wanted. There was no way that
Vlahakis could let the lawyers come up with some other scheme
that would prevent the museum from getting its money immediately.

"There's *no deal!*" she suddenly announced to every lawyer in sight. *"There is no deal as far as the museum is concerned!"* She let loose an expletive.

"Look," said Stuart Katz, one of Getty Oil's New York-based takeover lawyers. *"Something* happened in that board meeting tonight and we must come out with a press release. Let's finesse the dispute. Let's just postpone the issue."

Vlahakis agreed. But to make certain that she was not committing the museum to something irreversible, she took a copy of the press release that the lawyers were drafting by hand and scratched out an entire paragraph dealing with the museum's role in the transaction.

As finally written, the news release described a deal known in the trade as a "reverse triangular merger" because it involved two parties—the trust and Pennzoil—creating a *third* company that would end up merging with all of Getty Oil. All shareholders except Gordon would receive a total of $5.3 billion; Pennzoil would put up about half the money to buy out the shareholders, the new Getty Oil itself would borrow the rest. Gordon, although not spending any of the trust's money, was, indirectly, a buyer himself, since the trust would be the majority owner of the enterprise buying in the shares. Besides describing these terms, the press release would read:

FOR IMMEDIATE RELEASE
JANUARY 4, 1984

LOS ANGELES—Getty Oil Company, The J. Paul Getty Museum and Gordon P. Getty as trustee of the Sarah C. Getty Trust announced today that they have agreed in principle with Pennzoil Company to a merger of Getty Oil and a newly formed entity owned by Pennzoil and the trustee. . . .

The agreement in principle also provides that Getty Oil will grant to Pennzoil an option to purchase eight million treasury shares for $110 per share.

The transaction is subject to execution of a definitive merger agreement, approval by the stockholders of Getty Oil and completion of various governmental filing and waiting-period requirements.

Following consummation of the merger, the trust will own 4/7ths of the outstanding common stock of Getty Oil and Pennzoil will own 3/7ths. The trust and Pennzoil have also agreed in principle that following consummation of the merger they will endeavor in good faith to agree upon a plan for restructuring Getty Oil on or before December 31, 1984, and that if they are unable to reach such an agreement, then they will cause a division of the assets of the company.

By about midnight, everyone present had signed off on the press release, which would be issued early in the morning under the letterhead of Getty Oil Company. Vlahakis and the lawyers for the trust and Getty Oil left the building, walked past the just-darkened theater marquee and headed into the night on Third Avenue with the temperature falling below the freezing point.

The Baker & Botts lawyers whom Pennzoil had flown in from Houston remained in the law office, pondering the final four-party merger agreement that it was their assignment to write. At about 1:30 A.M., after an incredibly exhausting day, two of them decided to pack it up, leaving behind another two to continue the drafting. They had left all the other lawyers with the impression that the documents would be drafted first thing in the morning. But even if it took a little longer than that, the essential terms of their deal would be cast in stone once the news release went out in a few hours—a news release that eighteen months later would be stamped with a little sticker reading PLAINTIFF'S EXHIBIT #5.

· 13 ·

If Wall Street has a nervous system, it is the Dow Jones News Service. When a company announces a takeover, increases its dividend, fires its chairman, introduces a hot new product or does anything apt to move its stock price, the world usually learns of it first from the Dow Jones ticker, so named because of the thousands of teletype machines that spend all day clacking in investment offices of the nation. Just as fingers and toes twitch on electrical impulses from the brain, the stock traders of the nation jerk into action on the news spewing forth from the ticker.

Bruce Wasserstein, co-head of the mergers and acquisitions department of First Boston Corporation, was still slumbering in a Houston hotel room when the ticker carried the news of Pennzoil's agreement in principle at 7:43 A.M. Central Time on Wednesday, January 4. Wasserstein was visiting Houston from New York to arrange a possible deal that had nothing to do with Pennzoil or Getty Oil. But one of his astute colleagues back at the office recognized the Pennzoil-Getty announcement as something that Wasserstein would want to know about immediately. Among many New York investment bankers such as Wasserstein, the announcement of an "agreement in principle" was little more than a form of notice to the market that companies were *trying* to work out a deal; it hardly meant that a deal was done. After the phone rang in his room at the Four Seasons Hotel, it would be two days before Wasserstein ever got back to bed.

Wasserstein was a smirking thirty-six-year-old with a hurried manner and reddish hair—thin on top and long on the sides, giving him a resemblance to a certain well-known television host for children. A Harvard lawyer, a Harvard MBA, the recipient of a postgraduate degree in economics from Cambridge, Wasserstein had quickly established himself as one of the most inventive—and possibly *the* most aggressive—dealmakers on Wall Street. In 1977, at about the time he left his position as an M&A lawyer at the white-shoe Wall Street law firm of Cravath, Swaine & Moore, an internal report had glumly concluded that First Boston was "no factor at all" in the burgeoning takeover-service industry. Within five years, Wasserstein's flair for doing deals had helped to propel First Boston nearly to the top of the pack.

Wasserstein operated by the same philosophy that rules every automobile showroom and new-home development in the country: Close the deal quickly or it may go away. Wasserstein was credited, for instance, with the invention of the "lockup" option—the very deal-clinching mechanism that Hugh Liedtke of Pennzoil was employing in his deal with Getty Oil. While Texaco was letting the acquisition of Conoco slip through its fingers, Wasserstein's client, Du Pont, moved in quickly and scored. For Wasserstein's services, Du Pont paid First Boston a mind-numbing fee of about $15 million.

"We're paid to perform," he once explained. "If we don't perform, we lose out."

Four days before the Pennzoil-Getty announcement, First Boston had closed out a year in which it had handled $10.2 billion worth of takeovers, representing both buyers and sellers—a performance that ranked second only to Goldman, Sachs, the firm now representing Getty Oil. If somehow Wasserstein could find a buyer for *all* of Getty Oil—not just the 60% of it that wasn't owned by Gordon—he could equal his entire 1983 performance in a single deal. The sale of Getty Oil was probably worth upward of $10 billion. It would be the biggest acquisition in history. It could reap an astonishing investment-banking fee. Wasserstein did not want such a deal to go forward unless he were a part of it.

Immediately after getting dressed, Wasserstein collected one of his colleagues from another room at the Four Seasons, raced to a local office maintained by First Boston in Houston and began studying the Dow Jones announcement. *If there's a deal here, it's awfully sketchy,* Wasserstein concluded. Like every other deal-maker on Wall Street, Wasserstein had been following the tumult at Getty Oil for months, just in case a takeover opportunity arose. He knew the lineup of players as well as anyone: His friend Larry Tisch had just joined the board, and his friend Marty Lipton was acting as the museum's lawyer. Yet the news release didn't begin to take account of the complexity of the situation. *Who is buying the public shares? Who is buying the museum shares? Is this a merger or a tender offer? When is the deal supposed to close?* The press release said nothing of this. As far as Wasserstein was concerned, the bidding for Getty Oil had only begun.

Wasserstein, who thinks in terms of checklists, knew he had to do two things: (1) Establish that the deal was still open, (2) Find a client. He picked up the phone and called Lipton.

"Is there a deal here?" Wasserstein asked.

"There is no deal yet."

"If you had somebody who was willing to make a higher offer, would that be considered?"

Lipton had to be evasive. Wasserstein did not have a client, and although Lipton considered Wasserstein a friend, he could not be certain that Wasserstein wasn't fishing for inside information to give First Boston an edge if it wanted to take a big position in Getty Oil stock. Lipton continued politely dodging the question, but Wasserstein persisted. *Would the museum consider a higher offer for its shares?* he demanded.

There are ways to answer a question on Wall Street without actually answering it.

"I'd answer the phone," Lipton replied.

"How much time do we have before we can do something?"

"I don't know," Lipton said. "We're working with Pennzoil."

"Thank you very much."

Wasserstein also called his friend Larry Tisch, one of the new members of the Getty Oil board. Of course, Wasserstein had no way of knowing that Tisch had hoisted a glass of champagne with Ann and Gordon Getty to toast the Pennzoil deal. Nor was Wasserstein fully aware of the nagging anxiety that Tisch felt: Like all the directors of Getty Oil, Tisch was personally liable to shareholders' lawsuits because Geoff Boisi, Getty Oil's invest-ment banker, had refused to render an opinion that the Pennzoil deal was fair to the public shareholders of Getty Oil.

"Is there a deal here?" Wasserstein asked Tisch.

"No, not yet," the famous investor replied.

"Is there going to be a tender offer? A merger? What is the form of it?"

"Frankly, I don't know if it's worked out yet."

Just as he'd initially suspected, Getty Oil was wide open. Now all he needed was a buyer to represent.

From time to time over the past few months, Wasserstein had talked informally about the Getty Oil situation to one of his contacts at Texaco, Dick Brinkman, one of its finance men. Now it was time for a heart-to-heart talk.

The previous evening, instead of attending the madcap effort to draft a press release at the offices of Pennzoil's law firm, Geoff Boisi of Goldman, Sachs had gone home for his first real night of sleep in days. But the Getty Oil investment banker knew that a press release would be hitting the wires early that morning, and he resolved to reach the office early so that he could follow up on the phone calls he had placed to other potential bidders before they heard the news for themselves. After all, Boisi had promised to keep several other companies abreast of where things stood between Getty Oil and Pennzoil.

After reaching the office, Boisi placed one of the first follow-up calls to his acquaintance Al DeCrane, the president of Texaco. "There was a handshake on price," Boisi told him of yesterday's board meeting, "but there are a lot of other issues and they're working on a definitive agreement." The prior day, DeCrane had already had a team undertaking an initial investigation of Getty Oil—how much it would cost, how well it would fit into Texaco—but the studies were nowhere near completion. Boisi, however, left DeCrane with the distinct impression that if Texaco wanted to bid, it had better not dilly-dally.

"As things develop," DeCrane told Boisi, "let's keep in touch."

Boisi had one other housekeeping chore to attend to that morning. In an unfortunately timed action, the gravity of which he could not foresee, he presented his client Getty Oil with a bill for the $6 million minimum fee to which Goldman, Sachs was entitled under its retainer agreement. The presentation of the bill on the same day that Pennzoil's agreement in principle had been announced would not look good in a jury's eyes.

Up in White Plains, DeCrane summoned the members of the Texaco Financial Policy Advisory Committee for a status report on the analysis of Getty Oil. Dick Brinkman announced that he had heard from Bruce Wasserstein of First Boston and related the conversation: *The Pennzoil deal is not a done deal. Texaco can still get in. Getty would be a good fit for Texaco. Time is of the essence. If Texaco wants to bid it will need outside help to analyze the situation and to develop tactics.* And First Boston was available to work as Texaco's investment banker.

Chairman McKinley was ending his holiday vacation in Tuscaloosa and would be arriving on a Texaco jet in a few hours. He would have to make the decision about how to proceed. In the meantime, DeCrane instructed the team to continue the studies.

Completing the final paperwork for a $5 billion deal turned out to be anything but perfunctory. Even apart from the boxcar dollar amounts involved, this was no ordinary deal. Unlike most mergers, this one didn't involve two parties, but four: Pennzoil, Getty Oil, the Sarah Getty Trust, the J. Paul Getty Museum—not to mention the newly formed paper corporation through which the trust and Pennzoil would jointly acquire Getty Oil. The multiplicity of parties simply multiplied the number of pages that had to be drafted. Moreover, this wasn't a straight cash deal; there was a "stub" that had to be accounted for in the documents. In addition, when Gordon's lawyer Tom Woodhouse examined the draft in progress, he became concerned about a potential tax complication to the trust; the problem was disposed of, but it took time.

Finally there was the damnable matter of who would actually buy out the museum. The museum wanted its money *now*—that much was clear. But Getty Oil's lawyers had raised a fuss over permitting the company to buy the shares immediately, because doing so would put Gordon closer to majority control before the public had been paid. Pennzoil could certainly buy the museum shares without having to pay any more money, simply by having the company purchase the portion of the publicly held shares that *Pennzoil* was supposed to buy; it was a wash. Pennzoil, however, couldn't immediately buy the museum shares because the deal still had not been blessed by the federal government. *That* problem, however, could be surmounted if Pennzoil simply put the museum's money in an interest-bearing escrow account.

As far as Liedtke's lawyers were concerned, none of these issues mattered in the slightest, and a solution existed to each of them. But despite how quickly the main terms had been agreed on, it took time to develop solutions even for issues that the Pennzoil lawyers judged to be minor.

And of course the lawyers wasted a certain amount of time on pure banter and gossip, for lawyers like war stories as much as anyone. Moulton Goodrum of Baker & Botts recalls listening to tales of the battle with Petersen's people from Tim Cohler of Gordon's group, who took time to express his deeply held hope that things would be different between Liedtke and Gordon.

Fortunately for Pennzoil's draftsmen, reinforcements were already on the way. At 6 A.M. Wednesday morning, Goodrum had called back to Houston ordering an additional two of his underlings to pack their bags, rush to the airport and grab the 8 A.M. flight to LaGuardia. One of the new recruits was the redheaded Joseph Cialone, a company commander in Vietnam who had become a crack M&A lawyer at Baker & Botts and whose experience in complex deals included aiding Boone Pickens in his hostile assault on Cities Service. Since Pennzoil had announced its tender offer a week earlier, however, Cialone hadn't been involved in any of the documents work in connection with Getty Oil. Obviously, the paperwork was going to take a little longer than Pennzoil had expected.

When Bart Winokur, one of Petersen's people, had left the news-release-writing session at Paul, Weiss the previous evening, he, like everyone else, was certain that Liedtke's lieutenants would complete a draft of the final merger document and get copies to Getty Oil and the museum first thing in the morning. When Winokur arrived at his law firm's New York office at 8 A.M., the document was nowhere in sight.

While Winokur waited, he conferred with Geoff Boisi. Boisi said he had resumed his contact with Texaco; was it all right to assure Texaco that only an agreement in principle had been reached with Pennzoil? *No problem,* Winokur replied. *It's only an agreement in principle.* The draft of the definitive agreement, apparently, was still in the word processor a few blocks away at the Paul, Weiss offices.

At 10 A.M., when the draft of the final documents still had not shown up, Winokur decided to call Moulton Goodrum, Pennzoil's chief draftsman. *Mr. Goodrum is unavailable,* Winokur was told, *but they're working on it and hope to have it completed shortly.*

At noon Winokur again called the Paul, Weiss law firm asking for Goodrum, who was still unavailable. *The documents are nearly completed, but not quite yet.*

At 2 P.M. Winokur called again and was informed that he

should now be speaking with a Mr. Cialone. Mr. Cialone, however, was unavailable. *Not to worry, the documents are done, they're in word processing and they'll be out any minute.*

At 4 P.M. Winokur called again. *It just left with the messenger,* Winokur was told. "Great," he replied.

At 6 P.M. Winokur called again, and this time he actually got Cialone to take the call.

"I thought you sent a messenger over two hours ago. Your offices are only three or four blocks away. How can it possibly take two hours for a messenger to get four blocks?" Cialone replied that the messenger might have gotten lost.

Messenger schmessenger, Winokur thought. "Well, listen," he said. "Why don't I just send a messenger over to just pick it up? We'll just send somebody over to get a copy of it."

"No, you don't have to do that," Cialone replied. "We'll find the messenger and I'm sure you'll have your copy shortly."

Ten o'clock, 2, 4, 6—for the sake of symmetry, Winokur called back at 8 P.M.

"We still haven't received anything yet."

"Are you *sure*?"

"Well, let me check again."

Winokur walked to the reception area. Nope. No packages from Pennzoil had managed to sneak in.

"We absolutely don't have it. I mean, *I'll come over there and get it!*" A corporate lawyer really means business when he offers to pick up a document himself. "There's no reason for this. It's only four blocks!"

The first draft showed up at 8:30 P.M., twelve and one-half hours after Winokur had expected it.

The difficult task of drafting the final documents didn't spoil the fun for all of Liedtke's lieutenants. As far as Jim Glanville, the investment banker, was concerned, the deal was in the can. After spending the night on the town, he arrived at his office and talked on the phone with Marty Lipton, the museum's lawyer, who, as Glanville would later claim, heaped praise on him for helping to arrange "the greatest investment banking deal in the world." (Lipton doesn't recall saying so.) By noon Glanville was at the New York Yacht Club, attending to some business as part of the effort to help the U.S. regain the America's Cup from the Australians. After that he went home.

*   *   *

That afternoon, while Gordon was preparing for another get-together with his new partner, Hugh Liedtke, another member of the Getty family was heard from. Claire Getty, one of the three daughters of the deceased George Getty II, sent her lawyers to the Los Angeles County Courthouse, where they received a court order temporarily blocking Uncle Gordon from signing "any legally binding agreements"—with Pennzoil or anyone else—until a hearing could be held the following day. It was a blind-siding development.

It had been exactly two months since Claire had written Uncle Gordon reminding him that "there are a number of provisions of the Trust that are important to me" and requesting "some appropriate voice in Trust matters." Now she was being asked to sit idly by and watch from across the country as Uncle Gordon fundamentally altered the trust's relationship with Getty Oil—and as he fundamentally restructured the oil company in which the girls' deceased father had made his career. Claire wasn't necessarily opposed to the Pennzoil deal; in fact, it strengthened the family's control of the company. But the deal was complex. Too little time had passed for her to understand fully what it meant to her and the other beneficiaries. And from a distance it looked like Uncle Gordon, who had flunked out of Getty Oil, was now bargaining his way into the number-one position with the company—a position that Claire's father, after a lifetime of service to the company, had never lived long enough to attain.

"Gordon has been seeking to reorganize or dismantle Getty Oil since he became sole trustee some eighteen months ago," Claire's lawsuit said. It continued: "Gordon has been in a continuous struggle with the current management of Getty Oil. . . . Gordon refuses to give adequate disclosure of information to the beneficiaries, and acts in a clandestine manner . . . as if he personally owned the entire trust corpus. Gordon has used the trust to accomplish his personal objectives."

For his part, Liedtke was still thrilling in the afterglow of the Getty Oil deal. Even Howard Kaufman, the president of Exxon, had called to congratulate him. Liedtke wasn't too much perturbed by Claire's suit. Claire seemed intent on building up and preserving Getty Oil, which was exactly what Liedtke and Gordon wanted. Liedtke would dispatch his number-two man—Baine Kerr, whose giant intellect and soft-spoken manner made him a shrewd negotiator—out to California to mollify the beneficiaries.

And while Liedtke and Gordon talked the deal through once more this evening of January 4, they began planning a trip of their own to California. Gordon was anxious to be seen walking through the corridors of Getty Oil with his new partner, to assure the managers of Getty Oil that Pennzoil was a savior, not a raider. He was especially keen on winning the confidence of Bob Miller, Sid Petersen's number-two man—an oilman through and through. Liedtke, in fact, already had made some discreet inquiries into Miller's background and reputation in the oil patch and found that although Miller was very much one of Petersen's people, he was also a small-town Oklahoma boy, which meant he must be okay.

Liedtke did not, however, want to rush out to Getty Oil headquarters—at least not while Sid Petersen was technically still in command. Sid Petersen was history as far as Liedtke was concerned, but there was no point in rubbing salt into the wounds of his defeat by marching through the headquarters building before the guy had a chance to clean out his desk. Liedtke assured Gordon that they would make the trip together at the first opportune moment.

Passing up his first chance to plant both feet in the home office of Getty Oil would be a decision that Liedtke would live to regret—one of the biggest regrets, in fact, of his career. For one thing, it would have gotten Gordon out of New York. And for another, possession, as the saying goes, is nine-tenths of the law.

Despite Liedtke's calm reaction to Claire's lawsuit against the Pennzoil deal, the unforeseen event gave new purpose to the lawyers trying to complete the paperwork. The lawyers for Liedtke and Gordon began calling their counterparts with the museum and Getty Oil, exhorting them to the Paul, Weiss offices to begin reviewing the freshly typed and long overdue first draft of the merger contract. *No way,* they said. *We're not going to try to beat some court order. . . . We're tired. . . . Nobody's had a decent night of sleep for days. . . . We spent all day waiting for you to produce a draft and we haven't even read it yet. . . . We haven't even eaten yet. . . . We'll see you tomorrow.*

Six months later, while giving a deposition, John McKinley of Texaco would be asked a simple, direct question: "When did you first become serious about acquiring Getty Oil for Texaco?"

McKinley would sit silently, deliberating over the question as if he had been asked to define the Theory of Relativity. "I am puzzling over when *did* I first become serious," the chairman would answer. "I am usually serious about everything." It was vintage John McKinley. *Serious about everything.*

McKinley had never forgotten the long hours over Bunsen burners and labyrinths of glass tubing back in Professor Montgomery's chemistry lab at the University of Alabama. "I remember his emphasis on precise observation, on meticulous recording of data and on generalizing only after we examine our results," McKinley would recall. "That has been a pattern which has stayed with me for forty years."

When he arrived at Texaco headquarters in White Plains from Alabama late the afternoon of Wednesday, January 4, the executive suite was imbued with the kind of high-stakes atmosphere in which McKinley was unaccustomed to making huge decisions. Financial executives, economists and lawyers paraded through his office bearing press clippings, annual reports, pro formas, SEC reports—everything they had been able to put together on Getty Oil. McKinley grilled all who came before him, and pondered the documents like Tantric mandalas. As greatly as Texaco needed oil-in-the-ground—as much as McKinley might have wished to capstone his career with a bold stroke, as beautifully as Getty Oil's operations seemed to line up alongside Texaco's—nobody was about to rush McKinley into anything. Moreover, he would step into the Getty Oil fracas only with a personalized invitation; he would make no exception, not even in this case, to Texaco's long-standing policy against unfriendly acquisitions.

He took a call from John Weinberg, Geoff Boisi's boss at Goldman, Sachs. Boisi himself certainly had sufficient rank in the investment-banking community to place the call himself, but Boisi had heard that it was McKinley's style to keep up his guard with strangers. Weinberg, on the other hand, was an old golfing buddy of McKinley. With Boisi on the line, Weinberg assured his friend McKinley that Getty Oil would be delighted to receive a takeover proposal from Texaco. Even if Getty Oil had announced reaching an agreement in principle with Pennzoil, *it had no binding agreements with anyone,* Weinberg asserted. *Getty Oil, in fact, is seeking offers.*

McKinley then telephoned his houseguest of the prior few days, Bob Baldwin of Morgan Stanley, to see if that firm could accept an engagement as Texaco's investment banker. Alas, McKinley had called his friend too late; Morgan had already accepted an invitation to advise another oil company that was considering a

bid for Getty Oil; it was believed to be Chevron. He authorized his
subordinates to hire another investment-banking firm while continuing their studies. And shortly after dinner hour, McKinley left
the office for home, where he would spend the night pondering
little else.

Bruce Wasserstein of First Boston had not spent all day merely
crossing his fingers for an engagement from Texaco. He had
called Rawleigh Warner, the chairman of Mobil, inquiring
whether the nation's second-largest oil company might have an
interest in bidding for Getty, but alas it did not. Finally, in the
early evening—just about the time that the first draft of Pennzoil's
final paperwork was coming out of the word processor in New
York—Wasserstein was notified that Texaco would happily hire
First Boston for advice in connection with a possible bid for Getty
Oil. "Fine," said Wasserstein, who was still in Houston on the
unrelated deal. "We'll have an advance team up there in an
hour." The retainer agreement was worked out hastily on fairly
customary terms: First Boston would receive a fee of at least
$500,000 no matter what; if Texaco made an acquisition, First
Boston would be paid a percentage of the transaction value. If the
deal were worth, say, $10 billion in round numbers, First
Boston's fee, also in round numbers, would total $10 million.
    Wasserstein immediately made arrangements for a chartered
plane to fly him to New York. But before meeting his plane, he
undertook to do some hiring on his own. When it came to takeover
law firms, the undisputed leader on the offensive side was Skadden, Arps, Slate, Meagher & Flom—the firm in the perennial
neck-and-neck race with Marty Lipton's firm for distinction as the
number-one takeover law firm for *any* side. After spending the
day talking all over the oil industry and the investment-banking
community, Wasserstein was keenly aware of the competition for
Getty Oil—not just from Pennzoil but now from Chevron and
possibly others. Wasserstein had thus called Skadden, Arps earlier
in the day, asking that the firm give him the chance to sign off
before going to work for anybody else on the deal; within minutes
of securing the firm's retainer with Texaco, Skadden, Arps did
indeed get another call; Chevron, it appeared, was also closing in.
    A firm like Skadden, Arps, with more than four hundred law-
yers and a finger in upward of half the big takeovers going on at
any time, couldn't begin working for any client until it had absolutely ruled out a conflict of interest. But one partner's possible

conflicts remained unaccounted for by the time that Morris Kramer, one of the firm's senior lawyers, left his midtown office in a limousine that evening for White Plains. When he reached the entrance to the Texaco building about an hour later, he instructed the driver to stop at the guardhouse, where he called back to the office. The last partner had been accounted for; no conflicts existed.

For his part, Wasserstein wasn't about to get on a chartered plane for a four-hour flight without putting his two cents into the development of a $10 billion takeover strategy. As his "interim crew" was congregating in White Plains, Wasserstein was hooked into a speaker phone and began holding forth to a group of Texaco executives, sharing the insights he had gleaned into the various Getty Oil parties from his own dealings and from common market rumor. While he spoke, two Texaco executives—Al DeCrane, the president, and Patrick Lynch, the associate controller—took copious notes.

As the notes would reflect, Wasserstein told the Texaco team that there was resentment among Petersen's people over the inadequacy of Pennzoil's price. At the board level, Larry Tisch was the *key director;* he's got the public shareholders' interests at heart and is *looking for a better deal* (even though he had already voted to okay Pennzoil's). Goldman, Getty Oil's investment banker, was *mad, embarrassed* that its client was doing a deal at *less than fair value*—and *upset with the museum* for going along with it too. The museum and Marty Lipton *feel they are getting squeezed,* because if Pennzoil got enough shares in its tender offer it could *assume the leverage role with Gordon;* Lipton, by the way, says *there are twenty-four hours,* Wasserstein reported. Gordon . . . well, Gordon is *a buyer at one price and a seller at another*—and $125 a share may be the price he sells at.

This isn't an ordinary merger that Pennzoil has planned, Wasserstein told the assembled Texaco team. The museum shares are going into an escrow account, which could lock up the deal. But Pennzoil, he emphasized, *only has an oral agreement* at the moment.

By about 11 P.M. Wasserstein was off the squawk box and on his way to the chartered plane, but already a small cadre of other First Boston strategists were on hand at Texaco to continue the strategy session. Leading them was Joseph Perella, a baldheaded, black-bearded former encyclopedia salesman and accountant who was Wasserstein's equal as co-head of First Boston's M&A

group. Perella liked to consider the emotional elements in take-over strategy. ''You couple fear and greed, which are the two greatest forces at work in the market,'' Perella once said in another context. ''. . . On this matter I would say we have no equal. The path is littered with other firms that have tried to tangle with us.'' The Texaco executives continued taking notes as Perella and his First Boston colleagues picked up where Wasserstein had left off.

Like a drumbeater keeping time for oarsmen, Perella also emphasized the importance of quick action. *Timing is critical. . . . There are probably twenty-four hours. . . . We've probably got until noon tomorrow to get the ball rolling.* Texaco's offer should be made quickly to *stop the signing* of Pennzoil's final documents, to *stop the train.*

Pennzoil *did* have a ''lockup'' option for eight million Getty Oil shares, but fortunately, *they won't get it until their definitive agreement is signed,* the Texaco executives were told. Buying out that option at a profit to Pennzoil, however, might be a way to placate Pennzoil if it made any noises at having been beaten out of its deal. The option, in other words, *could be a way to take care of Liedtke.*

In any event, it would be wise to get the museum shares before they were gone for good to Pennzoil. Marty Lipton of the museum is the *key person.* The *ideal plan* would involve locking up those shares first and then going to Gordon.

Gordon was a special matter. An outside lawyer who had studied the Sarah Getty Trust documents explained to the Texaco group that Gordon *cannot sell except in a merger, to avoid a loss to the trust. We might have to go the tender offer route to merge him out or create concern that he will take a loss. . . . The problem is that there seems to be no way to get Gordon on base first.* Developing a strategy to get Gordon to sell was so crucial that Al DeCrane, the president, put three stars alongside this section of his notes. For extra emphasis he wrote ''Imp,'' for ''important,'' which he underlined twice.

It was a complex, delicate situation requiring all the guidance that Texaco could obtain. At about 4 A.M., after his chartered jet landed in New York, Bruce Wasserstein finally made it to Texaco headquarters.

Also at about 4 A.M., on the Upper East Side, Marty Lipton rolled out of bed to begin reviewing the first draft of Pennzoil's final merger agreement, which a messenger had left leaning against his door the prior evening. An hour before daybreak, while he continued studying the document by the light of a lamp in his apartment, Lipton's telephone rang. It was Wasserstein calling from Texaco headquarters in White Plains.

"Number one, good morning," the ebullient Wasserstein said. "Sorry to call you so early, but what I would like to know is have you signed an agreement yet?"

"No," Lipton answered.

"Well, you should know that we are working with a client," Wasserstein said. "I will be back to you to give you some sense of whether they have an interest in this."

· 14 ·

Turnabout is fair play. After waiting all day Wednesday to receive Pennzoil's draft of the merger papers, Petersen's people would spend today—Thursday, January 5, 1984—making the other side wait to discuss them.

At 10 A.M., Joe Cialone of Pennzoil talked by phone to Bart Winokur of Getty Oil, beseeching him to Pennzoil's document-drafting room at the Paul, Weiss law offices. *We haven't completed our review,* Winokur said. *We'll be over at noon.*

At noon they conferred again. *What's the rush?* Winokur asked. A restraining order had been imposed on Gordon and a hearing was scheduled for that afternoon. Cialone, however, was as insistent as ever on wrapping up the deal, and again implored Petersen's people to come over. *We're not through reviewing the document,* Winokur said, *but we ought to be over in about two hours.*

At 2 P.M. one of Petersen's people talked again to Cialone. *We've finished looking at the document. But we want to draft something additional on the ERC stub. We'll be over about four o'clock.*

At 4 P.M. Cialone again talked to one of Petersen's people. *We're still working on the stub. We'll be there at six o'clock.*

Cialone was getting awfully impatient. After barely a day on the scene, he had been left in charge of completing the paperwork by his boss, Moulton Goodrum of Baker & Botts. Goodrum, for his part, was spending the day with his client Hugh Liedtke, sitting in on a conference-call meeting of the Pennzoil board, basking in some of the limelight that was still shining on Liedtke. After receiving the 4 P.M. phone call, Cialone did not know what to do besides call Goodrum at the Waldorf.

"Look, unless you do something, I'm going to get a call at six o'clock saying they can't be here until eight o'clock," the tempestuous Cialone told his boss. Goodrum called his old friend Herb Galant, the New York lawyer representing Getty Oil, complaining about the apparent slowdown. Galant assured Goodrum that his colleagues weren't stalling, that they were simply caught up in some complex drafting issues involving the ERC stub; the Getty Oil lawyers would later insist that they were only trying to structure the stub in a way that potentially could add several dollars a share to what they otherwise considered an unfair deal, and that that was taking some time.

Nevertheless Galant assured Goodrum that the Getty Oil lawyers would be coming by shortly.

Goodrum called his subordinate Cialone back about 5 P.M. "Okay," Goodrum said. "They've said they'll come within the hour."

Great. Almost an entire day *wasted*.

Overnight, John McKinley had finally decided to reach for the brass ring. If his board approved, and if it could be accomplished on a friendly basis, Texaco would bid for Getty Oil—all of it. A meeting of the Texaco board was hurriedly scheduled for early afternoon.

With Wasserstein by his desk reviewing in advance what his client would say, McKinley started working the phones. One call went to Marty Lipton, custodian of the museum's nine million shares. McKinley knew Lipton from Texaco's ill-starred attempt to acquire Conoco and from another attempted acquisition that had not panned out.

"Good morning, Marty," McKinley said. "I understand you received a call this morning from Bruce. We are the client that he has been working with." McKinley explained that his board would be assembling shortly to approve making a bid much greater than Pennzoil's price—as long as the museum and everyone else connected with Getty Oil was free to deal. Lipton assured him that everyone was free to deal.

Wasserstein had impressed on McKinley the importance of speed; if Texaco wanted the trophy, it had to run like a thoroughbred, not like the "jackass" Texaco of yesteryear.

"If necessary," McKinley told Lipton, "we'll abbreviate the meeting to the degree we can. We do want to make this proposal to you, and we want an opportunity to be heard, before you sign the other agreement."

Lipton assured McKinley that the negotiations over Pennzoil's final merger documents would not likely be completed during the day.

Another call went to Tisch, with whom McKinley also had some acquaintance. "The transaction is still open," Tisch told McKinley.

McKinley also called Petersen, who was making no secret of his interest in obtaining a better price; indeed, when contacted by a representative of First Boston, Petersen had declared, "The fat lady hasn't sung." And Boisi of Goldman, Sachs, of course, had been broadcasting the message all over the oil patch that Getty Oil remained *free to deal*.

"Sid," said McKinley, who knew Petersen through the American Petroleum Institute, "I'm calling you up because I understand from your investment bankers that Getty Oil is desirous of receiving proposals and is free to receive them. You know that we only engage in friendly acquisitions."

McKinley had reached the identical, critical moment he had achieved more than a year earlier, when Texaco was all set to make a formal bid for Conoco—and stopped first to ask the permission of Conoco's chief executive, who said he would rather that Du Pont acquire his company. *If you don't want an offer*, McKinley told Petersen, *we'll walk away*.

Petersen assured McKinley that Getty Oil was not only wide open for dealing, but that it would welcome a bid. However, other companies were also eyeballing Getty Oil. Texaco, Petersen indicated, had better hurry.

John McKinley walked into the Texaco boardroom, ready to convene the most important board meeting of his career.

At Getty Oil headquarters in Los Angeles, a middle-level manager was busily writing a letter to the staff about the Pennzoil deal. When Pennzoil asked to see a copy before it was distributed, the employee-relations man consulted with a company attorney. "Don't send the letter to Pennzoil yet," the attorney instructed him.

* * *

A reporter from Reuters called Marty Siegel, Gordon's investment banker, to obtain a comment. "There is a rumor in the marketplace that some major oil company is about to make a bid," the reporter explained. Siegel called Glanville, his counterpart with Pennzoil, to see if Pennzoil had heard anything. When Glanville didn't call back, Siegel left a second message.

Glanville, however, had already moved on to other matters, spending the morning in Armonk, New York, trying to drum up some investment-banking business from IBM and the afternoon at a meeting on another deal. The Pennzoil deal, after all, was done, and any investment banker worth his pin-stripes is always looking to the next deal. Unbeknownst to Glanville, his neighbor and friend of twenty-five years—Frank Cary, the recently retired chairman of IBM—would be attending a special meeting of the Texaco board as an outside director only a few hours after Glanville had completed his sales call on IBM.

That evening, Glanville would take his family to dinner and the theater. He never would return Siegel's phone calls.

This would be the biggest acquisition in history, and there were a few things that the Texaco board wanted to know. Where was the price of oil headed? (At the moment, sideways.) How about interest rates? (Beginning to drift downward.) How much would Texaco have to borrow to buy Getty Oil? (Initially about $9 billion.) How much of Getty Oil's oil-in-the-ground was in politically unstable areas of the world? (Very little, compared with the great amount securely located in the U.S.) How much can ERC, the insurance company, be sold for? (The consensus was $1 billion to $1.4 billion.) How much is Getty Oil really worth? (As much as $140 a share, First Boston estimated.)

What is the status of the Pennzoil deal? *It is not a done deal,* the board was assured.

Bruce Wasserstein, wearing a pair of Sperry Top-Siders that he had not had time to change from since leaving Houston a day and a half earlier, was present for about an hour of the meeting. Wasserstein told the board that "the right approach" would involve seeking the museum's shares first, *then* going to Gordon Getty. It was a coincidence, Texaco would later say, that the resolution adopted by the board also listed approaching the museum first and Gordon second, and then obtaining the publicly owned shares.

The board authorized McKinley to bid as much as $125 a

share—$9.98 billion—for all of Getty Oil. Even at that kind of price, a premium of $12.50 a share over Pennzoil's price—Getty Oil was still a good buy, the board was assured.

McKinley finally left the boardroom at about 4:50 P.M., after about three hours, and turned to Wasserstein, who was waiting outside the door.

"What should we do now?" McKinley asked, as Wasserstein would recall. With his investment banker at his side, McKinley again began working the phones.

"Marty, our board has approved making a firm proposal with our conditions to you. The way we would like to handle this is Bruce and a team of our people will be arriving at your office to discuss this with you. . . . I assume that you have not signed anything yet? . . . I assumed you would not sign anything until you heard my proposal."

Again, another call went to Tisch, who told McKinley that Gordon could be reached at the Pierre. McKinley also called Geoff Boisi, Getty Oil's investment banker from Goldman, Sachs. McKinley said that Texaco would be approaching the museum first, and then Gordon Getty; Boisi reminded McKinley that there were still other possible bidders out there. "You better put your best shot forward," Boisi said.

Moments later, at twilight, a caravan of chauffeur-driven cars was readied at Texaco headquarters for the one-hour trip to Manhattan. James Kinnear, the vice chairman, and William Weitzel, the general counsel, would ride with Wasserstein to Lipton's office to begin dealing for the museum shares. Four financial executives got into another car headed for Skadden, Arps, Texaco's freshly engaged law firm, to begin preparations for any paperwork. A third car would take Al DeCrane, the president, to the Essex House Hotel, where the company had an apartment; DeCrane would remain there as a sort of utility infielder, waiting to see where he could do the most good. Chairman McKinley got into a fourth car bound for the Carlyle Hotel, where he would resume his effort to contact Gordon Getty.

The cars slipped out of the Texaco driveway on to Westchester Avenue and then to the New York State Thruway, driving against the headlight glare of oncoming commuter traffic headed for the deep suburbs of New York.

Patricia Vlahakis, the young lawyer working with Marty Lipton on the museum account, had spent most of the day holed up in her office with the telephone switched off, picking apart Pennzoil's proposed merger document. "It was completely unacceptable from the museum's point of view," she and her boss had agreed—not so much for what it included, but for what it was missing. She drafted language setting up an escrow account so that Pennzoil could pay for the museum's shares immediately, in advance of government antitrust approval. She put together a section guaranteeing that if Pennzoil or anyone else paid more than $112.50 for the public's shares—for any reason or at any time—the museum would receive the extra money as well. She worked on an additional section in which everyone connected with Getty Oil would agree to hold everyone else perpetually harmless for all the months of feuding. Some additional language also would make Pennzoil hold the museum harmless in case a shareholder of Getty Oil ever decided to sue the museum for throwing its support behind the Pennzoil deal.

Vlahakis sent the new sections by messenger to Pennzoil's law firm, where her presence—and that of the Getty Oil lawyers—was being anxiously awaited for a meeting on the final merger contract. Then, just after darkness had fallen on Manhattan, Vlahakis took a call from the Pennzoil lawyers assembled at the Paul, Weiss firm.

"Everybody is on their way," she was told. "Please come over."

"I'll be over just as soon as I can," she replied.

Vlahakis gathered her coat and her papers and stopped by Lipton's office to see if he had any instructions, which, as a matter of fact, he did.

"You can't leave," Lipton said. "I need you here."

Texaco, he said, had a team on its way into the city to discuss buying the museum's shares.

Liedtke had gathered some of his lieutenants in the Pennzoil suite at the Waldorf. *The definitive agreement is essentially agreed to,* he was told. *There had been mechanical delays in distributing one of the drafts, but certainly nothing to be concerned about.*

But some of Liedtke's lieutenants were disquieted by rumors they had picked up during the day. Mobil, Gulf Oil and Chevron were said to be gearing up to make bids for Getty Oil. These were strictly rumors, however, and rumors of all kinds always fly after a deal. What's more, the museum and trust shares—52% in all—were already committed to the Pennzoil transaction, and the Getty

Oil board had blessed the deal. As far as Liedtke's lieutenants were concerned, the deal was already done.

One of the Pennzoil people would also recall someone mentioning that Texaco's name was also in the rumor mill. But Texaco, the Pennzoil group decided, was one of the least likely oil companies ever to try sticking its nose into Pennzoil's deal. Texaco was just too big.

As Texaco's team penetrated the darkness of Manhattan, the late-afternoon sun was still shining brightly in Southern California. A horde of litigators marched down the long and brightly lit corridor to courtroom #40, where Los Angeles County Superior County Judge Richard P. Byrne suddenly had a gigantic case on his hands. Judge Byrne's twenty-four-hour order preventing Gordon from signing any "legally binding agreements" was about to expire; now the judge had to decide on the request of Claire Getty, Gordon's niece, to extend the order for twenty days. Claire wasn't necessarily opposed to the Pennzoil deal, her lawyer said. She simply wanted some time to study it in depth.

Pennzoil, which was as vitally interested in the outcome as Gordon's group, had dispatched two experienced litigators from Baker & Botts in Houston to observe the proceeding. The lawyers, John Jeffers and G. Irvin Terrell, watched in disbelief as the courtroom overflowed with lawyers representing the various legacies of J. Paul Getty—not just the born and unborn, but "the dead and undead," as they would recall. Three groups of lawyers represented the three daughters of the late George Getty. Ronald Getty and his children were represented. Another group represented the children by the first marriage of J. Paul Getty Jr. Still another turned out in the name of yet another lawyer who was the guardian for the child by Paul Jr.'s second marriage, Tara Gabriel Galaxy Gramaphone Getty. As for J. Paul Jr. himself, he was represented without actually being represented; his lawyer, anxious to avoid any action that would make her client subject to the jurisdiction of the California courts, announced that she was present to observe the hearing, but that she otherwise wasn't present at all. All told, some forty lawyers were present.

"What a zoo!" Pennzoil's Terrell would recall.

This particular hearing before Judge Byrne might have been an irrelevant event in the history of Getty Oil Company except for one thing: Gordon's lawyer, Tim Cohler, stood up to argue that but for the final signatures, the Pennzoil transaction was a done deal. In fact, he said, the deal could be wrapped up—all the paperwork completed, all the *i*'s dotted and *t*'s crossed—"in the very, very, very near future, perhaps this evening."

As Cohler explained: "There is presently a transaction *agreed upon* among Getty Oil Company, the J. Paul Getty Museum . . . the trustee . . . and Pennzoil Company. This is the first time there has been agreement among the four of those parties. Indeed, this is the first time there has been agreement among the museum and the trustee and the company.

"This is an agreement which has been entered into after *extremely careful consideration.* . . . There have been investment bankers from several of the well-known firms in New York, one acting on behalf of Getty Oil Company, another acting on behalf of the museum, another acting on behalf of Pennzoil, another acting on behalf of the trustee. There have been lawyers from law firms reputed to be expert and skilled in matters of this sort, generally known as mergers and acquisitions." Nothing, in short, had been entered into lightly.

Cohler assured the assembled lawyers that by increasing the trust's ownership of Getty Oil to four-sevenths—or 57%—from the current 40%, the Pennzoil deal fulfilled the intent of their clients' grandmother, great-grandmother and great-great-grandmother, the wish that the Sarah C. Getty Trust should at all times maintain the greatest possible control of Getty Oil. "This is the first time since Mr. J. Paul Getty died that the trustee, or trustees, of this trust have attained that control," Cohler said.

The lawyers for Gordon's fractious relatives were unmoved. Based on the press reports—*and that's basically all we have to go on*, said a lawyer for one of Claire's sisters—Getty Oil will be sold off, wiped out, or liquidated into oblivion if Liedtke and Gordon weren't getting along after a year. "That does not sound to me like what Sarah C. Getty intended," the lawyer said.

At about 6:20 California time, after enduring nearly two hours of argument, Judge Byrne announced that he would continue the hearing until the next morning. No action by the court had undone any commitments that Gordon had already made to Pennzoil—in the Dear Hugh letter, for example, or in the memorandum of agreement that spelled out the four-sevenths/three-sevenths deal. Judge Byrne expressed doubts about his authority to block the signing with Pennzoil, but said that until noon the next day, Gordon remained under court order not to sign any more documents with Pennzoil—or with anyone else, for that matter.

*   *   *

Marty Lipton had been pleased to get $112.50 from Pennzoil for each of the museum's nine million shares; even by Lipton's standards, raking in $1 billion for shares worth half that less than a year earlier was a big deal. Now, half the top brass of Texaco was seated around a conference table in his office offering an even higher price. But despite this seemingly happy circumstance, Lipton was throwing a small tantrum.

"Are you prepared to give us your price?" Lipton demanded.

"No."

"You know, there's no point in us discussing this unless you are going to give me a price."

"Well, we're not."

This was, to say the least, a rather odd way of negotiating a multibillion-dollar deal. But the Texaco executives had a few extremely good reasons for refusing to divulge their offering price. For one, they didn't know it. Only their chairman, John McKinley, had the authority to name the price, and he was sitting on the Upper East Side at the Carlyle trying to get Gordon Getty on the phone. Moreover, as one of the Texaco executives told Lipton, "Any price we give *you* just becomes a floor to talk with Mr. Getty. So we're not giving a price till we see Mr. Getty." And that was that.

Periodically, one of the Texaco executives would step into the hallway with Lipton to try to settle him.

"Calm down, Marty," said William Weitzel, Texaco's general counsel. But when the Texaco executives left the room at one point, their investment banker, Bruce Wasserstein, did his best to comfort his friend Lipton that Texaco had a big number in mind.

Nevertheless, Lipton remained annoyed. His client at the museum, Harold Williams, was celebrating his fifty-sixth birthday with his children and had authorized Lipton to undertake whatever negotiations with Texaco were necessary. And Texaco was refusing to negotiate until it gave its price to Gordon.

"Gordon Getty is going to be in *bed*," a Texaco executive would recall Lipton saying. "If you want to see him, you'd better get over to see him." Lipton would deny making the comment, but it seemed plain that if more money were to be paid for the Getty Oil shares held by the museum and everyone else, Lipton was entirely in favor of it.

Wall Street's tag-team takeover procedure was in its finest hour. Texaco's investment banker (Wasserstein) had notified Getty

Oil's investment banker (Boisi) that Texaco was ready to make offers to the museum and to Gordon. Getty Oil's investment banker (Boisi) hiked over to the offices of the museum's lawyer (Lipton). Lipton notified Gordon's investment banker (Siegel), who was eating Chinese food from a carton, that Texaco's people had likewise reached his office and would be presenting an offer. Siegel notified his client (Gordon) that Texaco would be in touch. Texaco's chairman (McKinley) called Gordon and set a meeting for 9 P.M. If Howard Cosell were handling the play-by-play on this one, he might be forgiven for concluding that Texaco had the momentum—and that Pennzoil was missing one scoring opportunity after another.

It was about 6 P.M. when five of Petersen's people finally showed up to go over the draft merger documents with the Pennzoil lawyers at Paul, Weiss. Joe Cialone, the lawyer from Houston's Baker & Botts left behind to quarterback the session, wanted to begin immediately. "Anybody got any comments on page one?" he asked.

For seven hours—while the California court hearing dragged on, while Texaco's team swept into the city, and while everyone else was connecting with Lipton and Gordon—the lawyers toiled over the proposed final contract page by page. Would Pennzoil and Gordon agree to share Getty Oil's final quarterly dividend with the public? Who would control the sale of the ERC insurance business—Liedtke's lieutenants or Petersen's people? What protection was Pennzoil willing to provide for Getty Oil's nearly 20,000 employees, who were noted throughout the industry for their loyalty and hard work? Something had to be discussed on nearly every page, if only a typographical error.

As the hours passed, Pennzoil's Cialone began noticing something strange. Petersen's people kept absenting themselves from the drafting room for phone calls. Even Tom Woodhouse, Gordon's lawyer, got up to take a call.

There was something mysterious about all these calls: Nobody said who they came from or what they were about, and the whole thing was interfering with completion of the final documents.

Then, at about 8:30 P.M., Woodhouse announced that he had to leave. Within a few minutes, Bart Winokur, one of Petersen's top people, made the same apology and excused himself. The Pennzoil lawyers continued their efforts with the somewhat thinned-out crowd of lawyers that remained, but by 1 A.M., the session was over. *Everyone* from Getty Oil had left.

Oh, well, the Pennzoil lawyers thought. *What few open points remain will easily be resolved once we get everyone together first thing in the morning.*

· 15 ·

John McKinley is hardly a threatening man. Although he had acquired a certain tough-guy image in his three years as Texaco's chairman—closing refineries left and right, turning up the pressure for new oil discoveries, peppering his subordinates with hard questions—he remained in fact a demurring, even shy executive by the standards of Big Oil. Gordon was immediately struck with McKinley's gentlemanly, Alabama-bred manner—and his assurances that Texaco refused to make takeovers on anything other than a friendly basis.

"I appreciate your seeing me," McKinley said.

Instead of playing to Gordon's psychological needs, as Liedtke had attempted five days earlier, McKinley broke the ice by acknowledging Gordon's musical interests, reminding him of Texaco's sponsorship of the Metropolitan Opera. Even Texaco's no-nonsense, high-pressure investment banker, Bruce Wasserstein, initially aroused Gordon's cultural interests over his economic concerns; still wearing his Top-Siders and having gone without sleep for nearly two days, Wasserstein was disarmed to hear Gordon express familiarity with the works of his sister, Wendy Wasserstein, an accomplished playwright.

"I'm here to discuss with you a possible offer by Texaco to the trust for your shares," McKinley said.

Yet amid all the happy talk, as McKinley and his associates

pulled their chairs closer to the coffee table separating them from Gordon, there was something threatening about the situation, if not the people. Gordon and his advisors knew that Texaco had already paid a visit to Lipton's office, and that the museum's 12% was as good as gone. The museum shares were like iron filings around a magnet; the highest price could pick them up in the blink of an eye. Unlike the family trust managed by Gordon, the museum faced no grandmotherly prohibitions on the sale of its stock. To the contrary, the law required the museum eventually to dispose of its shares. The museum's sole incentive was to obtain top dollar.

But if Texaco owned the museum's 12% and then bid for the public's shares, it would own a majority of Getty Oil. If Gordon did not sell under these circumstances, he faced the inescapable fate of holding a 40% interest in a company 60% owned by someone else—the worst imaginable scenario. The value of the trust shares would plummet; there would be absolutely no market for them. If Texaco wanted to play rough, it could exercise its majority control and forcibly buy out the trust's 40% on whatever terms it chose.

A comment by J. Paul Getty about his drawn-out takeover of Tide Water seemed more meaningful than ever: "I had very vivid memories of the disadvantage of being a minority stockholder." Even the warning issued by Petersen's people two years earlier, in the fateful meeting at the Bonaventure Hotel, seemed apt. *Somebody's gonna get squeezed, Gordon, and you could be the juice.*

"Texaco would be interested in acquiring the Getty Oil Company."

Nothing in McKinley's voice suggested malice; all the pressure in the room was in the minds of the beholders—Gordon and his advisors.

"Texaco played hardball," Gordon says today. "But they weren't crooks." But that did not change the issue in Gordon's view, and in that of his advisors. They perceived a threat no matter what McKinley said.

However, this was not an altogether unhappy occasion. If Texaco truly offered a premium price, Gordon thought, it would be tempting to take under any circumstances. Under the complex plan he put to the museum in London, Gordon was either a seller, as trustee, or a *buyer,* as the controlling shareholder of Getty Oil, depending on the price. Price had been a factor in deciding whether he should be a buyer or a seller under the plan by Petersen's people to keep Pennzoil at bay for ninety days. Certainly, at Pennzoil's price of $112.50, Gordon was a buyer.

Which was Gordon now—a buyer or seller?—with his newfound partner Hugh Liedtke getting ready to crawl into bed at the Waldorf, and with John McKinley seated in Gordon's suite. Under the deal with Liedtke, Gordon had preserved the heirloom status of the shares, yes—but this heirloom was not a priceless one. This was not Great-grandpa's pocket watch or the oil portrait of Aunt Bess. These were shares in a major corporate enterprise subject to the whims of the marketplace. This was hard economic reality. When Gordon took over as sole trustee, the shares were valued by the market at $50 each. Now Texaco would be offering $120—maybe $125. This kind of offer might never come around again.

Can you hear a number? McKinley asked.

Aye, there's the rub. *Am I free to sell?*

As the Texaco executives would recall, Gordon conferred with his lawyer, Tom Woodhouse, who was seated at his side on the sofa. It is unusual that the sellers in multibillion-dollar transactions confer among themselves in the presence of the buyer; Gordon, in fact, would later insist that he and his lawyer could not have been overheard.

But they were.

The legal documents governing the Sarah Getty Trust prohibited the sale of the family's stock, except "to save the trust estate from a substantial loss." Surely the trust was now in precisely that danger; holding a minority interest in a company that was majority owned by someone else constituted the potential for economic loss. Moreover, what if years and years passed and no one offered as much money, even after Gordon and his brothers had died and the trust *were* free to sell? Wouldn't the trust be accepting an "economic loss" if it let Texaco walk away?

"I feel that I would like an offer from you, and I see no reason why I shouldn't," Gordon finally said. McKinley's men watched Gordon turn to his lawyer for counsel. "Do you agree with that, Tom?" Tom agreed.

There was one other small issue. What about Pennzoil?

About the time the subject came up, Gordon's other lawyer, Tim Cohler, telephoned Gordon's suite at the Pierre—from the offices of *Pennzoil's* Los Angeles law firm. His purpose was to report on the hearing that had just concluded before Judge Byrne, before whom Cohler had proclaimed that there was "a transaction presently agreed upon" among Gordon, the museum and Pennzoil. In fact, Cohler had just filed an affidavit with the court, telecopied to Los Angeles by Gordon's investment banker Siegel, who swore that the board had "approved a corporate reorganization" and that the "principal terms of the transaction" with Pennzoil were included in Getty Oil's press release of a day earlier.

At the Pierre, Gordon's lawyer Tom Woodhouse took the telephone call from Cohler, with McKinley and the rest of the Texaco team seated a mere fifteen or twenty feet away, enabling them to hear one side of the conversation. It was apparent to the Texaco people that Gordon's lawyers were discussing whether the Dear Hugh letter that Gordon had signed three days earlier prevented their client from undoing the Pennzoil deal and accepting an offer from Texaco.

Remarkably, out on the West Coast, two of Liedtke's lieutenants who had attended the court hearing walked into the office where Gordon's other lawyer was holding up his end of the conversation. Cohler waved the Pennzoil lawyers out of the room, indicating that the conversation was private. The Pennzoil lawyers left the office, unaware that a group of executives from Texaco was privy to the other side of the conversation.

If the slightest doubt existed in the minds of Gordon's lawyers about whether Pennzoil already had a done deal, it was not apparent to Texaco. After concluding his call from California, Woodhouse returned to the sitting area of Gordon's suite and produced a copy of the Dear Hugh letter. Where Gordon had promised to support the Pennzoil plan and opposed any other proposal, the words "subject only to my fiduciary obligations" had been added by hand. Where Gordon had promised even to throw out the Getty Oil board if necessary to complete the Pennzoil deal, the same words had been written in. Gordon had a gut feeling that it was better for the trust to sell its shares at $125 than to buy in shares with Pennzoil at $112.50. Thus, didn't Gordon have a "fiduciary obligation" to accept the best available offer? Maybe those handwritten words on the Dear Hugh letter had saved the day.

"It's my understanding that you would be free to, and indeed should, receive any higher offers," the Texaco people heard Woodhouse tell his client.

"That's my understanding, too," they heard Gordon reply.

Woodhouse handed a copy of the letter across the coffee table to William Weitzel, the general counsel of Texaco, who thought to himself that the "subject to" handwriting indeed preserved Gordon's freedom to do a deal with Texaco. That left only the price.

Nobody wants to make the first move in the bargaining over a used car, much less over a $10 billion corporation. An almost comical Kabuki dance ensued, with Texaco gently trying to prod Gordon into naming his number and Gordon's group trying to goad a price from the Texaco team. *I sure would like to hear your offer, if the price is right. . . . I hope you have a big number in mind . . .* Gordon's group said. *Would you like cash or securities? . . . We would pay materially more for the public shares than Pennzoil had agreed to pay*, the Texaco team said. These were some of the most noted negotiators in the country—even the world—and some of the best-educated men in New York. Of the eight people in the room, three—Weitzel of Texaco, Wasserstein of First Boston and Woodhouse of Gordon's group—were Harvard lawyers. Siegel of Gordon's group was a Harvard MBA. (Wasserstein had that pedigree also.) Jim Kinnear of Texaco had studied strategy at the U.S. Naval Academy. John McKinley of Texaco held two honorary doctorates. Gordon's friend Mark Leland was a lawyer working for the U.S. Treasury. And yet in their discussions with Gordon, an English literature grad from the University of San Francisco, not one could get a dollar sign on the table.

Marty Siegel, an investment banker noted for always making the deal, watched this fencing and sparring go on for several minutes and decided that the situation called for a catalyst. He tracked down Larry Tisch at a formal birthday party at the restaurant Lutèce and begged his presence. Siegel knew that despite his confrontation with Gordon during the marathon board meeting, Tisch still held Gordon's deep respect, and had even toasted the Pennzoil deal with Gordon and Ann. A few minutes later, after knocking on the door of Gordon's suite, Tisch strolled in in all his baldheaded and black-tie glory. Tisch—a Wharton MBA, but, alas, a teenage dropout from Harvard Law School—urged everyone to do the Getty Oil board a favor and come to terms on a price that the directors could feel comfortable with. And with that, Gordon's group asked for the chance to caucus.

The Texaco team sat in an elevated sitting area of the Hotel Pierre lobby, beneath the pastoral, turn-of-the-century etchings of Central Park scenes, wondering just what would happen. They bantered to their boss, John McKinley, about what price Texaco should pay, and about whether the court order in California would prevent Gordon from committing to sell the trust shares to Texaco. After about fifteen minutes they were surprised to see Marty Lipton, the museum lawyer, strolling through the lobby. Like Tisch, Lipton had been summoned to Gordon's suite to help get the negotiations going.

Everybody had a different view of exactly what was motivating Marty Lipton during these first five days of 1984. After functioning as the go-between in Pennzoil's dealings with the Getty Oil board, Lipton—whose client owned only 12% of Getty Oil—was now jumping into the dealings between Gordon's group and McKinley's men. Was he acting out of goodwill, simply as an experienced takeover lawyer called in to help bring a major transaction to fruition, as Lipton himself would suggest? Was he acting out of ego, trying to prove that nothing big could happen unless *he* were involved, as some of his own friends would later explain? Or had he rushed over to the Pierre to put some heat on Gordon Getty, to act as an enforcer—to make sure that the indecisive heir capitulated to Texaco and thus assured a higher price for the museum? This would be Pennzoil's theory.

Lipton would recall that before heading to the elevator, he stopped to chat with McKinley and his men. "It wasn't clear to them that Mr. Getty desired to receive a proposal from them, and they were concerned about what Mr. Getty's position was, whether he would be receptive to a proposal," Lipton says. "They seemed confused by the meeting that had taken place and weren't quite clear in their own minds what they wanted to do next or what Mr. Getty would like to have happen."

Lipton said he would see what he could do and headed for the elevator.

While Lipton and Tisch discussed the Texaco deal with Gordon, Liedtke's lieutenants were wrapping up their night-long document-drafting session with the thinned-out group of lawyers representing Getty Oil Company and Gordon. Except for the absence of a few key lawyers on the other side, they were pleased at how smoothly the discussions seemed to be proceeding.

After less than thirty minutes alone with Gordon and his advisors, Lipton and Tisch re-emerged in the lobby of the Pierre to summon

McKinley and his men back up to the suite. Tisch, a Getty Oil board member, and Lipton, the museum lawyer, remained as vitally interested as ever in the price that Texaco would offer; after all, the same price paid to Gordon would also be paid to everyone else connected with Getty Oil.

While the group ambled past the front desk to the elevator, Lipton put on the full-court press. "What price are you going to offer Gordon?" he demanded.

McKinley wasn't about to offer the maximum price that his board had authorized—$125 a share—only to find Gordon trying to bargain him higher. McKinley said about $122.50.

"That just won't do it," Lipton said. "He won't accept that. He is just not going to bargain with you. He is not going to go back and forth. You've got to put your best price forward, and the price that will be required to make this deal, in our opinion, is one hundred and twenty-five dollars." The difference of $2.50 would be worth $23 million to Lipton's client, the museum. It would be worth nearly $100 million to the public shareholders represented by Tisch and the rest of the Getty Oil board.

Tisch chimed in. "At one hundred and twenty-five dollars, I can assure you, John, that the Getty board will wholeheartedly support the Texaco proposal and enter into a merger agreement with you."

With the half-dozen others and an elevator operator riding with him, John McKinley turned the numbers over in his mind. As he walked through the labyrinthine upper-story hallway, as he entered the suite—even as he prepared to talk—McKinley remained uncertain about what price he would offer to bring off history's largest corporate transaction. The ensuing conversation would be played out over and over before the jury two years later, and understandably so. For despite McKinley's spontaneity, there was a faint air that the entire exchange had been preordained, that the transaction already was a fait accompli—that Gordon's acceptance had been procured by someone outside of Texaco before McKinley himself ever re-entered the room.

McKinley told Gordon that he had given a great deal of thought to the price, and that he had expected to offer something like $122.

A smile crossed McKinley's face.

"But I have gotten some other indications here that there is another price that would be more agreeable to you, and I am prepared to offer—"

"I accept," Gordon blurted, and as he caught the surprised look of the Texaco people he quickly added, "Oh! You're supposed to give the price first!"

Laughter swept the room, breaking the tension of the moment. "I think we can probably do business," Gordon recalls saying, "if the courts okay it and the price is what I have heard."

In their mirth, no one mentioned the champagne toast that Gordon, his wife and Larry Tisch had hoisted in the same room only two days earlier after the Getty Oil vote on the Pennzoil deal, or an affidavit filed with the California court only a few hours earlier, in which Gordon's investment banker had sworn that the *transaction had been approved.*

The problem of the pending injunction imposed by Judge Byrne in California was handled with dispatch by Marty Lipton of the museum, a lawyer noted for ingenious solutions to the most difficult complications. By longhand—for even a full-service hotel like the Pierre had no typists available at midnight—Lipton drafted a letter for Gordon's signature that would function as Texaco's equivalent of the Dear Hugh letter to the extent that it committed him to see the deal to completion. Through his remarkable skill at writing in the negative, Lipton created a kind of takeover Newspeak that committed Gordon to the Texaco deal while specifically absolving him of any commitment. As long as Gordon remained committed in some fashion, Texaco's generous price of $125 a share was locked in for the museum and everyone else.

> I regret that an order of the Superior Court in California prevents the trustee of the Sarah C. Getty Trust from entering into any legally binding agreement in any way concerning the stock or assets of Getty Oil Company. Therefore I as trustee cannot commit to sell the Getty Oil Company shares held by me as trustee to Texaco pursuant to the offer of $125 per share being made by Texaco to all shareholders of Getty Oil Company. I believe that I have a fiduciary duty to seek to accept the offer by Texaco. It is my intention to request the court to lift the order. . . . As soon as I am able to do so, I intend to agree to sell or tender the shares to Texaco. . . . I will request Getty Oil Company to approve the Texaco offer. . . .

In one final effort to make sure that he was free to sign the letter, Gordon had his advisors undertake a feverish effort to locate Moses Lasky, his father's old lawyer. After defending the Sarah Getty Trust *against* Gordon in the dividend wars of years earlier, Lasky probably knew more about what Sarah Getty really intended than any other man alive. Unfortunately, Lasky at the moment was known only to be dining at a Chinese restaurant somewhere in San Francisco, and locating him proved no easier than finding your best friend at a 49ers game without knowing his seat number. When Lasky was finally reached, he was read the letter and raised no objection to its signing.

Ann Getty, who had made several appearances during her husband's negotiations with Pennzoil, remained in the bedroom of the suite throughout the evening that her husband was dealing away the company altogether. As a result, Gordon's investment banker, Marty Siegel, had to go down to the lobby when he could wait no longer to use the facilities.

When he returned, the deal was done. The "I regret" letter was signed by Gordon and dated January 6, 1984, for midnight had just passed. *After all these twists and turns, I can't believe I missed the final act,* Siegel thought.

By the wee hours of the morning, Marty Lipton's thirty-fifth-floor offices on Park Avenue were teeming with about two dozen of the nation's top lawyers and investment bankers. And among the issues confronting them was Hugh Liedtke of Pennzoil, who at that moment was sleeping fitfully right next door on the thirty-eighth floor of the Waldorf.

Nobody in Lipton's office—not Texaco, the museum, the members of Gordon's group, nor any of Petersen's people—*really* thought that Pennzoil had a binding contract with anyone. Some of Petersen's people, in fact, had just arrived from the drafting session with Pennzoil's lawyers eight blocks away in the Paul, Weiss law offices on Third Avenue. As far as they were concerned, the evening-long meeting with Pennzoil had failed to resolve several "open issues," such as the continuance of benefit plans for Getty Oil employees, the formula for dividing Getty Oil's last quarterly dividend, how to deal with the injunction in California and how the sale of ERC could be structured to increase the value of the stub. Burning issues. *Essential terms.*

You had to give Pennzoil credit, though. Liedtke had been ingenious, cultivating Gordon the way he did and *almost* bringing off a masterful deal. Too bad for Liedtke that his lawyers hadn't been faster with the paperwork.

There was absolutely no doubt in anyone's mind. Of course Pennzoil didn't have a contract! Certainly, everyone knew that Pennzoil, Gordon and the museum had signed a memorandum of agreement, but that didn't mean a thing. The deal was contingent on approval by the Getty Oil board and on the signing of a *definitive* agreement. And as everyone knew, the Getty Oil board had so far approved only Pennzoil's *price*. Pennzoil's "deal" would hold up in court about as well as a cease-fire in Beirut, if it ever came to that.

But you couldn't be too careful.

Throughout the morning hours, the nagging idea of a lawsuit by Liedtke, an old lawyer and boardroom infighter, came to the forefront of the museum's contract negotiations with Texaco. Lipton, in fact, had broached this concern to some of McKinley's men even before anyone had met with Gordon during the evening. The museum wanted to walk away with its $1.16 billion and never have to worry about another thing—not Pennzoil, not even the months of bad blood with Petersen's people and Gordon's group.

The solution was an indemnity.

In the thrill of the moment, after several sleepless nights, no one in Marty Lipton's office could foresee how large the indemnity would loom two years later. Indemnities, indeed, are usually nothing more than a plain paragraph of legalese, a little boilerplate routinely included in major commercial transactions. The word comes from the Latin *damnum*, meaning "loss" or "hurt," as in "damn." The party granting the indemnity agrees to hold harmless the recipient for any legal problem associated with the transaction. Authors indemnify publishers against the risk of libel or copyright judgments: publisher is sued, author pays the legal expenses and damages, if any. If Unethical Pharmaceutical Company agrees to a takeover by MegaGiant Corporation and some of Unethical's shareholders claim they were ripped off, MegaGiant's directors, not Unethical's, defend themselves in court.

But Lipton wanted something different than a plain-vanilla indemnity. If anybody found a reason to sue the museum for its role in the consents and the "stockholders agreement" with Gordon that foisted new bylaws on the Getty Oil board, Texaco would have to cover it. If someone had an excuse to sue the museum over Lipton's ill-starred truce among the warring Getty parties, this liability would also befall Texaco. And finally, if anyone— namely Pennzoil—tried to pony up a breach-of-contract suit against the museum, that, too, would be Texaco's problem, not the museum's.

And to make it absolutely clear that the museum was accepting no risk—even if the risk were infinitesimal—Lipton wanted one further thing in the museum's contract with Texaco. Many sales agreements, whether for a house or a secondhand turbine blade, contain a section of "representations and warranties" in which the seller gives his word that he really owns what he's selling and has the authority to sell it. The museum was certainly willing to promise that it actually owned the shares it inherited from J. Paul Getty, "except," Lipton would write, "that no representation is made with respect to the standstill agreement, the consent, the stockholders' agreement or the Pennzoil agreement."

McKinley's men were taken aback by Lipton's demand. Lipton showed them all the documents in question, including the Pennzoil memorandum of agreement, and assured them that the worst thing Texaco could possibly face was the risk of some legal fees. Period. The Getty Oil board had not approved the Pennzoil deal, despite the company's announcement that it had reached an agreement in principle. The Texaco representatives satisfied themselves, by looking at all the documents, that Pennzoil didn't have a binding contract, but still they balked at Lipton's demand. "You just don't understand the psyche of my client," he finally explained. Harold Williams—museum president, former SEC chief—had had it up to here with all the feuding and pressure. He wanted to wash his hands entirely of the problems at Getty Oil. "That's just a non-negotiable point," Texaco was told.

Texaco's lawyers finally relented on the indemnities. After all, Lipton indisputably was one of the world's outstanding authorities on the law of mergers and acquisitions. He had been present for every minute of the Getty Oil board meeting. If anybody knew for certain whether Pennzoil had a done deal or not, surely it was Lipton.

Lastly, granting the indemnities was the only way to lock up the museum shares—and at the moment, locking up the museum shares was the surest way to prevent the Pennzoil deal from coming to fruition. "If we wanted to make the deal," says Weitzel, Texaco's top lawyer, "we were going to have to give up these terms, and we did."

The museum was not alone in its demand for an indemnity. When the time came to write a final agreement to buy Gordon's shares, he too wanted the same protection.

Texaco's representatives had heard Gordon say many things during the meeting at the Pierre. *I am free to commit to a sale under the trust instrument. . . . I will be free to sell once the court injunction is lifted.* And indeed, a lawyer for Claire Getty quickly gave his consent to the Texaco deal, freeing Gordon to sign a binding agreement with Texaco. After all the discussions about what Gordon actually had the freedom to do, all of Texaco's representatives—its chairman, its vice chairman, its general counsel—would later swear that they heard Gordon repeatedly state that he was *free to deal.*

Texaco, alas, must have got it wrong, Gordon says today.

Whatever Texaco heard or overheard, Gordon's recollection is that he "never represented that I was free to deal"—and never would have done so. Where Pennzoil was concerned, "the whole idea was that Texaco, not I, was taking that risk by indemnifying me." Gordon was not sure at the time whether Pennzoil had a contract; it was beyond his expertise, he says, and he remains unsure to this day. "The lawyers thought I was free to sell," Gordon says, "but they weren't positive *how a court would rule*" if ever Pennzoil sued.

"That's why we wanted indemnification." And Gordon got it.

At about 5 A.M. Bruce Wasserstein, the Texaco investment banker, was satisfied that the deal was close enough to fruition that he could leave for some sleep; he had not been to bed for nearly forty-six hours. Al DeCrane, the president of Texaco, showed up at 5:30 A.M. to oversee completion of the museum agreement and to begin finalizing the contracts, which were well under way, with Gordon's group and Petersen's people.

Lipton—"pursuant to telephone authority" from his client, Harold Williams—signed the museum's contract with Texaco at 7 A.M., twenty minutes before daybreak in Manhattan. The news release would go out in an hour, about ten minutes after Liedtke's lieutenants had gathered for breakfast at the Waldorf.

WHITE PLAINS, N.Y., Jan. 6—John K. McKinley, chairman of the board of Texaco Inc., announced that Texaco today signed an agreement with the J. Paul Getty Museum, owner of 11.8% of the outstanding common stock of Getty Oil Company, to purchase all of that stock for $125 per share in cash.

Mr. McKinley also stated that the board of directors of Getty Oil Company will be meeting later this morning to consider a

business combination proposal from Texaco. Further information will be released later today concerning Texaco's proposal,
which is being recommended to Getty's board by its management and also has the support of Gordon P. Getty, the trustee of
the Sarah C. Getty Trust, which owns 40% of the outstanding
Getty common stock.

It was still the middle of the night in Los Angeles when Sid
Petersen called another Getty Oil executive to relay news of the
Texaco deal.
"The fat lady has sung," he said.

· 16 ·

Liedtke's lieutenants were expecting to witness the second eruption of Vesuvius. "I am *not* going to be the messenger," one of them had said.

But instead of exploding, Liedtke sat in stony-faced silence, listening as one of his public-relations aides played aloud a recording of the Texaco press release that had been recited over the telephone by a staff member back in Houston.

What's happening here? Liedtke thought. The situation was confusing; nothing made any sense. The museum had broken ranks—that much was clear. But Gordon? How could Texaco possibly have obtained his "support"? In the Dear Hugh letter, Gordon had promised not only to see the Pennzoil deal through to completion, but also to oppose all other deals. And how could anyone ask the Getty Oil board to approve a deal with Texaco? They had already voted fifteen to one to make a deal with *Pennzoil.*

Liedtke's lieutenants watched and waited for a reaction.

The capstone of Liedtke's long career in the oil patch—a once-in-a-lifetime deal, his dream of creating a giant oil company before going into retirement—lay in ruins. *Those bastards!* he thought. *What deceit! What treachery! They all know they had a deal with us. Texaco must have known they had a deal with us. Hell, our deal was all over the newspapers—just yesterday morn-*

*ing! Those bastards have stolen this company away from me!
Texaco is like a big gorilla,* he thought—*a big gorilla that sleeps
wherever it goddamn wants to. They've made off with a billion
barrels of my oil!*

Finally he spoke. "We're going to sue everybody in sight."

He grabbed a legal pad and began composing a Telex to the
Getty Oil board:

Gentlemen:

We expect you to comply with the terms of your agreement
with Pennzoil Company approved by your board by a 15-1 vote
only three days ago, including specifically the option granted to
Pennzoil to purchase eight million Getty treasury shares at $110
per share.

If you fail to keep your agreement, we intend to commence
actions for damages and the shares against Getty Oil Co., your
individual board members, the Getty Trust, the Getty Museum
and all others who have participated in or induced the breach of
your agreement with us.

Liedtke paused, thinking back to a meeting he had had at Sutton
Place with J. Paul Getty twenty years earlier. The Old Man had
offered to sell Getty Oil's West Coast refining and marketing
operations to Pennzoil, a tiny company, because Exxon (then
Jersey Standard) couldn't qualify to make the purchase under
antitrust laws. How could anyone justify Texaco purchasing that
refinery? Liedtke wondered. *Texaco is playing a huge Monopoly
game here,* Liedtke thought. *They're buying up every house and
hotel in sight!* Liedtke continued to write:

In your evaluation of your course of action, we trust that you will
consider not only your obligation to us, but also the significant
antitrust issues presented by the Texaco offer. . . .

Very truly yours,
J. Hugh Liedtke

Liedtke made arrangements to cancel a quail-hunting trip.
There was no time for shooting birds now. Liedtke had to plan a
gorilla-hunting expedition.

No one could believe that John McKinley's Texaco had pulled it
off. The company whose headquarters had been nicknamed Port
Fumble had scored with an interception.

McKinley and his men had moved with "daring," with "a speed worthy of Secretariat," *The Wall Street Journal* wrote. *Newsweek* called it a "surprise attack." *Petroleum Intelligence Weekly* marveled at Texaco's "surprisingly swift" action. *The New York Times* hailed Texaco's "lightning" moves. "McKinley's deal for Getty appears to clinch his turnaround of Texaco," *Business Week* added. Texaco itself indulged in a little self-promotion, issuing an employee newsletter which said that McKinley had put the deal together "with a speed that dazzled industry analysts." Nobody, it appeared, was as surprised by the deal as Texaco itself.

Texaco's alacrity was all the more stunning when compared with Pennzoil's halting involvement in the deal. After Texaco's deal with the museum was done, Texaco's president (DeCrane), general counsel (Weitzel) and vice chairman (Kinnear) had become personally involved in the drafting of final agreements with the trust and Getty Oil itself. McKinley himself got on the telephone with Sid Petersen to personally negotiate a few elements of the deal.

By contrast, neither Hugh Liedtke of Pennzoil nor any of his top executives had met with a single officer or director of Getty Oil, although Liedtke had spoken to his old friend Harold Berg, the retired chairman. Why bother when the museum and the trust had already signed a memorandum of agreement committing 52% of the company's stock to the Pennzoil deal? The offers and counteroffers between Pennzoil and Getty Oil had been mostly transmitted through Liedtke's investment banker (Glanville) to Liedtke's lawyer (Liman) and then either to the *museum's* lawyer (Lipton) or the trust's investment banker (Siegel) before ever reaching the Getty Oil board. One of Getty Oil's counterproposals—a handwritten one at that—had even been slipped under a hotel-room door at 4:30 A.M. Moreover, Liedtke had delegated all the document-drafting work to his lawyers, who never gave him any reason to believe that their efforts were proceeding anything but smoothly.

The Texaco people did not call a break in the action for a champagne toast or for a celebratory dinner at "21," choosing to burn the midnight oil instead of making late-night toasts. In the midst of completing the deal with Gordon's group, Al DeCrane didn't think twice about letting his opera tickets go to waste—and shrewdly made note of this sacrifice when Gordon personally showed up at the drafting session. DeCrane also made a point of

helping Gordon's lawyers promote the virtues of the Texaco deal
to the lawyers for contentious family members, convincing them
to have Judge Bryne dissolve the order that had prevented Gordon
from signing any legally binding documents.

In addition, Texaco's deal was simpler, cleaner, more straightforward. It was taking *everybody* out: Gordon, the museum and
the public. Texaco had quickly knocked out deals with each group
in separate, two-party agreements; Pennzoil had tried to do the
entire deal with four-way documents, in what one lawyer would
call a "quadripartite" agreement that exponentially increased the
complexity of the deal.

Texaco's deal also did not present the complication of Gordon's
continued ownership of the company—an element of the Pennzoil
deal that reawakened all the fear and loathing toward Gordon that
had built up in Petersen's people. The months of bad blood at
Getty Oil spilled all over the Pennzoil deal; not a drop fell on the
Texaco deal. Michael Schwartz, a partner of Marty Lipton, would
later observe that when Liedtke's lieutenants sat down with Gordon's people, Petersen's and the museum's, "there were *five* people at the table, including the ghosts of transactions past."

Petersen's people were delighted to work as quickly as necessary to see the Texaco deal to completion. They had deeply resented Pennzoil's bear-hug approach to the company and were
outraged at Gordon's readiness to cast his lot with the enemy.
There were no phone calls to Texaco every two hours saying *we'll
be over in another two hours*. There were no protestations against
drafting documents in the face of a California court order preventing Gordon from signing them. By contrast, Petersen's people
were happy to remain seated at the drafting table with the Texaco
team until the deal was sewn up.

Besides the goodwill of Getty Oil, Texaco had another advantage that was never available to Pennzoil—the advantage of going
second. Texaco made certain to keep abreast of Pennzoil's deal at
every stage. Texaco's advisors knew from Getty Oil's advisors
that only quick action could *stop the train, stop the signing*, as
Texaco's executives had recorded in their handwritten notes, and
that Pennzoil *only had an oral agreement*. It knew from the museum's Lipton and Laurence Tisch of the Getty Oil board that the
company's directors desperately wanted a better price. The inter-
lopers also had the advantage of walking through the paper trails
already blazed by the negotiators for Pennzoil and the Getty
group. When the time came to draft the final merger documents

with Texaco, Petersen's people reached into their briefcases for their draft copies of language they had been working on only a few hours for the Pennzoil deal. The museum had already drafted language creating an escrow account so it could sell its shares immediately while Pennzoil awaited antitrust approval to buy them; "the very next day," Pennzoil would later claim, "the museum and Texaco did little more than change the name of the purchaser on the documents."

Texaco, of course, had not done it alone.

Bruce Wasserstein and other investment bankers for First Boston remained at Texaco's side during the entire seventy-nine hours between the moment they were hired the evening of January 4 until the last of Texaco's merger documents was signed at about midnight on January 7. If Wasserstein had allowed the deal to fall through, First Boston would have been paid about $500,000. Completing the deal, however, earned First Boston a fee of $10 million—which worked out to $126,582 *an hour*. Petersen's investment banker—Geoff Boisi's firm, Goldman, Sachs—would have collected a fee of about $9 million under the Pennzoil deal. Under the Texaco deal, Goldman, Sachs was paid a little over $18 million.

The total investment-banking bill, including payments by the trust and museum: $47.1 million.

Although John McKinley of Texaco recognized that the deal might not have been possible without the takeover professionals, he nevertheless felt a deep sense of personal satisfaction. In barely twenty-four hours the company's finance department had raised $9 billion in short-term borrowings from banks in the U.S., Europe and Asia—an extraordinary feat. In one fell swoop, he had *doubled* Texaco's holdings of U.S. oil-in-the-ground, elevating the company from the ninth-largest owner of worldwide oil reserves to the second-largest (behind Exxon), from an also-ran to a leader, restoring Texaco's long-lost dominance. *And we did it with a good, friendly merger,* McKinley thought. *We didn't have to resort to any of this tough-guy stuff.*

In the hours after the emergence of the Texaco offer, a few additional legal details had to be attended to.

Out on the West Coast, Gordon's lawyer Tim Cohler had signed an affidavit swearing that Pennzoil's "agreement in principle" had been "approved by the Getty Oil board of directors." Although the affidavit had been filed with the court clerk just that morning, Cohler grabbed a photocopy of the document and began adding some words by hand. Now, Pennzoil's agreement in principle had been "approved by the Getty Oil board of directors, *as to certain terms, but was changed to a new proposal.* . . ." Cohler's addition did not say whether this "new proposal"—the $112.50 counterproposal to Pennzoil—had also been "approved" by the Getty Oil board. Clumsily, Cohler also scrawled some additional words on the last page of his affidavit and drew an arrow to indicate where he meant to insert them, making the sentence read: "I declare under penalty of perjury that each of the foregoing statements, *including the handwritten correction on page 5, on lines 9–11,* is true and correct." The edited photocopy was submitted to the court as a "corrected" affidavit.

The emergence of the Texaco offer also presented some unfinished business when Getty Oil's sixteen directors and a nearly equal number of advisors hooked up by telephone-conference call for a hurry-up board meeting at noon Eastern Time. The directors had not forgotten that they had ended their previous board meeting only sixty-five hours earlier by presenting Pennzoil with a $112.50 counterproposal. But even though Pennzoil had notified the board of its acceptance, the directors now voted to "withdraw" the counterproposal. Once that matter had been laid to rest, the Getty Oil board unanimously voted its consent to the Texaco deal.

After the telephone meeting, the general counsels of Getty Oil and Texaco—Dave Copley and Bill Weitzel, respectively—turned their attention to the threatening missive that had been issued by Hugh Liedtke of Pennzoil. It was ridiculous, of course, to think that Pennzoil could claim a breach of contract: "that is the most absurd thing I've ever heard," Copley said. But there was no telling what this guy Liedtke was capable of. Copley told his Texaco counterpart that it would be prudent to "clear the air," to make certain "that there aren't any shadows or problems." So at 6:40 P.M., just as the last of the Texaco–Getty Oil merger documents were being prepared for signatures, Getty Oil's lawyers filed a five-page lawsuit in a state court in Delaware. It was a request for "declaratory judgment," a determination by the court that Pennzoil had never had a contract—"or," the suit added, "that any alleged agreement is no longer valid."

The first round had been fired in court. And Getty Oil itself had pulled the trigger.

* * *

Two days after the late-night meeting between McKinley and Gordon, Marty Siegel, the investment banker, showed up at the Pierre to pay a visit to his client. But when he reached Gordon's suite, he was surprised to find that Gordon and Ann had checked into another suite—a much larger one.

"Gordon," Siegel joked, "you haven't gotten the money yet, so don't start spending it."

"A really *big* hitter has come to town," Gordon explained. The Sheikh of Oman and his very large family had arrived in New York, engulfing an entire floor of the hotel, including Gordon's suite, forcing Gordon to relocate to another suite in what had once been his own father's hotel. Somehow, Siegel would recall, "it put everything in perspective."

Hugh Liedtke wanted his deal back.

His first attempt began January 10, four days after Texaco had torn it apart, when Pennzoil filed a suit of its own in Delaware against Getty Oil, Gordon Getty, the Getty Museum and Texaco. It had taken that long not just to prepare the lawsuit but to figure out where to file it. Delaware might not be the most favorable forum, but at the time it seemed the most logical one. Texaco was based in New York, Getty Oil in Los Angeles and Pennzoil in Houston—but all three were incorporated in Delaware. In addition, the Delaware courts were accustomed to moving as quickly as most any hot-and-heavy takeover fight required.

Pennzoil's nineteen-page suit demanded "specific performance"—the law's term for a court order requiring a party to live up to its obligations. Pennzoil claimed that the memorandum of agreement with Gordon and the museum, the Dear Hugh letter in which Gordon promised to see the deal to completion and the fifteen-to-one vote by the Getty Oil board conclusively proved that a contract had come into being—that the parties had demonstrated their *intent to be bound*. And the press release announcing the agreement in principle—issued the day after the board meeting on Getty Oil's letterhead, with the approval of the museum and Gordon's group—made it plain that *all essential terms* had been agreed to: the price ($112.50), the "lockup" option, the structure of the transaction (a merger between Getty Oil and a company to be formed by Gordon and Pennzoil) and the plan for operating the company (liquidation after a year if Gordon and Liedtke couldn't get along).

A few days later, a collective "eureka" erupted among the Pennzoil lawyers.

They had obtained copies of the merger documents Texaco had filed with the SEC—documents showing that Texaco had indemnified Gordon and the museum, agreeing to hold them harmless for any breach-of-contract lawsuit from Pennzoil. Moreover, Gordon and the museum had specifically excluded any claim by Pennzoil in the language in which they "represented and warranted" the true ownership of their Getty Oil shares. To Pennzoil's lawyers, it was as if Texaco had knowingly bought a house for which the seller refused to promise a clear title!

Suddenly, in Liedtke's mind, it was clear why Gordon and the museum had broken the deal. "Just sign with us, boys, we'll hold you harmless," Liedtke imagined Texaco saying. "Lie, steal, cheat, do whatever you want to do. We will hold you harmless. You don't have to make any representations to us. We will accept this, even though we know you don't have any right to sell it." As Liedtke saw it, Texaco had judged the threat of litigation to be a small risk for the benefits of buying $10 billion worth of oil-in-the-ground. All Texaco would have to do is "pay damages to these little squirts," he thought.

Pennzoil quickly refiled its Delaware lawsuit, adding a claim against Texaco for "tortious interference" with Pennzoil's contract.

A tort is a wrongful act, as in "torture"—a legal claim usually reserved for lawsuits involving defective steering mechanisms, surgical errors and other personal-injury cases. Claiming tortious interference with a *contract* was an unusual application of the law, but one that had been successfully employed on occasion ever since 1854. Johanna Wagner, an opera singer, had a long-term contract to perform at an establishment owned by an impresario named Benjamin Lumley. One day, however, another theater operator named Frederick Gye induced the vocalist to abandon her employment—unaware that she was bound by contract. Wagner could have sued the singer for breach of contract. Instead, he went after his competitor, doubtless believing that he could collect a bigger judgment from him. Wagner sued *in tort,* meaning that Gye's ignorance of the contract didn't matter, for in tort cases wrongdoers may be held liable for their acts simply out of negligence. In a divided judicial opinion, Wagner won, and the concept of tortious interference with a contract was born.

As the law evolved, it became necessary to prove that the

inducer *knew* about the contract and set out *deliberately* to interfere with it. But with those differences, it remained fair game to
sue the interloper instead of—or in addition to—the reneger.
"Texaco knew of the existence of the valid contracts between and
among Pennzoil, the trustee, the museum and Getty," Pennzoil
said in its new Delaware lawsuit, "and intentionally procured the
breaches of those contracts, to Pennzoil's detriment."

But the courts were not the only place that Liedtke could try
clawing his way back into the deal.

Jim Glanville, Pennzoil's investment banker, had walked into his
office the morning of January 9 to find a message that John
McKinley of Texaco had called.

It wasn't at all unusual that McKinley should call Glanville.
They were neighbors, tennis partners and occasional drinking
companions in the bucolic residential area of Darien, Connecticut.
Glanville and his wife would often drop in on John and his wife,
Helen, if they were walking the dog and saw the lights on at the
McKinley home. The McKinleys and Glanvilles had been
friends—close friends—for years.

But this morning Glanville knew that McKinley was not
calling to set up a mixed-doubles date. Glanville immediately
called his client Chairman Mao, who said it was all right to return
the call—as long as Glanville let McKinley do the talking. As it
turned out, McKinley simply wanted to tell his friend that he bore
no hard feelings toward Pennzoil, that Texaco was only looking
out for its interests. Glanville, however, said that they should get
together that evening to talk.

Glanville and his wife walked down the street to the
McKinleys', where both couples had some snacks and a drink. As
McKinley recounted it, when the talk eventually turned to the
deal, Glanville pointed out that Pennzoil had "significant influence in Washington" that could be brought to bear—one way or
the other—when Texaco's acquisition of Getty Oil came before
the Federal Trade Commission for antitrust approval. "He
thought it would be a good idea if we thought about the idea of
joining with Pennzoil, or letting them join our arrangements,"
McKinley would later testify. Glanville would recall no discussion of Pennzoil's political influence, but would recall saying that
he thought Pennzoil could absorb some "antitrust lightning" if it
were somehow brought into Texaco's deal with Getty Oil.

Both Liedtke and McKinley had to be in Washington two days

later for oil-industry ceremonies with the premier of China. McKinley was scheduled to eat lunch with the premier, but Liedtke—the former business partner of George Bush and the first oilman invited to drill in China, where Bush had been a diplomat—was to join the premier for dinner at the White House.

Liedtke and McKinley flew out of New York on separate airplanes in one of the worst snowstorms of the year.

The Chairman met the Chairman in the Texaco suite at the Hay-Adams Hotel, only a few hundred yards from the White House. The meeting was a turning point, an event that would cast a long, dark shadow over the events to come. For although both men maintained a reasonably civil tone, Liedtke would leave the meeting feeling more outraged, wronged and obsessed than ever. McKinley, for his part, would become convinced that Pennzoil was unwilling to play fair.

"John, I feel like what you have done is highly illegal," Liedtke recalls saying.

"I have been very thoroughly convinced by our lawyers that we have a very sound legal position," McKinley countered. But he added: "We'd like to see the matter disposed of." As a means of settling the litigation, McKinley suggested that Texaco might be willing to sell at market price some of its interest in the Hueso field jointly owned by Texaco and Pennzoil.

Liedtke couldn't believe his ears. *That's nothing in a deal this big,* he thought. *They've made off with a billion barrels of our oil and they think they can buy us off with a puny little offer like that? These guys obviously stole our deal thinking that they could buy their way out of it cheap.*

Liedtke tried to regain his equanimity before answering.

"John, you must understand that *hueso,* in Spanish, means bone. You're talking about throwing us a bone. . . . The only basis we'd consider settling on is close to what we'd bargained for . . . a little less than three-sevenths of Getty."

Liedtke's determination to get back his deal seemed boundless. After starting the ball rolling barely two weeks earlier at the "generous" price of $100 a share, after permitting himself to be "chiseled" up to $110, after being further "chiseled" to $112.50, he was now willing to pay Texaco's ample price of $125 a share for three-eighths, rather than three-sevenths, of Getty Oil.

"Well, we have no interest in that," McKinley calmly said to Liedtke's proposal.

As McKinley and his men would recall, Liedtke also offered Texaco a little incentive. Liedtke seemed to go out of his way to mention his friendship with George Bush, according to Bill Weitzel of Texaco. "Mr. Liedtke was quite outspoken with regard to the influence that he felt he had—and would and could expect in Washington—in connection with antitrust matters and legislative matters," McKinley would say in a deposition. "This idea that Pennzoil was not without political influence that could adversely affect the efforts of Texaco in completing its merger."

"That's a bald-faced lie!" Liedtke would later respond. "The political-influence thing isn't true. I don't have any and McKinley knows it!"

In fact, Pennzoil *did* tell Texaco that it would lobby against the merger unless it were cut in on the deal. "It's always more difficult to get approval of something if you don't have opposition to it," Baine Kerr, who attended the meeting as Pennzoil's president, explains today. But as for claiming political influence? Absolutely not, Kerr says—and no one from Pennzoil invoked the name of George Bush.

The day before the meeting, an aide to Senator Howard Metzenbaum disclosed that he had already discussed the antitrust issues in the Texaco-Getty deal with representatives of Pennzoil. That was not, however, what anyone in Washington would call high-level political influence. Pennzoil did go on to lobby other federal officials, but in the end to no avail.

Pennzoil would have to find another way to defeat Texaco's union with Getty Oil.

In Los Angeles, as Getty Oil employees awaited word of their fate, the tension was only beginning to mount. A manifesto sweeping through the company office buildings along Wilshire Boulevard went through one photocopier after another, and before long almost everyone had seen it:

AN OPEN LETTER TO GORDON GETTY

We want to let you know that we harbor no hard feelings because you destroyed the company your father and grandfather had built up over half a century of hard work and dedication. . . .

We are sure you thought long and hard about the dedicated employees who make up Getty Oil (as it used to be known),

employees who worked 12 hours though they were paid for
eight. . . .

Naturally, these minor concerns had to be placed aside in view
of the one overriding one: to increase the value of the Sarah
Getty Trust from its mere $2 billion to a higher sum. As working
people, we understand how hard it must be for you to make ends
meet with a large family on only $28 million-a-year interest from
the trust. Given your lifelong struggle against poverty, you did
the only thing you could do.

As we move on to other employment, we want you to know
that we will never forget you. Never.

A Soon-to-be-Former Getty Employee

But the employees weren't the only members of the "Getty Oil
family" to attack the Texaco deal. There was also Ronald Getty,
the disinherited son of J. Paul Getty. Judge Byrne's California
courtroom would again figure in the action on the East Coast.

Ronald himself didn't have any financial interest in the takeover
of Getty Oil, because the legal documents limited his income from
the Sarah Getty Trust to a mere $3000 a year. But it meant all the
difference to his son and three daughters, ages eight to twenty-six.
They were "remaindermen" under the trust, meaning that they
would share with their cousins—all of them grandchildren and
great-grandchildren of J. Paul Getty—in dividing the assets of the
trust when their father and the last of their uncles were dead.
Under the Pennzoil deal, the trust would have gone forward as the
owner of four-sevenths of Getty Oil, an enterprise that with any
luck would have grown in value for decades to come by reinvest-
ing most of its profits for the future. Once the trust received its
nearly $4 billion under the Texaco deal, however, the cash would
go straight into Treasury bills, with the annual income paid out
directly to the income beneficiaries—of which Ronald Getty decidedly was not one.

Ronald Getty's law firm, Irell & Manella, engaged a former
dean of the UCLA Law School, William D. Warren, to serve as
the guardian for Ronald's children. Pennzoil's lawyers agreed that
if Warren could somehow undo the Texaco deal, Pennzoil would
go forward with its original deal—even increasing its price to
$125 a share from $112.50, so that the public shareholders would
not be treated any worse than under the Texaco deal.

On January 18 Warren filed a suit demanding that Judge Byrne
order Gordon "not to consummate the sale" to Texaco "and
instead to enter into and consummate certain agreements with

Pennzoil Company"—in short, to do exactly the opposite that Gordon's niece Claire had demanded only two weeks earlier. "Gordon Getty's execution of the Texaco Agreement was a flagrant and willful breach of trust," it said.

Texaco, however, wasn't about to let Pennzoil get the benefit of a family feud.

Within hours of hearing about the suit, Al DeCrane, Texaco's president, climbed on a company jet with five lawyers from the company and the Skadden, Arps firm, flew to Philadelphia to pick up Bart Winokur, the Getty Oil lawyer, and began winging his way to Los Angeles. During the flight, Winokur walked DeCrane through the Getty family tree so that by the time he landed, DeCrane would know just which branch was cross with which, and who stood to gain or lose under various scenarios.

The following morning, a Saturday, DeCrane and his army of lawyers met with a throng of Getty family lawyers. Quickly, DeCrane smelled another purpose building among the lawyers: to extract more money from Texaco. One of the family lawyers said that his clients opposed the Texaco offer because they didn't wish to see the Getty family name removed from the company's street-corner gasoline stations—but it was implicit in the comment that a little more money would help sever their emotional ties. Another lawyer bluntly suggested that if Texaco would agree to pay the trust $140 a share instead of $125, the whole matter might blow away. *And by the way,* one of the family lawyers told the Texaco president, *we've got a planning session scheduled with some people from Pennzoil first thing Monday morning.*

DeCrane was incredulous. This was everything but blackmail! Every additional $1 a share squeezed out of Texaco would cost the company $80 million, because the museum and the public would also have to be paid the same increased price. DeCrane resolved not to leave Los Angeles until the matter was disposed of and Texaco had washed its hands of the Getty family feud—for good.

"If this has any value, it's nuisance value," DeCrane told the family lawyers. He offered to sweeten the pot by one dollar a share. The lawyers held out for more.

Finally DeCrane acquiesced to an additional three dollars. Texaco's $9.98 billion acquisition had just become a $10.12 billion deal; ridding itself of the Getty family had cost Texaco nearly a quarter of a billion dollars.

After spending almost an entire night completing the paperwork, some of the family lawyers wanted to go home to shave, shower and change from their weekend sports clothes before showing up at the courthouse to enter the settlement. DeCrane, who by this point had no idea what to expect next, would have nothing of it, hauling the entire bleary-eyed and scruffy-looking group into Judge Byrne's courtroom, where the settlement was formally entered. Later, Ronald's lawyers at Irell & Manella would ask the court to approve a fee, to be paid by the trust, of $5 million for the firm's role in coaxing a higher price.

At six o'clock the evening of the settlement, the Texaco president, still wishing to leave nothing to chance, personally accompanied Gordon Getty into a walk-in vault at Security Pacific National Bank—and walked out with the Sarah Getty Trust's sheaf of ownership certificates for 31,805,088 shares of Getty Oil Company.

Delaware . . . Washington . . . Los Angeles. On January 17 Pennzoil took its crusade against the merger to a federal court in Hugh Liedtke's home town of Tulsa. The additional advantage of filing there was the recent takeover by Occidental Petroleum Company of the Tulsa-based Cities Service Company, whose half-completed headquarters building—along with thousands of jobs—was cast in a pall of uncertainty. Although that takeover had been set in motion by Liedtke's friend and protégé T. Boone Pickens, Liedtke's lawyers figured that the loss of independence by one of Tulsa's biggest and most-loved companies might dispose the judiciary unfavorably toward Texaco's megadeal.

The issue in Tulsa was antitrust. "This acquisition and the merger fever it will touch off is likely to result in a substantial lessening of competition in this vital sector of the economy, to further concentrate economic power," and so on, the suit said. One of Liedtke's lawyers at Baker & Botts in Houston—John Jeffers, whose expertise was antitrust—considered the suit a long shot. But at a meeting with Jeffers four days earlier, Liedtke had made his interest plain. "I won't get another deal like this," he declared. *"We want those reserves"*—and decoupling Texaco from Getty Oil was Pennzoil's only hope of ever accomplishing that. Over the years, Liedtke had had more than his share of deals scuttled on antitrust grounds, such as his proposed takeovers of Wolf's Head Refining Company and Asarco Incorporated. "Now you're telling me a major oil company can do what I couldn't do?" Liedtke demanded.

Before the Tulsa case went to trial, Liedtke told local reporters that, unlike the Pennzoil deal, the Texaco deal would harm Getty Oil employees in Tulsa, who numbered more than one thousand. Donald Schmude, Texaco's top executive in Tulsa, was outraged. "Pennzoil's frustration at being outbid is understandable," he countered, "but it does not warrant a deliberate attempt to frighten Getty employees and their families as part of an effort to delay the implementation of Texaco's signed contracts with Getty."

Before long, after a five-day trial, U.S. District Court Judge James Ellison refused to block the Texaco-Getty deal. But Pennzoil had Texaco on the run. No sooner had Texaco's lawyers returned to New York than they got news that Pennzoil had filed a motion for an expedited appeal of the Tulsa decision in the U.S. Court of Appeals in Denver. The Texaco plane turned around and took off for yet another trip west. Alas for Pennzoil, that effort too would fail.

The last chance to undo the merger on antitrust grounds occurred in a federal court in Providence, Rhode Island. Fairlawn Oil Service, a little heating-oil and gasoline distributor that purchased most of its supplies from Getty Oil, filed a lawsuit to block the Texaco deal on the ground that as a small business it could never *buy* products from a Texaco subsidiary and expect to compete in the resale market with Texaco itself. The owner's argument was undeniably interesting, but when the local press asked him how he could afford to take on a battery of high-priced Texaco lawyers, he said it was easy: Pennzoil had given him a stipend—later established to total $50,000—to finance the litigation.

"It's lucky for me that they could foot the bill," he explained.

Although this suit, too, would go nowhere, it would foreshadow the intensity of the courtroom dramas to come. On the day that the federal judge in Providence was scheduled to rule on Fairlawn's demand for a restraining order, Texaco at last was all set to take physical possession of the Getty Oil shares that had been tendered by the public—its final step in completing the takeover. A Texaco lawyer commandeered the only pay telephone in the courthouse, where he opened a line to Texaco headquarters in White Plains. The judge took the bench and began reading from his opinion, which in the beginning made it plain that he would allow the merger to proceed. Immediately, a Texaco lawyer joined a horde of reporters and Wall Street professionals who were dashing from the courtroom in the kind of indecorous display rarely seen in the federal courts.

"Madame marshal, secure the courtroom!" the judge bellowed, and with that the uniformed woman blocked the exit, a large-caliber handgun on her hip. Unable to relay the message to begin "lifting" the Getty Oil shares, Texaco's lawyers squirmed in the courtroom for what seemed an interminable time, while the poor man in the phone booth kept turning away anyone else who required the use of the phone. The experience would prove an invaluable dry run for events yet to come.

Liedtke and his lawyers knew all along that their best shot at undoing the Texaco deal was in the court system of Delaware. Proving that Pennzoil had had a contract to buy three-sevenths of Getty Oil—and that Texaco had knowingly and willfully destroyed it—would be a lot easier than crying foul over antitrust in the age of Reaganomics. Pennzoil's lawyers wasted no time in blanketing Texaco and everyone else in the case with subpoenas for depositions and documents.

When the documents began arriving, the Pennzoil lawyers did everything but dance in the streets.

From Getty Oil arrived a sixty-eight-page set of typed notes recording the action of the marathon meeting of the Getty Oil board. "Mr. Williams"—the same Harold Williams whose lawyers were now claiming in Delaware that Getty Oil approved no contract—"then moved that the board accept the Pennzoil proposal . . ." the notes read. And later: "Mr. Siegel"—the investment banker to Gordon, who was also contesting Pennzoil's contract—"advised the directors that Pennzoil had indicated to him that it would accept *the counterproposal presented by the board.*"

The Pennzoil lawyers couldn't believe it. It seemed to them that Getty Oil had just delivered the proof of a binding contract.

Texaco's documents were every bit as good.

Will you look at this stuff? You ought to see this one! Can you believe it! In the boxes of documents produced from Texaco's files was a pad of yellow legal paper with the handwriting of Al De-Crane, the president. They appeared to be notes taken when Texaco was plotting strategy with Wasserstein and other advisors from First Boston—and they made it clear that, through a pipeline into the boardroom of Getty Oil, Texaco knew everything about the status of Pennzoil's negotiations. "Only have an oral agreement," DeCrane had written. *That's "actual knowledge" of our contract!* In some notes about Gordon, DeCrane had written:

"Might have to go the tender offer route to merge him out, or create concern that he will take a loss." *Evidence that Texaco was putting the squeeze on Gordon,* the Pennzoil lawyers exclaimed.

They sifted further, coming upon another legal pad with notes by Patrick Lynch, an associate controller of Texaco who had also attended some planning sessions. "Timing—24 hours until noon tomorrow," the notes said. "Meet with Marty at noon." *Of course,* the Pennzoil lawyers thought. *Lipton was the enforcer, the guy who put the arm on Gordon to sell to Texaco so the museum could get a higher price for its shares.* "Stop train," the notes went on. "Stop signing."

Liedtke, a lawyer himself, was keenly interested in one passage of the notes by Texaco's president. They referred to the "lockup" option that Pennzoil had received in its agreement in principle— shares that Texaco, it appeared, had considered buying back from Pennzoil at a price that would permit Liedtke to walk away from the deal with a small profit. "Option to PZ," the notes read. "Could be a way to take care of Liedtke."

Liedtke vowed a fight to the death.

Pennzoil's hearing before Judge Grover Brown of the Delaware Chancery Court on January 25, 1984, made a few things very plain. Hard feelings abounded, especially on the Pennzoil side; Liedtke's Houston law firm of Baker & Botts—the same firm that had failed to get the definitive agreement signed—would also handle the courtroom work, and they were willing to play rough. In addition, it was clear that the battle had been transformed into a full-employment case for lawyers; some thirty were in attendance, with the total legal bill already approaching $1 million. The hearing also proved to the doubtful that greed had become a permanent fixture in the case, for the courtroom was brimming with Wall Street arbitragers intent on making a quick buck from any information they might glean from the proceedings.

Pennzoil's case was presented by John Jeffers of Baker & Botts, a forty-one-year-old Yalie and University of Texas Law graduate with dark hair, eyebrows long enough to comb and a perpetually knitted brow. With his understated, almost muted tone, Jeffers had never been famous for oratory among Houston trial lawyers. But it was clear to anyone who listened that he had a keen intellect—and this morning everyone was listening.

After reading the reams of affidavits filed for the hearing by Texaco, Getty Oil, and trust and museum lawyers involved in the deal, Jeffers knew he had to prove that Pennzoil's contract did not depend on the signing of a definitive agreement.

"The principles of contract law are as ancient as Anglo-Saxon commerce, and they are universal," Jeffers said. "If it is proven by objective and hard evidence that the parties intended to be bound and that they executed writings showing their intent to be bound, then they *are bound,* regardless of what further documentation they contemplated."

Jeffers hoisted poster-sized enlargements on an easel, something not commonly done in the Chancery Court of Delaware. One by one they went up: the memorandum of agreement, the press release announcing Getty Oil's and Pennzoil's agreement in principle, the statements by Gordon's lawyer in the California courtroom, the clumsily amended affidavit by Gordon's lawyer, the notes of the Getty Oil board meeting. Jeffers discussed how Gordon Getty and Hugh Liedtke made plans to visit California together, how Gordon invited Liedtke over for a glass of champagne. *Proof of intent to be bound. Proof that all essential terms had been agreed to.*

"That's our contract," Jeffers said.

He turned his attention to Texaco. "They just took the Pennzoil papers and changed the names," Jeffers said. He called the court's attention to handwritten *stop the train* and *stop the signing* notes. All of it, he said, was evidence that Texaco knew Pennzoil had a deal. Jeffers pleaded with the judge to discount the depositions and affidavits filed by the lawyers and other participants on the other side of the deal. Pennzoil's evidence was *hard,* Jeffers said, while the others' consisted of "after-the-fact self-serving statements." Don't hold it against Pennzoil that Marty Lipton of the museum was one of the nation's most respected lawyers, Jeffers said; Lipton got his young associate, Pat Vlahakis, to pony up an affidavit to cover up his duplicitous actions. "I don't accept what she says there at face value," Jeffers said. As for a certain remark in the deposition of Harold Williams, the former chief of the SEC, "I don't think that's very credible," he added. Don't believe the testimony of Getty Oil's investment banker: "Mr. Boisi of Goldman, Sachs makes nine million dollars with Pennzoil and eighteen million with Texaco."

In so many words, Jeffers said everyone else involved in the deal had a reason to lie.

"There's too much at stake; the pressures are extraordinary. The people on the other side who are arguing for 'no contract' and 'no interference' are people with indemnification, people speaking after the fact. They are people who lured the museum and the trustee away from meetings where they were supposed to be signing final documents, and they are people who left those meetings without advising why—all to stop a train and stop the signing.''

Jeffers took a parting shot by noting that the secretary's notes of the Getty Oil board meeting had been typed from handwritten notes—and that the original notes had been destroyed three days after the litigation had begun. No evidence existed that even a scintilla of valuable information was lost with the original notes, but their destruction was an indiscretion that Jeffers exploited for all it was worth. "You have people on that side—lawyers!—destroying evidence after suit was filed. The original evidence went into the trash bins and sewers of Philadelphia. For those vile and dishonorable acts, the individuals responsible should be excoriated and sanctioned.

"But Pennzoil only wants to have back its deal.''

The other lawyers were incredulous. That was the kind of legal argument usually confined to a criminal court. John Jeffers had just called some of the nation's most prominent lawyers and business figures liars!

Michael Schwartz, an intense, chain-smoking trial lawyer for Lipton's firm, was unable to contain himself at the accusation that Marty Lipton's young associate Pat Vlahakis had sworn to a perjurious affidavit. During a break following Jeffers' presentation, Schwartz approached Pennzoil's lawyers in the hallway and demanded an apology. "Buzz off,'' said a trial lawyer from Baker & Botts. Schwartz then found Perry Barber, the general counsel of Pennzoil, and demanded that he instruct the company's lawyers to issue an apology.

"Fuck off,'' Barber snapped.

During the afternoon session, the lawyers on the other side made powerful presentations of their own. Besides pointing out that Pennzoil had only an "agreement in principle, *subject to* a definitive agreement,'' they showed the judge the blank signature line where no one from Getty Oil had signed Pennzoil's $110-a-share memorandum of agreement. "That document was dead and buried'' long before the Getty Oil board meeting was even halfway over, said Matthew Broderick, representing Getty Oil. What had the Getty Oil board approved by a fifteen-to-one vote? Only a price of $112.50—not a complete deal. Why hadn't the paperwork been completed? Because Pennzoil's lawyers were *slow*.

The people whom Jeffers had so severely defamed were people of impeccable integrity, said Rodman Ward of Skadden, Arps, representing Texaco. "No one in this room is unaware of the name and reputation of Martin Lipton. . . . Laurence Tisch is one of the most respected businessmen in the United States. . . ." The unwavering belief among everyone on the Getty-Texaco side was that no deal existed, while all Pennzoil has to go on are some flimsy documents, Ward said. Ward dismissed the *stop the train* evidence produced from the files of his client Texaco as "a few indecipherable notes which were taken during the initial confused meeting with Mr. Wasserstein when they were just beginning to get into the matter." Gordon's champagne toast meant absolutely nothing. "I must say that if I were in Mr. Getty's position I would have a little more champagne than I do."

After working himself up over the assaults on his partners, Mike Schwartz of Lipton's firm, representing the museum, delivered the most succinct reply to Jeffers: "Your Honor, the whole situation was, I think—if you look at it as a contract—laughable. If you look at it as *I* think it ought to be looked at—as people *trying* to get a contract—it begins to make sense."

After a full day of arguments, Judge Brown dismissed the lawyers, promising to render a decision as soon as possible. "This is a massive task," he said.

· 17 ·

When Hugh Liedtke needed time alone, he found it in the privacy of his cattle barony in the backwaters of Arkansas. Although Liedtke loved the thrill of deep-sea fishing in the Gulf of Mexico, in reflective moments he preferred dropping a hook and line into the tranquil river flowing through the spread. While most Fortune 500 executives—John McKinley of Texaco, for instance—spent their free time on the golf courses, tennis courts and country clubs of exclusive vacation communities, Liedtke preferred pulling on his boots and clapping the shoulders of his cattle or paying a visit to his two pet buffalo, named Butt and Snort.

Liedtke ruminated at the ranch while awaiting the decision of Judge Brown in Delaware. The suit for ''specific performance'' remained his best shot at regaining his lost prize—that one billion barrels, a unique asset, the kind of oil-in-the-ground he had been coveting through his entire thirty-year relationship with the Getty Oil Company. The opportunity would certainly never present itself again, at least not in the three years remaining before he was scheduled to give up the reins and retire from Pennzoil. *A once-in-a-lifetime deal,* he kept thinking.

But in nearly a month of hauling Texaco, Getty Oil, Gordon and the museum from one courthouse to another, it was clear that Liedtke might never get his deal back. If he could not, then *Texaco would pay for what it had done.*

Years later, some of the people closest to Liedtke would confide that he might have let the matter drop had it been Exxon, Shell or almost any other major oil company on the other end. But it was Texaco, the company once infamous for refusing to sublease idle oil lands to independents, the company reputed to squeeze the last nickel from every deal, the outfit that had become a giant on a sweetheart deal with the cronies of Huey Long. Liedtke's harshly negative views were all the more emboldened by his directors, business associates and friends, many of them oil patch veterans who in the past had themselves come to swords' points with Texaco. And even one member of the Big Oil fraternity—Howard Kaufman, the president of Exxon—had rallied to Liedtke's defense.

"Hugh, I want you to know that I am shocked," Kaufman had said. "Exxon would never do business this way."

"Well, I've been in this field for years," Liedtke had answered, "and the attitude is quite different at Exxon than it is at Texaco."

It didn't matter to Liedtke that under John McKinley, Texaco had begun holding out its hand to the small companies of the oil patch. Jim Kinnear, McKinley's vice chairman, had become a hunting buddy of Hugh Liedtke's brother, who ran POGO Producing Company in Houston. In the pursuit of a nice-guy image, Texaco had offered to make an investment in POGO, to inject cash into the company when it was trying to resist a hostile takeover by Northwest Industries Incorporated of Chicago. Texaco had even done a deal with Liedtke's friend Boone Pickens. For its part, Pennzoil itself was now a partner in Texaco's Hueso field.

None of that mattered now, for January 6, 1984, when Texaco took Pennzoil's deal, lived in infamy. In Liedtke's mind, the "Texas Company," as he still often called it, had once again proved that it was willing to squash whatever small-fry happened to get in its way, that it had earned anew its distinction of the 1960s as "the meanest company in the world." Gordon had promised. The museum had promised. The Getty Oil board had promised. And what had Texaco done? Enticed them with a pot of gold and some indemnities to protect them in case Pennzoil refused to roll over. The reference to Pennzoil's lockup option in the handwritten notes of Texaco's president kept echoing through Liedtke's mind: "Could be a way to *take care of Liedtke.*"

This wasn't merely corporate honor at stake. This was personal honor.

To some, it was also a physician-heal-thyself case.

Over the years, few people in the oil patch had played tougher than Liedtke. He had left hard feelings in the wake of his entry into South Penn Oil, the company that had become Pennzoil. He had practically invented the hostile takeover as practiced by modern-day corporate America. He had battled his way into control of the nation's largest pipeline. He had been charged with insider trading. His company had been accused of trying to milk a public-utility subsidiary. That subsidiary had been accused of *breaking contracts*—contracts involving the sale of fuel on which jobs and petrochemical production and the air conditioners of the South had depended—even though those contracts had been entered into long before Liedtke's time. His company had been accused of sowing political benefits with Nixon campaign money.

But Liedtke's scrapes with judges and regulators had become things of the past. Most of the controversies were forgotten by those who even knew about them. For years Liedtke had not only kept his nose clean but had amassed a reputation as an executive who took care of his shareholders. American General Insurance and other investment titans had put money behind Pennzoil and had done exceedingly well. Liedtke had become something of an elder statesman of the oil patch. As his lawyer and close friend Joe Jamail would claim: "The difference between Texaco's reputation and Hugh Liedtke's reputation is the difference between chicken shit and chicken salad."

Jamail, a blue-eyed, blue-tongued Lebanese who stood 5-feet-9 in boots, had spent several evenings in January drinking beer with Liedtke and watching his friend become more consumed by the day. Time after time Liedtke relived the deal and its undoing, expressing incredulity over how the lawyers and investment bankers on the other side had been so ready to deal on both sides of the street. And although Jamail, a personal-injury lawyer, was out of his element in the high-stakes takeover game, his friend's story shared many elements of an intersection collision or an on-the-job injury, except that instead of a lost limb this case involved a lost billion barrels of oil.

This case, Jamail had thought, *would play beautifully in front of a jury.*

Now, in late January, Jamail had joined Liedtke at the ranch, watching his friend pull one fish after another from the water while he caught none. As they commiserated in the middle of a boat, both men were painfully aware that as a matter of law, jury trials were unavailable in the Chancery Courts of Delaware. Pennzoil certainly was entitled to a bench trial on damages from Getty Oil, Gordon, the museum and Texaco, even if the judge refused to block the Texaco deal. The problem was that judges rarely make the kind of boxcar-sized awards that juries do, especially for punitive damages—the damages intended to *punish* wrongful behavior. Jury or not, the Delaware Chancery Court doesn't even allow punitive damages.

But where there's a will there's a way—and there was a lot of will in that boat, and back in Houston at the Baker & Botts firm.

"You really wanna do this thing?" Jamail asked.

"You're goddamn right," Liedtke answered. And with that they motored ashore for a drink.

A week later, on February 6, 1984, Delaware Judge Brown handed down a forty-nine-page opinion with the last paragraph reading: "The application of Pennzoil for preliminary injunction is denied. It is so ordered." For Hugh Liedtke, Getty Oil was gone for good.

But what the judge wrote on the preceding forty-eight pages would send a shock wave through the law.

In 1976, SCM Corporation, the maker of Smith-Corona typewriters, decided that it wanted out of the photocopier business in Europe. A New York investor named Norman Muller offered to buy the operations for $9 million. Soon, SCM announced reaching an "agreement in principle," which was "subject to a definitive agreement." As negotiations wore on, a final draft of the agreement was typed, hands were shaken and someone involved in the negotiations expressed gratitude that "the deal is done." Before it could be signed, however, SCM discovered it had understated the value of the subsidiaries; other complications also arose. SCM announced that the deal was off, and Muller sued for breach of contract.

Four years later, a federal judge in New York awarded Muller $1,062,000 for breach of contract. The evidence that the parties expected to sign the final drafts was sufficient to prove their intent to be bound. Merely because the agreement in principle was "subject to" something else and merely because certain terms remained unresolved did not prevent the agreement from ripening into a binding contract. The opinion in the SCM case quoted the distinguished jurist Edward Weinfeld, who said, "The day is long past when a red ribbon and seal is required. . . ."

While evaluating Pennzoil's claim that it had a contract, Judge Grover Brown in Delaware read carefully through the SCM opinion and measured it against the documents on his desk. He picked apart the notes of the Getty Oil board meeting, compared the terms in Pennzoil's memorandum of agreement with those contained in the press release, and took note that Sid Petersen had renewed his golden parachute and that Goldman, Sachs had presented a bill before Texaco ever made an offer. He pondered the affidavits filed by Gordon's group in the California case and considered the refusal of the museum and the trust to "warrant" that they were free to sell their shares. Finally he wrote:

. . . Under the law of New York, the existence or nonexistence of a contract is a question of the intention of the parties . . . to be determined objectively, based on their expressed words and deeds as manifested at the time. . . .
I can only conclude that on the present record Pennzoil has made a sufficient *preliminary* showing that in all probability a contract did come into being between the four parties.

The judge then turned his attention to whether Texaco was legally responsible. "The point is a close one," he said.

I have no doubt that Texaco deliberately set out to use what are said to be its superior financial resources in an effort to wrest the acquisition of an interest in Getty from Pennzoil, and to acquire all of Getty for itself. But on the present record, I cannot conclude that Texaco did so with full knowledge that a contract had already been entered into. . . .

But despite everything else the judge had to say, he ultimately refused to thwart the Texaco merger and award three-sevenths of Getty Oil to Pennzoil. Pennzoil could obtain an injunction only by proving that it had no other way to make itself whole. And Pennzoil did have a recourse, the judge said. It could present the case all over again at a full-blown trial, right in Delaware, and try to convince a judge to award it money instead of oil-in-the-ground.

Seldom do judicial opinions circulate as quickly as did Judge Brown's. Take that old SCM case. Judge Brown did not know that while he was writing his opinion in *Pennzoil v. Getty*, a federal appeals court had quietly reversed the very lower-court decision on which Brown's had relied in his finding that Pennzoil probably did have a contract. SCM, the appeals court found, had reserved the right "not to be bound" before a definitive agreement had been signed.

"There are still situations," the appeals court said, four days before Brown published his opinion, "where the absence of a signed, formal agreement is fatal."

Hugh Liedtke wasn't the only person connected with Pennzoil whose pride was on the line in the Getty deal. There were also the partners of Baker & Botts, who were already hearing and reading stories about how New York's community of takeover professionals was joking over Pennzoil's performance in the Getty deal. The rap was that the big, prestigious Houston law firm had gotten clobbered when it tried to do battle on the turf of the New Yorkers, that Baker & Botts had dropped the ball, that the hayseed lawyers from Houston couldn't even get people to sign on the dotted line. Although it was Baker & Botts' corporate lawyers who had handled the deal in New York, the trial department vowed to restore the firm's reputation—partly by showing that whatever other problems they might have experienced, the Houston lawyers drafting Pennzoil's papers had been the victims of a slowdown so that the other side could shop for a better deal.

At Baker & Botts, no one was better suited to the job than Irv Terrell.

Eventually *The American Lawyer* would dub him Pennzoil's "Hit Man." Thirty-seven years old, tall and slim, with a rapidly rising hairline and deep cleft in his chin, Terrell took delight in a fight. His partner—John Jeffers, who had argued Pennzoil's case in the Delaware hearing—was usually courtly and understated. But Terrell was animated, full of nervous energy and, as events would soon demonstrate, adroit at making himself every bit as acerbic as a hostile witness required. Terrell had succeeded in making the switch to low-tar cigarettes, but only by tearing the filters off.

On January 6, the day that Texaco took away Getty Oil, Terrell was in Los Angeles with Jeffers after attending the hearing in Claire Getty's lawsuit against Gordon. The next day, Terrell flew back to Houston, picked up a tuxedo and flew to Atlanta, where he was married the following day. After the wedding he threw himself into the war on Texaco, and would not take a single day off until he began his honeymoon—twenty-seven months later, in April 1986.

Terrell had undertaken the daunting assignment of devising a way to take Liedtke's tale of woe before a jury of his peers. But Judge Brown's denial of Pennzoil's demand for an injunction did not free Pennzoil to go running into the Texas courts; Pennzoil had started the lawsuit in Delaware, and by all legal protocol was bound to see the case through to completion there. Pennzoil *could* ask the Delaware judge for permission to dismiss the case there and file it in a state court elsewhere. But Texaco and the Getty group were certain to put up a fierce fight to remain in the familiar jurisdiction of Delaware. Moreover, there was no clear jurisdiction for suing the entire Getty group in Texas: Gordon lived in San Francisco, and the museum and the company were headquartered in Los Angeles.

But while studying the unfamiliar (to him) terrain of the Delaware courts, Terrell had come upon a seldom-used rule: If someone fails to file a formal answer to a lawsuit, the lawsuit may be dismissed at any stage of the case—without even asking the judge for permission. Getty Oil had filed a response. So had the trust and the museum. *Texaco hadn't.* As near as Terrell could tell, Pennzoil was now free to drop Texaco from the Delaware lawsuit without even having to ask the judge for permission, then refile the case anywhere in the country where both companies had operations. That, of course, meant Hugh Liedtke's Houston.

While awaiting the decision on the injunction motion, Terrell discussed the move to nonsuit Texaco with the Pennzoil local law firm in Delaware, which agreed that the move just might succeed. However, special security measures were required. The Chancery Court of Delaware is a genteel court, where local custom requires the lawyers on one side of a case to notify their adversaries in advance of any filings. Doing so in this case would obviously permit Texaco to hustle up a piece of paper formally answering Pennzoil's suit.

"Maintain silence," Terrell told his cohorts in Delaware. "We're not gonna follow this gentlemanly rule."

At midnight the day that Judge Brown finally dashed Pennzoil's hope of restoring its deal, Terrell conferred with his partner Jeffers and with Liedtke, who were in Washington appearing before the Federal Trade Commission in yet another effort to thwart the Texaco deal. (It too would fail.) Liedtke gave the go-ahead. Early the following morning, Pennzoil gave the Delaware court clerk a $25,000 check to cover its costs in the case and filed a one-sentence document: "Please take notice that plaintiff, Pennzoil Company, hereby dismisses this action without prejudice as to defendant Texaco Inc. pursuant to Chancery Court Rule 41 (a) (1) (i)."

Fifteen minutes later, Terrell filed the case of *Pennzoil Company v. Texaco Incorporated,* Cause #84-05905, in the District Court of Harris County, Texas. In the last paragraph of the twenty-page document, Pennzoil made the greatest damages demand ever seen in Harris County, and probably anywhere in the world: "Pennzoil respectfully prays that upon trial by jury, this Court enter judgment against Texaco in such amount as is proper, but in no event less than $7,000,000,000 actual damages and $7,000,000,000 punitive damages."

The amounts would later be increased to $7.53 billion each, after Pennzoil had more fully refined its case.

On the instruction of his client, Terrell identified Liedtke's friend Jamail as the lead lawyer on the pleadings. Terrell had never met Jamail, but when he reached him by telephone out of town, he readily obtained his new partner's consent to scratch the signature of Joseph D. Jamail Jr. on the lawsuit.

The "King of Torts," as *Newsweek* dubbed him in 1973, flunked torts at the University of Texas Law School. His professor, Clarence Morris, urged Jamail to do everyone a favor by discontinuing to take up space at the school. Jamail resolved to apply himself, and did so well in a class called "death action seminars" that within a few years he convinced a jury that the city of Houston had negligently created an "optical illusion" by placing a tree in the middle of a boulevard—a tree that happened to get in the path of his client's husband, who was drunk, while he was driving home with an order of fried chicken. "I read with interest about the Glover case," Professor Morris, who by then was dean of the University of Pennsylvania Law School, wrote his former student. "I told you you'd never make a lawyer, and I meant it."

When Hugh Liedtke had been denied for good his billion barrels of oil-in-the-ground, he knew that Jamail was just the lawyer to drill him a gusher in the courtroom.

He was born as Joseph Dahr Jamail Jr. on October 19, 1925, his middle name, drawn from an old branch of the family, meaning "the strong side of the mountain." He grew up comfortable in a family of greengrocers who had immigrated from Lebanon at the turn of the century, Anglicizing the family name from Gemayel. Joe's fourth cousin was Amin Gemayel, the president of Lebanon.

After a year in pre-med at the University of Texas, Jamail

decided he couldn't tolerate the smell of formaldehyde, dropped out, joined the war for the Pacific as a Marine and returned home a party boy. His father shook him out of a hangover one day at 6 A.M. "You're worrying your mother," Joe Sr. said. "I don't wanna hear any bullshit war stories because I fought in the *real* war," meaning World War I. "Here's two hundred dollars. You be outta here by noon."

Jamail headed for New Orleans, where $200 just after the war could buy all the fun a nineteen-year-old veteran could handle. But Jamail never made it past Lafayette, Louisiana, about two-thirds of the way. "I ran into this big-titted bar lady," he would recall. "Boy, I was enamored of her. I had a hell of an evening." But he was also nearly out of money.

While spending the last of it on a few more drinks at the Buckhorn Bar, he made the acquaintance of a lawyer named Kaliste Saloom, who convinced him to enroll at the local University of South Louisiana, home of the "Ragin' Cajuns," and to embark on a career in the law. Jamail did just that, but after a year of studying English in French-speaking south Louisiana, he yearned to rejoin his fellow Longhorns at the University of Texas. In order to complete his last term at USL in time to begin the next semester at UT, Jamail convinced his English professor to render an early oral exam over a few beers, back at the Buckhorn Bar. Jamail scored a B.

For a young man whose chosen profession was steeped in procedure and protocol, Jamail would abide by little of it. He talked his way into UT Law School without having to take the entrance exam. Then, before completing his second of three years, he was sitting around Hillburg's Bar in Austin drinking beer with some recent graduates who were bemoaning the upcoming bar exam. *Cut your whining,* he told them. "Okay, mouth," one replied, "we'll bet you a hundred dollars *you* can't pass that sonufabitch." Jamail persuaded a lady judge sitting on the bar association admissions committee to let him take the exam a year before graduating, and passed by one point. "We all went up to the lake and got bombed on my hundred bucks."

Then, as now, the Houston bar was dominated by three firms: Vinson & Elkins, Baker & Botts and Fulbright & Jaworski, as they are known today. Jamail started out briefly at Fulbright—a firm that would become his archnemesis in the aftermath of the Texaco trial—and almost immediately tired of sitting in a row of desks. He became an assistant district attorney, but after two years

of keeping Texas safe from mule sodomy and other sundry infrac-
tions, Jamail again searched for a greater challenge, one that
would make better use of his gifts.

He found it in personal-injury law, a specialty where success
belongs to the lawyer who can stir a jury to passion over lost limbs
and eyesight, who can arouse a jury's anger over a sore back or a
whiplash, who can move a jury to the edge of tears over the death
of a spouse, who can transform an intersection collision into a
morality play. And in the postwar buildup of Houston into the
offshore-drilling and petrochemical-refining capital of the world,
there was no shortage of clients.

Jamail would take pride in his unwillingness to engage in gro-
tesquerie, refusing to pass artificial limbs through the jury box or
to have women bear their breasts to show the effects of a wayward
surgical scalpel. Jamail's only courtroom weapons were words.
He seized on current events: In a fairly recent case involving a
little boy who had lost a hand touching an electrical wire, he
reminded the jury that the Ayatollah Khomeini had just threatened
to cut off the hands of the American hostages in Iran. He had a
perverse sense of history: "In your zeal for perfection," he once
asked a German engineer in a case involving a Volkswagen car,
"do you still maintain that you build the best ovens in the world?"
If necessary, he mocked his own client: While representing the
famous and fabulously wealthy defense lawyer Percy Foreman,
who had sustained a minor injury in a rear-end collision, Jamail
appealed to the jury simply to provide some mental comfort to his
aging client. "He's over the hill," Jamail said. "He's bordering
on senility." In open court, Foreman leaped from his chair, proclaiming his own lawyer to be a "goddamn son of a bitch!"

Jamail seized the moment. "See what I mean?" The jury
awarded his client $75,000.

Jamail was certainly never afraid to use props, as he proved in
the case that earned him a place in the *Guinness Book of World
Records*. Fourteen-year-old Will Coates was holding a rifle in a
deer blind when it accidentally discharged, paralyzing his father.
The boy's story—that the gun misfired while he was sliding the
safety lock into position—sounded improbable, but Jamail was
convinced. Sure enough, in a series of tests, the gun accidentally
discharged once in every ten tries.

Remington Arms vigorously denied that any fault lay with the
weapon, until a company official sat down in front of a camera for
a videotape deposition, which Jamail intended to play for the jury.
"Come close," Jamail beckoned, the gun in his hands. "You see

this weapon is on safety. You see my fingers. They are not touching the trigger.'' With blank rounds in the gun, Jamail then nudged the safety button—and the rifle exploded, sending the Remington man diving for cover. Rather than go to trial, Remington agreed to write Coates a check for $6.8 million, the largest settlement payment ever received by an individual.

Which bespoke another point about Jamail: that after winning damages for misplaced trees and all manner of unlikely hazards, his reputation had reached to the point at which fewer and fewer of his cases ever had to go to trial. Jamail's signature on a lawsuit gave the case immediate settlement value; in the 1970s it was said around Houston that his name added $100,000 to the value of a case. In 1981, for instance, his $3.5 million settlement for a man hurt in a grain-elevator explosion exceeded the amounts paid for some deaths in the same case, proving that Joe Jamail could get more money for injuries than other lawyers could get for killings.

Like most personal-injury lawyers, Jamail worked on a contingency basis, for a piece of the action. If he failed to obtain something for his client—which was exceedingly rare—he charged no fee. Of what any client got, Jamail usually kept one third, although his practice was to reduce or occasionally waive fees in cases involving children. Winning as much as $20 million a year for clients, as he once claimed he did, was sufficient to make Jamail one of the wealthiest lawyers in a state full of wealthy lawyers. A single $1 million judgment could bring Jamail more than some lawyers earn in nearly a decade of scratching out a living—and by the time Texaco came along, Jamail had won more than forty settlements of that amount or more.

He could have retired long before middle age and never begun to spend half of what he had, but Jamail thrilled at the power, the glory, the satisfaction of sustaining his record of victories. It was almost like keeping score: In the seventies he would boast that he was able to take on only one of 75 cases offered him. By the early eighties it was one in 200, by the mid-eighties—after *Pennzoil v. Texaco*—one in 300. Jamail had attained such celebrity that in May 1984 one of his clients mailed a postcard from Juneau, Alaska, with the address:

> Joe Jamil
> "Houston's Flamboyant Attorney"
> Houston, Texas
> U.S.A.

The Postal Service delivered the card, misspelled name and all, straight to Jamail's office.

Jamail's reputation made him a string-puller, a meddler, a renderer of favors—someone whom *judges* often called on for help or advice. No other lawyer shared his influence and reputation with the Texas judiciary, and none made the kind of campaign contributions he made—which by the 1980s could render his signature on the bottom of a pleading often worth a great deal more than $100,000.

A morning in Jamail's office is like watching triage on *M*A*S*H*. Jamail blows through the reception area and down the hallway to his office while two sons working with him fall into formation with a young lawyer from Baker & Botts. As he turns into his office he is joined by his secretary Denise Davidson, who ticks off the messages, receives his orders on how to handle them and takes down some rapid-fire dictation. Between sentences he asks for reports on the boys' cases and holds forth on everything he hears. To a son: "That judge is a fucking vodka head." To Denise: "Get Mrs. Jamail and me a hunting and fishing license, would ya?" To another son: "Here's your pipeline check; you can put it in your trust fund or just have a good time. I don't care." To the visiting lawyer: "How's everybody over at Baker & Botts?" To a woman judge on the phone: "Helloooo, darlin'."

To an interviewer: "I'm pretty fuckin' active."

A brass telescope stands in the window facing the Hyatt Hotel. Next to the couch is a volume of Twain, which Jamail often picks up when he lies down. "A lot of it rubs off on me." On an end table rests *The Prophesies of Nostradamus,* bound in black velvet. "That sonufabitch really had sumthin'," he says of the medieval astrologer. Like many trial lawyers, Jamail has a superstitious streak, refusing, for instance, to wear polka-dotted shorts ever since his zipper broke just before his summation in the Case of the Misplaced Tree, exposing his polka-dotted glory to the jury. Jamail has even looked to the stars to break a tie between two candidates in jury selection. "I'll take a Libra over a Scorpio any day."

But for all the trappings of his success—the ten thousand-square-foot home in the exclusive community of Tanglewood, the beachhouse on Galveston Bay, the Jaguar with the cellular phone, his friendships with country singer Willie Nelson and former

Governor Mark White and the billionaire Cullen family (whose leader, Corbin Robertson, had caused such a stir at Getty Oil)—there remained an unpretentiousness about Jamail. He would make a point to bag a few groceries while he talked to the troops at his cousins' *haute cuisine* grocery. He truly enjoyed the other side of the tracks, whether rubbing elbows with Houston's low-level executives in a topless bar called Rick's Cabaret or sitting as a judge in a bikini contest called Mary Belle's Miss Wharf Rat Contest at one of his favorite haunts, a fishermen's bar along the bay.

Joe Jamail was a good ol' boy, leather boots and all. And Hugh Liedtke didn't mind that in a man.

They had met a decade earlier, first through their children, who attended Houston's most exclusive prep school, and later at a party through their mutual acquaintance with a senior partner of Baker & Botts. Jamail and Liedtke amused each other; they were two lawyers who could get together without having the slightest reason to discuss the law. Jamail, who smoked exactly four cigarettes a day, might blow smoke in the face of the ex-smoker Liedtke. They cruised to Hong Kong and Peking together with their wives and the Cullens. Jamail accompanied Liedtke to the Indianapolis 500, where Pennzoil sponsored the Penske racing team, and sometimes conducted business from Liedtke's yacht. On the credenza behind Jamail's desk, amid the forest of family photos, sat a framed black-and-white photograph of Liedtke and Jamail horsing around.

Jamail had handled a few commercial cases in his day—a fraud lawsuit against Browning-Ferris Industries, the nation's largest garbage hauler, a battle over the estate of a colorful oilman known to all as Silver Dollar Jim, and the probate case involving his friends the Cullens. But never a takeover case, never a securities case. *Pennzoil v. Texaco* was about as far afield from *Coates v. Remington Arms* as a lawyer could get. Obviously there was a huge contingency fee in it—probably a gargantuan fee, assuming that Jamail could get anything from Texaco.

But to anyone who asked, Jamail's refrain through the entire proceeding never changed: "I'm doing this for a friend."

On February 17, 1984, Texaco presented Gordon Getty with a check for $4,071,051,264.00 for the Getty Oil shares held in his grandmother's trust. Other than one for a slightly larger amount written by the U.S. Treasury in the refinancing of postwar debt, it was the largest check ever written. The mere anticipation that the trust would reinvest the money in government securities caused a momentary increase in interest rates in the entire market of U.S. Treasury bills.

When Pennzoil sued Gordon in Delaware for breach of contract, Gordon received word through Fayez Sarofim, the Pennzoil investor acquainted with Gordon and Ann, that Liedtke intended no personal offense in the suit. "It brought a smile to my face," Gordon says of the message.

Gordon, however, would not be smiling for long. For despite doubling the value of the family estate in the time he was sole trustee, the travails of the Getty family had only begun. Even though Getty Oil had cosponsored Tara's suit to unseat Gordon as sole trustee, the litigation would live long after the company had been dissolved into Texaco. It became a focal point for the family's displeasure with Gordon, particularly for "George's girls," Claire, Caroline and Ann, who laid out their grievances in one court filing after another.

"In the first twenty months Gordon P. Getty acted as sole trustee of the trust, he lost the ownership interest in Getty Oil that the trust had held for fifty years. . . .

"He created instability and uncertainty in the management of Getty Oil, and fomented open hostility between himself and the management of Getty Oil. . . .

"The trustee still possesses the same personality traits that annoyed and were documented in numerous letters by his father . . . ones that played an important role in the takeover of Getty Oil Company. . . .

"There is substantial, increasing friction and hostility between Gordon P. Getty and many trust beneficiaries. . . ."

Much of the "friction," however, had less to do with Gordon's personality that with the common affliction of the superwealthy: taxes. The sale of the trust's stock imposed a capital-gains liability of $1.1 billion on the trust. The two beneficiary classes—income beneficiaries and remaindermen—would be fighting over who would pay Uncle Sam until long after *Pennzoil v. Texaco* was under way. "When you've got people with an interest in a hell of a lot of money," Moses Lasky would comment, "they find a lot to fight about."

In a remarkable irony, *Pennzoil v. Texaco* would ultimately only further poison the bottomless well of the Getty family fortune.

*   *   *

The museum, having satisfied its yearning to sell out for top dollar, got a check for $1,165,042,500.00, plus a few days' interest on the amount, raising its total assets to about $2 billion and elevating it into the world's largest charitable trust behind the Ford Foundation. The money was divided among eleven portfolio-management firms, which poured it into one of the greatest bull markets since just before the Depression. With a curatorial staff of about sixteen and fewer than one-tenth as many visitors as the National Gallery of Art, the museum's 1984 earnings would eclipse the combined operating budgets of the Metropolitan Museum of Art, the National Gallery, the Smithsonian Museums and the National Endowment for the Humanities.

Like the Sarah Getty Trust, the museum would remain a defendant on the back burner of the Pennzoil breach-of-contract case in Delaware. And in yet another remarkable irony, in 1986, the museum would have cause to initiate a suit of its own—against *Texaco*.

The twelve thousand public shareholders of Getty Oil were showered with computer-printed checks, many of them promptly signed over to money-market funds or cashed into certificates of deposit. A great deal of the money—particularly the windfalls to trust departments and professional money-managers—went right back into oil stocks. Gulf Oil was a popular investment now that T. Boone Pickens' attack on the company was gathering momentum. The New Era Mutual Fund took some of its $16 million and put it in Pennzoil. "The company has good management," a fund manager explained.

The consumer economy also received a momentary shot in the arm. The 118 shares that Clifford Smith of Gary, Indiana, inherited from a rich aunt bought him a secondhand Oldsmobile Cutlass. Kenneth Herman sunk his $12,500 into an apple farm in upstate New York. Theodore Harder took part of his $100,000 and made a down payment on a truck stop near Salina, Kansas. Edward Trofino, a steel-mill supervisor in Bethlehem, Pennsylvania, had bought ten shares in 1957 and a few more along the way because he believed in J. Paul Getty. He put his profits behind Lee Iacocca at Chrysler Corporation.

Sid Petersen unfurled his golden parachute for about $2.5 million (representing three years' salary and bonus), cashed out his management stock options for another $1.8 million and went to

Sarajevo, Yugoslavia, for the 1984 Winter Olympics. But Petersen spent little time celebrating in the aftermath of the company's sale. When Goldman, Sachs spent part of its $18 million fee on a sumptuous dinner at Jimmy's in Century City for key Getty Oil employees, Petersen toasted the occasion by saying, "We didn't achieve our goal; the odds were too great against us. But at least we got a fair price." As one participant would later remark: "It was like a wake." Nearby, all work had been halted on the internal improvements to Getty Plaza, where the "Imperial Chairman" was to have watched over his empire. "It looks like the San Fernando Valley's tallest building will become its largest white elephant," the *Los Angeles Herald-Examiner* commented.

In a letter to a friend, Petersen complained that he had to subject himself to an "interview" with McKinley and pledge fealty in order to obtain a seat on the Getty Oil "interim" board of directors. "Typical of our experiences over the past year," Petersen wrote, "the final events of the merger found someone else in control of our situation."

But clipping coupons gave Petersen no solace from the ache in his heart for the nearly twenty thousand employees of Getty Oil. When he would discuss their fate even two years later, tears would well up in his eyes.

Some one thousand two hundred and forty of them would be thrown out of work, an additional five thousand spun off with ESPN, ERC and the refining operations that the FTC forced Texaco to divest. Petersen had bargained hard with McKinley for severance. The benefits package that Petersen obtained was better than most victims of oil megamergers were receiving at the time, but Petersen knew it wouldn't go far for families thrown out of work without medical insurance into a job market glutted with unemployed geologists and other merger casualties—just as oil prices were hitting the skids. Petersen knew of a marriage that split up in the process. A Getty executive quickly developed a drinking problem and enrolled at the Betty Ford Institute.

And the worst was the waiting—for weeks to see which operations the Federal Trade Commission would force Texaco to dump, for months to see which jobs Texaco would eliminate. Stress-management seminars were held. Someone ripped the letter "G" from the Getty Oil logo on the fifteenth floor of corporate headquarters. The *esprit de corps* gave way to boredom and busywork. STOP TEXACO bumper stickers went around. A Texaco employees' newsletter headlined ANATOMY OF A MERGER was defaced to read

ANATOMY OF A MURDER. Obscene cartoons depicting Gordon and Petersen were photocopied and distributed anonymously. Someone in the Tulsa office embellished a photograph of a hooded, ax-wielding executioner identified as McKinley with the caption: WE WELCOME THE SURVIVING GETTY PERSONNEL WITH OPEN ARMS. TEXACO CERTAINLY HAS NO AX TO GRIND.

For all the complexity of the deals, the intensity of the action and the bitterness all around, one Getty Oil employee managed to put his tongue in his cheek about the time the companies were merged. The account of this anonymous employee received widespread distribution, and it told everything that had happened—and what would happen—in a parable that made it all clear. It was titled "The Lost Book of the Gettians."

And it came to pass in the fifth year of the reign of Sidney, son of Peter, that Gordon the Terrible brought civil war upon the land of the Gettians. . . .

And Sidney, son of Peter, didst incite Tara Gabriel, the Galaxy Gramaphone, nephew of Gordon the Terrible, to unseat him as sole trustee. And Gordon the Terrible didst conspire with Hugh of Pennzoil to defeat Sidney, son of Peter, and didst make a pact with him to betray the land of the Gettians for 112 pieces of silver.

And upon hearing this news, Sidney, son of Peter, was filled with fear. And he summoned unto himself his Investment Bankers and said unto them: "Find for us a White Knight that we may withstand the onslaughts of Hugh of Pennzoil" . . . And lo, out of the East came a White Knight, John, powerful chief of the Texacites. . . .

And Gordon the Terrible didst betray Hugh of Pennzoil and didst also enter into the pact with John, chief of the Texacites, for 128 pieces of silver, and didst tender the land of the Gettians unto him for that price. Agreeing in this was Harold [Williams], the Antiquarian, who didst likewise tender.

And Hugh of Pennzoil didst rant and rage at this.

J. Paul Getty and his firstborn, George II, whose death was ruled a probable suicide.
REX FEATURES LTD.

Gordon Getty, who emerged from his father's disfavor to control of the family enterprise—and then dealt it away.
THE NEW YORK TIMES

The Getty Oil Board in peaceful times. Left to right in front: Stanford Miller and Henry Wendt. Left to right in back: John P. McCabe, Gordon Getty, Willard S. Boothby Jr., Harold Berg, Sidney Petersen, Robert Miller, Harold Stuart, Chauncey Medberry III, C. Lansing Hayes Jr., Frederick Larkin Jr., Dr. Norman Topping.
GETTY OIL

Getty Oil chief Sidney Petersen, "the Imperial Chairman": "It's amazing how naive we were."
THE NEW YORK TIMES

Texaco chairman John K. McKinley: "I am usually serious about everything."

Pennzoil chairman J. Hugh Liedtke, "Chairman Mao": "We used to say that in the oil industry, business was done on a handshake. Should it now require handcuffs?"

DAN FORD CONNOLLY/PICTURE GROUP

Getty Oil lawyer Bart Winokur, "Back-Door Bart": "Somebody could get squeezed," he told Gordon, "and you could be the juice."
STANDARD PHOTO

Getty Museum president Harold Williams, the former SEC chairman who found himself in the position of kingmaker. Takeovers, he said, "feel different when you are in the middle of it."
© 1984 LOS ANGELES TIMES

"The two Martys."

Superlawyer Marty Lipton for the museum: "The whole situation was so bizarre." M

Marty Siegel, Gordon's investment banker. With events heating up, he broke off a Virgin Islands vacation and chartered a jet to New York.
JOHN MAGUIRE / MATAR STUDIO

Geoffrey Boisi, the "$18 Million Man," who advised Getty Oil, then brought Texaco into the picture: "You better put your best shot forward."
TED KAPPLER

Arthur Liman, Pennzoil's chief negotiator. "Do we have a deal? Can I go home? Can we get it over?" THE NEW YORK TIMES

Bruce Wasserstein of First Boston, who did not sleep until Texaco had grabbed Getty Oil from under Pennzoil's nose. "Is there a deal here?...How much time do we have?" THE NEW YORK TIMES

Judge Anthony J. P. Farris, "Tough Tony": "I want you to always, always, always, always, always, always, always, in a yes-or-no question, answer yes or no if you understand it."
© 1985 BY STEVE BRADY

Judge Solomon Casseb, Jr., the last-minute substitute. "I am at a disadvantage...."　　　　DAN FORD CONNOLLY/PICTURE GROUP

The Pennzoil legal team: Joe Jamail, center, flanked by Irv Terrell (left) and John Jeffers (right) of Baker & Botts. "We were two good trial lawyers," says Terrell, "with one extraordinary trial lawyer."

Texaco's Richard Miller, who pitted his "little chickenshit law firm" against the juggernaut of Jamail and Baker & Botts. "Most people around town are afraid of me, and that's the way I want to keep it."

Jamail, the "King of Torts," implores Judge Casseb to sign the $11 billion judgment. "We're hung out to dry, judge!"

· PART 3 ·

THE CASE

Home territory does help.
   —Judge Solomon Casseb Jr., April 2, 1986

· 18 ·

William Liedtke Sr., the "boy orator" who had espoused populist views at the Oklahoma Constitutional Convention of 1907, would have been proud.

The occasion was the 1984 Pennzoil annual meeting of stockholders, the subject was honor and the speaker was Hugh Liedtke. So many hundreds had poured from Pennzoil Towers that they clogged downtown traffic along the three-block walk to the Sheraton Hotel ballroom. Liedtke gave a brief discourse on the prosaic topics of oil prices and national energy policy, the standard fare for any annual meeting in Houston. "Finally," he said, turning to the subject at the forefront of everyone's mind, "I might just say a few words about Pennzoil's litigation against Texaco, because of the continuing interest in the matter."

After outlining the legal details of Pennzoil's claim, he said:

There is perhaps a greater question involved. It turns on the crucial point of integrity in our industry.

It's one thing to play hardball. It's quite another thing to play foul ball.

Conduct such as Texaco's is not made legal simply by protestations that the acts involved were, in fact, legal. All too often such assertions go unchallenged, and so slip into some sort of legal limbo, and become accepted as the norm by default. In this way, actions previously considered amoral somehow become clothed in respectability.

Pennzoil's litigation challenges this mindless slip into acceptability. We seek to test the acceptable standards of behavior in our industry.

A contract is a contract. We used to say that in the oil industry, business was done on a handshake. Should it now require handcuffs?

. . . We believe that integrity is more than just a word. It is a standard of conduct in a world perhaps gone slipshod. Our industry was built on that standard, and Pennzoil will continue to make every effort to see to it that this standard is upheld.

Thank you.

The shareholders and employees sprang to their feet, applauding madly. Hugh Liedtke had set the moral tone of *Pennzoil v. Texaco*. Single-handedly, their company, their chairman, would restore honor to a *world gone slipshod*.

Dick Miller got a good laugh out of Liedtke's speech, but he appreciated that Liedtke had made it. As Texaco's newly hired Texas trial lawyer, Miller now gauged that Pennzoil would try to turn a routine tortious-interference-with-a-contract case into a matter of honor. He could see it coming: Joe Jamail swaggering before the jury box, pleading for the ladies and gentlemen of the jury to "put the honor back in a handshake." There was only one word for it, a word that Miller would use over and over to describe Pennzoil's case: "chickenshit."

The whole thing reminded Miller of the movie *Boom Town*, in which Spencer Tracy and Clark Gable decided the ownership of an oil company on the flip of a coin. *This handshake stuff won't get you to first base in New York*, Miller thought. *Let them try to collect fifteen billion on a handshake. Let them try to come up with a single person from Getty Oil who actually shook hands with Hugh Liedtke*.

But Miller knew that the facts would become secondary in this case. Miller was in for the fight of his life . . . and he relished the prospect.

At fifty-seven years old, with brawny biceps, short-cropped hair, penetrating blue eyes and blood vessels showing through his temples, Miller looked every inch the ex-Marine that he was. He was Houston's most ferocious corporate trial lawyer and was considered by some to rank among the best in the nation. Only trial lawyers, he believed, were the true "artists" of the law. He viewed all other lawyers as Bartleby-like figures who toiled over paperwork without ever stepping into the field of battle. They were "assholes," in his judgment, "scissors-and-paste experts." Miller was no settler, no peacemaker, no negotiator. He specialized in "bet-the-company cases," in a go-for-broke legal strategy in which absolutely nothing was conceded to the enemy. To him, a trial was nothing short of war, a settlement nothing but surrender.

"Most people around town are afraid of me, and that's the way I want to keep it," Miller would say.

He had collected dozens of books on the world's most famous trials, from *The Bloody Assizes* to *The Trial of Oscar Wilde*. He was a nitty-gritty detail man, one who exploited every tiny shred of favorable evidence and who attacked even the smallest fact put forward by his opponents. In this respect he was the direct opposite of Pennzoil's Joe Jamail, who painted with a broader brush, whose vision of a case was more far-sighted than near-sighted. While Jamail kept a brass telescope in his office, Miller had a brass magnifying glass.

Although he considered Pennzoil's case laughable, Miller was acutely aware of the strength of Pennzoil's lawyer, himself an ex-Marine. Miller and Jamail were friends; they liked each other's brashness and irreverence and they respected each other's courtroom skills. Only a year earlier Miller had actually represented Jamail, defending him against charges that he had slandered a state senator accused of taking a bribe from Browning-Ferris Industries, the waste-hauling company. Miller quickly got the case against Jamail dismissed, and when Jamail had mailed Miller a payment of $5000, Miller had voided the check and returned it to Jamail with a note:

Joe,
   You've got to be kidding! I donate check #40649 to the "Fund to Get BFI."
   It was an honor to represent you.

If Jamail weren't opponent enough, Miller would have to contend with Jeffers and Terrell of Baker & Botts—two lawyers he had spent years helping to train, whom he considered to be protégés. And therein rested the reason that *Pennzoil v. Texaco* would explode not only into the highest-stakes courtroom in the history

of the law, but also into one of the most intense, personal battles
ever played out in a court.

Richard Beatty Miller was born on March 6, 1926, and, like Hugh
Liedtke, grew up in Tulsa. Thereafter, however, the similarities
ended. Liedtke grew up in an all-American household, the child of
a prominent hostess and a leading oil lawyer. Miller, whose
mother died in childbirth, was reared by an aunt, and his father
was a transplanted Dust Bowl farmer who had dreamed of being a
lawyer but who had to settle for selling animal feed. Liedtke lived
south of Twenty-first Street—the dividing line between the right
and the wrong side of town—and Miller far to the north of it;
Liedtke went to the Cascia Hall prep school, Miller to Will Rogers
High. Liedtke lived next door to the Skelly family, who introduced him to J. Paul Getty over cocktails; Miller lived within
three hundred yards of the cinderblock residence where J. Paul
Getty lived during the war, when he was managing Spartan Aircraft. Miller, however, never laid eyes on the great Getty.
  Miller's background bore more similarity to that of his
courtroom adversary, Jamail: two ex-Marines with the foulest
mouths in town, yet men who could turn on the charm in an
instant and talk their way into anything. Miller, for his part, quit
high school midway through his junior year to join the Marine
Corps, which to his dismay initially assigned him to foreign-
language school. "I wanted to kill Japs," he would recall, "not
talk to them." Finally Private Miller cajoled himself a position in
the Raiders Battalion, then into Scouts and Snipers School and
finally into the Fifth Division, which fought the battle of Iwo
Jima. His inability to rise beyond the rank of sergeant convinced
him that a high school dropout had little chance in the world, so he
talked his way into the University of Tulsa without a diploma and
after two years succeeded into transferring to Harvard College.
Following his junior year there, he talked his way into Harvard
Law School—again without a diploma.
  "You're telling me three years in the Marine Corps isn't worth
a year in Harvard College?" he asked the admissions director.
Miller spent his Harvard years studying in the coal bin at his
boardinghouse, and graduated from Harvard Law in the top tenth
of his class. And when Baker & Botts sent a recruiter to campus in
1952, Miller was scheduled for an interview by a placement director who mistook Miller's Oklahoma drawl for a Texas accent.
  Although Houston was fifty miles from the Gulf of Mexico, a

ship channel clawed out by its early boosters had made it the nation's third-largest port, with a refining and shipping industry dominated by multinational corporations that needed local counsel. Unlike Dallas, Denver, Kansas City and even Los Angeles, Houston experienced such growth in demand for legal services that eventually it boasted of having three of the nation's biggest law firms. The city's largest was Vinson, Elkins, Searls, Connally & Smith, which handled local work for Texaco—and whose partners included Hugh Liedtke's friend John Connally. Next was Fulbright, Crooker & Jaworski, where Joe Jamail had quit almost immediately after starting. And nearly as large was Baker & Botts, where an unduly high proportion of the lawyers—70%— did at least some trial work. Dick Miller decided that this was the firm for him. "I never had any ambition—still don't—but to be the best trial lawyer in the world."

Shortly after Miller joined Baker & Botts, the firm was just beginning to help three young oilmen out in West Texas—George Bush and the Liedtke brothers—build up the Zapata Petroleum empire. Over the years, as Zapata became Pennzoil and moved to Houston, the Liedtkes' company became one of the largest and most valued accounts at the law firm, and Miller could not help establishing close ties to it. He became a close acquaintance of Baine Kerr, who left Baker & Botts eventually to become the president of Pennzoil, and forged a strong family friendship with Perry Barber, who left to become its general counsel. On occasion he socialized with Bill Liedtke and his wife. Once, when Pennzoil was being sued for breach of contract in Louisiana, Miller represented Hugh Liedtke at a deposition.

But in twenty-five years at the firm, while establishing a reputation as its star litigator, Miller struggled against the bureaucracy and internal politics that accompany the growth of any law firm. After becoming the head of the trial department, he railed against the increasing practice of assigning several lawyers to a case by borrowing the motto of the Texas Rangers: "One riot, one Ranger," the state police said. "One lawsuit, one lawyer," Miller ordered. As the firm's managing partners changed, Miller became convinced that Baker & Botts was getting soft; indeed, over the years small-time personal-injury lawyers in Houston came to delight into going head-to-head with the giant clients of Baker & Botts, because the firm was considered a soft touch for small settlements, in the range of twenty thousand dollars.

"I had planned to leave that goddamn law firm from the day I got there," Miller would later say. "I didn't realize how good people thought I was. I didn't realize my commercial value." On the outside, with his own small firm, Miller figured he could quadruple his salary at Baker & Botts, where senior partners could easily make $500,000 a year. So in 1983, Miller struck out on his own, enlisting a half-dozen of Houston's brightest legal stars to form a firm called Miller, Keeton, Bristow & Brown. He liked to call it "my little chickenshit law firm."

As soon as Pennzoil nonsuited Texaco in Delaware and pulled its surprise switch into Texas, a Texaco divisional lawyer in Houston, Charles Irvin, began canvassing local law firms and found several willing to take on the case against the Jamail–Baker & Botts team. Irvin asked everyone he contacted what they thought of Miller's new, hotshot law firm; the praise was unanimous, even from those firms that hoped to receive the engagement for themselves. Irvin tracked down Miller in Midland—the old stomping ground of Hugh Liedtke, where Miller was representing Philip Anschutz, an oil magnate ranked by *Forbes* among the nation's one dozen billionaires, in a property dispute with Mobil. Would Miller like to take the Texaco case?

Initially Miller declined. How could he take a case against his old law firm, against his friend and onetime client Jamail, against his friends who worked at Pennzoil? "I knew they would regard my entry into the case as an act of hostility," Miller would recall.

And all the more so because of the nature of the case—a case in which the performance of Baker & Botts in the takeover deal was bound to become an issue. With $15 billion on the line, Miller would have to play as rough as he knew how, and playing rough would mean launching a frontal assault on his old law firm. Miller knew the partners and the internal politics of his old firm as well as anyone in Houston; he sensed that the lawyers working on the Getty Oil deal had purposefully kept their number small so that they wouldn't have to share credit in what had clearly started out as a brilliantly conceived transaction. "I could see the Baker and Botts corporate lawyers had fucked up," Miller says.

When Texaco continued pressing Miller to take the case, he took the matter up with his new partners.

Although Miller questioned whether his "little chickenshit law firm" had enough staff to handle so large a case, his partners considered Texaco's offer a godsend. *We gotta be crazy not to take this case,* they said. *We're a new, small firm. We're trying to make a name for ourselves as a full-service litigation firm. We have to take this case!* When Miller expressed his reservations about fighting the firm where he had spent twenty-five years building his reputation, his partners reminded him that he had left his partnership interest, worth about $100,000, at Baker & Botts to assure his freedom to compete with the firm.

On a Saturday morning in February, as Miller was debating the issue with his new partners, the phone rang. It was his former partner John Jeffers from Baker & Botts, who had heard that Miller had been offered the case against Pennzoil. Jeffers said he believed that Miller's firm was disqualified from representing Texaco.

"Why, for Chrissake?" Miller demanded.

Jeffers said that one of Miller's new partners—Daryl Bristow, also formerly of Baker & Botts—had briefly represented Pennzoil in connection with Getty Oil just before leaving the firm.

"That's absolute bullshit!" Bristow told Miller.

The next day, Miller called his old friend Perry Barber at Pennzoil to ask him if there was any reason he shouldn't take the case. According to Miller, Barber replied, "No, it's just two big companies fighting each other."

Miller was relieved. Maybe he could take the big case after all.

Then Barber called back.

"Listen," Barber said. "I want to be completely candid. I just want you to know all your friends at Pennzoil are going to be plenty pissed off if you take that case."

That did it. *I don't owe those bastards anything,* Miller told himself. He took the case.

On March 6, 1984, the day after he officially filed a response to Pennzoil's lawsuit, Miller turned fifty-eight. That morning he walked into his office and was greeted by a stack of depositions, more than any man could easily carry. Pennzoil had scheduled depositions with twenty-nine potential Texaco witnesses—all to occur during only the last two weeks of the month. Moreover, they were to be held in five cities spread over both coasts: in New York, for all the investment bankers, the museum's Marty Lipton and others; in Philadelphia, where some of Getty Oil's lawyers resided; in Houston, for certain Texaco executives; in San Francisco, for Gordon and his group; and in Los Angeles, for Sid Petersen and all his people.

In a $15 billion case, the juggernaut of Jamail and Baker & Botts was going to push Miller's tiny, newly formed "full-service litigation firm" to the limit—in only its first week on the case.

When Miller called Jamail to complain about this "chickenshit" move, Jamail suggested they get together to talk. But after Miller's supposed friends had inflicted what he viewed as professional discourtesy on him, he wasn't about to try negotiating a more reasonable schedule for depositions and other "discovery" proceedings. A judge had just been assigned to handle all pretrial matters. Miller would demand a court order *forcing* Pennzoil to consent to a more reasonable discovery schedule.

The majority of states elect their local judges, but every Texas judge from the county bench to the state Supreme Court must withstand the onslaught of partisan elections. The reason is rooted in the state's Reconstruction Constitution of 1869, in which the carpetbaggers gave the governor the power to appoint some ten thousand political officials around the state. When Texas finally bade farewell to the Yankees, the locals saw to it that virtually every important public job in the state—including every judge from the county level up—was controlled by the voters.

In modern times, conducting a judicial re-election campaign posed little problem in the numerous rural burgs of Texas, where little was required of the incumbents beyond some handshaking at the livestock show or the county fair. But in a town like Houston, the fourth-largest city in America, voters on a single election day might face choices for three Supreme Court races, fifteen appeals-court races, thirty-seven state-district court races and seventeen county court races. The only hope of assuring re-election was name recognition, and in a county with so many candidates and so many voters that could be obtained only through money, most coming from lawyers and law firms. Big money.

State District Judge Anthony J. P. Farris, a curmudgeonly former U.S. attorney with a long white mustache and rather mournful eyes, was facing no opposition in the 1984 Republican Primary but expected to have a Democratic opponent in the fall. And his campaign chairman was fretful over the dearth of campaign funds raised to date. Even Joe Jamail, a big-time judicial donor who sat on the steering committee of the judge's campaign, had donated a paltry $100.

On March 5, 1984, Judge Farris was assigned jurisdiction over all pretrial legal matters in *Pennzoil v. Texaco*. Two days later, he received a campaign check for $10,000, four times greater than the largest amount he had previously received. It came from Joe Jamail.

"A princely sum," Judge Farris noted of the contribution in a letter to his campaign chairman. Judge Farris duly recorded the contribution on a campaign-disclosure form for the Harris County clerk, but it would not be subject to public filing for four months.

When he held the first pretrial hearing in *Pennzoil v. Texaco*, Judge Farris had distinguished himself as an underachiever of the Harris County bar. During the war he had amassed nine decorations as a Marine gunner aboard dive bombers and torpedo bombers, but in civilian life he had mostly failed to regain his glory. After graduating from the University of Houston Law School, he ran for the U.S. Congress—unsuccessfully, as most Republicans in Texas did. He operated a general law practice—handling civil, criminal, divorce and plaintiffs' personal-injury cases—from the Sterling Building, a structure teeming with small law firms in the 1950s and 1960s. In an era when Baker & Botts and the other major firms had not yet engulfed the Houston legal community, the small-firm lawyers of the Sterling Building operated in a close-knit club; in the period when they dominated the Houston Bar Association, the group's activities consisted mainly of golf outings and stag parties. During this period, Tony Farris and Joe Jamail forged a strong acquaintanceship, according to a lawyer who worked with them in the Sterling Building.

As GOP lawyers had always been a rarity in Houston, Farris became the U.S. attorney in Houston when Nixon held the White House. He became widely known for his administrative skills, but his tenure as U.S. attorney was also marked by a controversial plea-bargain arrangement in which the kingpin in a celebrated political scandal got probation and a mere $5000 fine. Farris' election to the county bench in 1980 brought him a step closer to his longtime dream of an appointment to the federal judiciary, and initially he threw himself into the job with boyish enthusiasm—putting in long weekends at the office, for instance, something that judges nowhere are famous for doing. But after assuming the bench, Judge Farris had collapsed in his chambers with a serious coronary attack, developing a cardiopulmonary condition that required continual medical supervision and severe restrictions on his workload.

He acquired a reputation for sardonic behavior that sometimes bordered on outright rudeness. In the midst of the Pennzoil case he would write a caustic letter to Houston Mayor Kathy Whitmire, complaining that the city's "little meter maids" seemed to be going out of their way to issue parking tickets to judges. "Tough Tony," a columnist of the *Houston Post* christened him. Among

nineteen state-court judges included in a 1985 survey by the
Houston Bar Association, he would be ranked as the thirteenth
hardest-working, as the seventeenth most impartial and dead last
for courtesy and attentiveness to attorneys and witnesses.

It was before Tough Tony, the meanest judge on the state-court
bench in Harris County, that Dick Miller took his plea to relax the
hurry-up "discovery" schedule that the lawyers from Pennzoil
had imposed.

"You realize," Miller told the judge, "that these people are
seeking more damages than the national debt before World War
Two started!"

"And we reserve the right to amend," Jamail said—meaning
that Pennzoil might ask for even *more*.

"I'll see if they have nerve enough to raise it," Miller said.

"We do."

"Just sit back," Miller snapped. "You'll get your turn."

Miller told the judge of the "animosity and vindictiveness ema-
nating from Pennzoil" because of the failure by Baker & Botts'
lawyers to get Getty Oil's signature on a contract. "They're under
a lot of heat, and they know it and I know it," Miller said. "They
don't want me to be ready. . . . It's a shame people feel like they
have to practice law like that."

After listening to both sides argue the issue for about an hour,
Judge Farris made his first ruling in the pretrial phase of the case.
"Texaco's motion for protection is denied," he finally said.

Dick Miller was as familiar as anyone with the large reputation
that his opponent Joe Jamail enjoyed with the local judiciary.
Even though Judge Farris' official campaign records so far reflected only a $100 contribution from Pennzoil's lead lawyer,
Miller was concerned. Jamail, he learned, was sitting on the
judge's re-election steering committee. By itself that suggested
nothing untoward; lawyers try to help their favorite judges get re-
elected the same way congressmen do for presidents. In fact,
Miller's own partner, Richard Keeton, would sign a political ad
endorsing Farris' candidacy. But this was a gigantic case. Out of
an abundance of caution, Miller decided that he had better hire
another member of the judge's campaign committee as Texaco's
counterweight to Jamail.

With Texaco's permission, Miller talked to his close friend Don
Bowen, whose law partner, George Pletcher, also served on the
Re-elect Judge Tony Farris Steering Committee. Pletcher was in

Maui at the time, but readily assented over the phone to joining the big case on Texaco's side. Now, *both* sides were represented by a member of Judge Farris' committee.

For a few days, anyhow.

When Pletcher returned from Hawaii, he had several messages waiting from Joe Jamail. Pletcher and Jamail had been friends for thirty years. Their families had been having Christmas breakfast together ever since their kids, now adults, had been in diapers. Their beach houses on Galveston Island were a block apart. Jamail was the godfather of one of Pletcher's sons and Pletcher was the godfather of one of Jamail's sons.

"George," Jamail told Pletcher, "this case is gonna be mean and dirty."

Pletcher was confused; he had understood that Baker & Botts was taking the lead for Pennzoil. His friend Jamail, he thought, would be playing only a small role in the case.

"Baker and Botts doesn't have me," Jamail said. "*I've* got Baker and Botts!"

That changed things entirely. Although the partners in Pletcher's firm argued strenuously for continuing the engagement with Texaco, Pletcher visited his old friend Judge Farris to request permission to withdraw. The judge urged Pletcher to remain with Texaco; "it would help balance things," Pletcher recalls the judge saying. But, he adds, "I just decided a big case wasn't worth a friendship."

Jamail too would insist that he only wanted to preserve one of his closest friendships. Texaco's Miller, however, would say he learned something valuable from the Pletcher affair. Not only did Pennzoil want "one-sided name recognition" with Judge Farris, he would say, but "we knew then that we were up against a results case, not a methods case," meaning that Pennzoil wanted results, no matter by what method.

The lawyers were back before Judge Farris twice more in the next two months, each time fighting over documents that the other side was refusing to produce to Pennzoil. Getty Oil alone had produced twenty-five thousand pages of subpoenaed documents, but finally threw up its hands and said "no more" when Pennzoil, among other things, demanded every document in any way connected with the preparation of Getty Oil's tax returns; to comply, the company would have to search one hundred million pages of documents scattered throughout five hundred individual offices.

Judge Farris ordered Texaco to haul someone from Getty Oil to Houston to negotiate a compromise with Pennzoil.

"Do it within a week," the judge told one of Miller's partners.

"I'll do my best," he replied.

"That's not a firm enough answer," Judge Farris said. "I cannot do anything about Getty, but I can do something about Texaco."

Pennzoil also hauled Texaco into court when First Boston, Texaco's investment-banking firm, refused to provide confidential internal studies about takeover candidates. First Boston had even run into a New York courtroom for an order to quash a subpoena sanctioned by the Houston court—an action to which Judge Farris did not take kindly.

"Mr. Miller," he told Texaco's counsel, "I'm beginning to consider that the prestigious First Boston people believe us to be a colony down here."

Another theme was emerging. *Pennzoil v. Texaco* was rapidly becoming *Texas v. New York*—even *Rebel v. Yankee.*

Despite the extraordinary speed with which Pennzoil had attempted to bring Texaco to justice, enough time had passed since the complex events of early January that memories were already fading. The depositions were all the more complicated because they were the second set of interrogations conducted in the case; both sides had taken copious depositions in the case in Delaware. "I am not sure what I remember from remembering it directly, or what I remember from reading my deposition of what I remembered at the time, in January," Gordon's investment banker Marty Siegel said during his second third-degree by Pennzoil.

Some witnesses could recall little more than their names. Richard Howe, the head of public relations and communications for Pennzoil, managed to communicate virtually nothing about the planning sessions that occurred before the tender offer for Getty Oil:

"Did you know what the subject of the meeting was going to be?" a Texaco lawyer asked.

"I don't recall."

"How long did that meeting take place?"

"I don't recall at all."

"Was it a short meeting or a medium-sized meeting or what?"

"I can't remember."

"No recollection at all?"

"Not at all."

"No notes, I guess?"

"No."

"Have you ever had any notes of that first meeting?"

"No recollection of notes."

"What did you discuss at the first meeting?"

"I don't remember. All I remember is a series of meetings."

"Well, you would think you would remember the first meeting best, wouldn't you?"

"Not necessarily."

"Which meeting *do* you remember the best?"

"I don't remember any specific meeting."

And so on.

As the deposition phase of the case continued, with the Pennzoil and Texaco lawyers following one another on commercial flights from city to city, their emotional baggage got heavier on each stop.

It was only a few minutes into the deposition of the austere John McKinley, for instance, when one fight erupted. Pennzoil's Jamail asked the Texaco chairman whether his responsibilities included the power to "bind" the corporation with contracts. Miller's top partner, Richard Keeton, instructed McKinley not to answer. Jamail then asked whether McKinley understood the question. Keeton wouldn't let him answer that question either.

Jamail decided to stonewall.

"I think I will look at this watch," he said.

"You are browbeating!" Dick Miller barked.

"Sit down," Jamail shot back. "I haven't *started* the browbeating."

"Cut that crap and get on with it."

When Jamail asked McKinley whether he believed an "agreement" represented a "meeting of minds," McKinley was again instructed not to answer.

"I am asking his opinion as an expert," Jamail explained.

"He is not an expert."

Jamail turned to McKinley. "Your counsel is claiming you don't know how to do your job."

"You are so unfair," Miller said disgustedly.

Jamail again addressed McKinley. "Are you going to take the Fifth Amendment on that?"

"That is such a chickenshit remark."

"I do not care what you think."

Before long, Terrell of Baker & Botts jumped into the fight against his old boss. "You have a short fuse, Dick."

"Shut your mouth," Miller said. "You better be careful."

"You better *make* me. You have a hell of a short fuse."

When Texaco took depositions from people on Pennzoil's side, it chose not to videotape, and in a few cases it was just as well. Bob Brown, another partner of Miller representing Texaco, had to enter the enemy territory of Baker & Botts' offices in Houston's One Shell Plaza in order to take the deposition of Moulton Goodrum, the lawyer who had been in charge of completing the Pennzoil's documents in the Getty deal. After four wearying days of examination, tempers finally flared in a discussion of the Dear Hugh letter.

"Look," Terrell told Brown, "you work on quit being rude."

"I have *not* been rude."

"Work on asking good questions."

"I've been asking real good questions. Why don't you go out and instruct him to answer a question for once."

"No," Terrell said. "I am instructing *you* to be polite or get out."

"Well, I don't take instructions from you."

"Be polite or get out!"

"If you would like to invite me out, I will go out with you."

"Well, look, I am telling you to leave the law firm, if you won't be polite. If you are not polite I am going to throw you out of here."

"I challenge you to do that."

"All right."

"I don't believe you intend to. I don't believe you are physically able to throw me anywhere."

"What is your next question, tough guy?"

"Well, I don't feel tough at all. But I'm not intimidated by you or your law firm."

And so it went.

History's biggest civil case had become one of its least civil cases, and the trial was still a year away.

Dick Miller of Texaco received a call in late August from a lawyer friend who happened to work in the same office building. "Did you know that Joe Jamail had contributed ten thousand dollars to Farris?" the friend asked.

Miller was stunned. He assembled a battery of researchers to paw through the campaign-finance records filed by fifty-one local judges between 1980 and 1984; Jamail, they found, had contributed a total of $32,710 to nineteen judges—with nearly one-third of that amount going solely to Judge Farris. Another $10,000 of the amount had gone to an administrative judge who had authority over Farris. In Miller's mind, suddenly everything began to make sense.

Miller began roughing out a new court motion, adding up everything he knew. Here was Jamail, a dyed-in-the-wool Democrat, suddenly contributing an extraordinary sum to a Republican candidate he had never before supported—and doing it two days after the judge was given jurisdiction! Jamail, a sore-back-and-broken-arm lawyer, handling a takeover lawsuit that Baker & Botts was perfectly capable of handling on its own—surely he was in the case only for his sway with the judge, Miller thought. That's Pennzoil's style, all right; in fact, it probably explained why Pennzoil escaped through the back door of the Delaware court, Miller thought. It was "in keeping with its practice of attempting to bring political influence and pressure." Pennzoil probably had "reimbursed" Jamail for the ten grand, he wrote. And every single pretrial ruling that Miller could think of had gone Pennzoil's way.

Miller refused to recognize that Pennzoil had plenty of other reasons to hire Jamail, namely his persuasive powers with juries and his friendship with Pennzoil's chairman. But there was no disputing the poor appearance of the $10,000 gift. Miller wound up his motion with a flourish: "If one assumes that Judge Farris was ingenuous enough not to recognize the true purpose of the gift, nevertheless, the Jamail-Pennzoil contribution *appears* to have succeeded, *appears* to have affected the outcome of the hearings, *appears* to show bias, and *appears* to the bar and the public as an affront."

After consulting with Texaco, Miller filed his complaint as a motion to have Judge Farris pitched out of the case.

Dick Miller was fond of quoting Sherlock Holmes' line, "Watson, I don't believe in coincidences." Miller was convinced that one such coincidence occurred in late October, when the Honorable E. E. Jordan of Amarillo was assigned as a visiting judge to hear Texaco's motion to throw out Judge Farris. Texaco would later claim that the Visiting Judge Jordan had been designated for the hearing by the administrative judge who had received the *other* $10,000 contribution from Joe Jamail. Not so, says Thomas Stovall, another administrative judge in South Texas. Judge

Stovall says he appointed the special judge—and no one ever accused Judge Stovall of taking campaign money from Joe Jamail. In any case, the hearing to get rid of Judge Farris was held on October 25, 1984.

As officers of the court, lawyers are always presumed to be telling the truth while standing in the hallowed halls of justice. But after Dick Miller had briefly laid out Texaco's case against Judge Farris, Joe Jamail called his old friend to the witness stand and subjected him to the humiliation of a swearing-in.

"Tell us your name, please," Jamail then said.

"Richard B. Miller."

"And you claim to be a lawyer?"

Jamail proceeded to ask Miller the basis for his claim that Pennzoil was behind the $10,000 contribution.

"It's based on what I know about this case and what I know about Mr. Liedtke and his past." Miller also took note that Jamail's formal answer to the charge never denied that allegation.

"Sir," Jamail whined, "that is so cowardly."

"Wait a minute!" the visiting judge cried. "I'm not going to tolerate this. You can curse one another all you want to—somewhere else, but not in this court."

Jamail apologized, but explained that he was only trying to show that the campaign contribution "could not have influenced Judge Farris."

"I don't believe it anyway," the visiting judge said.

"That settles that," Jamail said. "I have no further questions, judge."

Miller stepped down from the stand and made his closing argument, which concluded: "If the public does not feel that there is going to be equality under the law—and if there is not at least one branch of this government which is above suspicion, which is above graft, which is above handouts, which is above influence-peddling—then we are in bad shape." Miller then took his seat.

Jamail then rose for his closing argument. "It is a disgrace for an officer of the court to come, and with suspicion and surmise smear the judicial system. I have contributed money, Your Honor, to almost *every* judge. . . . To say that I have never given a contribution of ten thousand dollars before is a *lie*. I have. And he knows it. . . . Those of us who have been successful at the bar, unlike Mr. Miller, are called upon to do more. . . .

"It's a long whine and pout by Mr. Miller. . . . Mr. Miller became surly because he couldn't have his way. . . . Mr. Miller's problem is that he can't run over Judge Farris. . . ."

Besides, Jamail explained, Judge Farris was assigned to handle only pretrial matters. "Your Honor, *Judge Farris won't try this case*. This will go on the central docket and it will go back and be reassigned."

The real reason for Texaco's "despicable" and "peevish" charge, Jamail continued, was that Texaco wanted to delay the case in Texas—so that it could find a way to get back into Delaware. Jamail looked at Miller. "He does not want to go to trial. We don't have enough cowboys in Texas to whip him with, to make him come try the case."

After letting the lawyers harangue for several more minutes, Visiting Judge Jordan spoke his peace. "The lawyers associated in this case have become so acrimonious that, well, it's just disgraceful in my opinion."

The Texas Constitution provided that a judge could be removed from a case only if he had a direct financial interest in the outcome or if he were related by blood, marriage or by prior legal service to one of the litigants. "Mere bias," under Texas case law, was insufficient grounds to disqualify a judge.

"The Constitution of the State of Texas . . . means the same thing today that it meant the day it was written," in 1876, Judge Jordan continued. And even though a canon of the Judicial Code of Ethics could be construed to suggest that Judge Farris should be thrown off the case, he said, "the Constitution still controls. The Code of Ethics does not. Number two, there is not only insufficient evidence, there is *no* evidence in my mind to justify recusing this judge.

"It's so ordered."

Jamail asked the court stenographer to make him a copy of the transcript so he could file it with a complaint against Miller to the Bar Association grievance committee. Miller made the same request.

In early 1985 the district courts of Harris County retroactively changed their rules. Judges who handled pretrial matters would now handle the trials as well.

"Judge Farris won't try this case," Joe Jamail had exclaimed three months earlier. But he would after all.

"It takes thirteen men to steal a horse."

Dick Miller used the expression to describe any case in which he doubted the judge's impartiality. There were always twelve other people to appeal to for justice in a case—the jury. "A judge

can fuck us and cheat us all he wants, but that ain't enough,"
Miller believed. He swore he would sway the jury to Texaco's
side.

But Miller's initial preparation convinced him that although the
case was clearly winnable, it was hardly open and shut. Using
videotaped depositions and friends and partners playing live witnesses, he staged one-day mock trials before three "juries" selected to represent a cross-section of Houston's demographics,
then watched each panel deliberate through one-way windows. In
two cases, won by Texaco, the jury had decided that a party has
the right to see all the fine print in a contract before being bound
by it—an outcome that told Miller he should play up the "open
issues" that remained in Pennzoil's dealings with Getty Oil and
the museum.

But in the mock trial that Texaco lost, "there was sympathy for
the concept that a deal is a deal," Miller found.

In none of the three trials, however, was the outcome over-
whelmingly for or against Texaco. "It made it crystal clear to the
client," Miller says, "that this wasn't any jelly roll—that we had
a fight on our hands."

Miller's trial strategy would have many elements. Besides
pointing out the obvious evidence—that no one from Getty Oil
ever signed Pennzoil's original memorandum of agreement, and
that Getty Oil had only reached an agreement in principle "subject
to" a definitive agreement—Miller would lay out the following
arguments:

First, *Pennzoil, not Texaco, was the heavy in the Getty deal.*
Using high-pressure tactics and by pandering to Gordon's supposed insecurities, Pennzoil had sought an unfair deal from the
Getty Oil board, leaving it with no choice but to run into the arms
of Texaco. By the time he got through describing Pennzoil's
tactics, Miller would try to make Liedtke wish he had never heard
of tender offers—let alone invented them. And then, by laying out
the extent of Liedtke's harassment of Texaco, including the political threats to John McKinley that Liedtke had purportedly made
after the deal had fallen through, Miller would see to it that the
jury wouldn't award Liedtke a buck, let alone fifteen billion of
them.

Second, *Pennzoil knew it never had a contract.* Why else would
it have kept its original, hostile tender offer outstanding after
Pennzoil had supposedly secured board approval of the deal?
"The deal moved faster than Pennzoil thought," Miller believed.

It was the only explanation for the tardiness of Baker & Botts in completing the first draft of the merger documents. "The Pennzoil people began worrying that they'd overlooked something; they weren't sure if the chicken they had grabbed was really a fox." What Pennzoil was passing off as mere drafting sessions between lawyers were actually roll-up-the-sleeves negotiations in which all sides were still striving to obtain maximum advantage—hardly evidence that all *essential terms* had been agreed to.

Third, *Texaco did absolutely everything possible to satisfy itself that Pennzoil had no contract.* "The real villains"—if there were any villains at all—"were the Getty people," Miller thought. And Pennzoil was suing them in Delaware.

Thus, finally, if the judge allowed, Miller would clinch his case by announcing that *Pennzoil had a second bite at the apple coming.* "I would tell the jury that these thumbsuckers don't have a case against anyone, but if they do, it's in Delaware. They've already brought suit against Getty in Delaware. You can decide this case for Texaco and realize that you haven't hurt Pennzoil's case in Delaware one whit." There was no way to lose with that.

Pennzoil, for its part, had far less obvious strategy. Pennzoil would try Texaco not just for *its* "crimes," but for those of Getty Oil Company.

Judge Farris arrived fifteen minutes late for Day One of *Pennzoil v. Texaco* on the morning of Monday, July 1, 1985—and his glasses were broken. Bailiff Carl Shaw saved the day by repairing them.

It would take Judge Farris and the lawyers all week to get ready for jury selection. But neither side minded taking the time; in any trial, many of the most significant events occur before the opening arguments are conducted and the first witness sworn.

More so than most, it would turn out, in this trial.

The housekeeping matters consumed two days. Both sides and the judge agreed that New York contract law would apply in the case because it was there that Pennzoil claimed its contract came to life. Judge Farris set the hours for the case to be from 9 A.M. to 4:15 P.M. with an hour and a half for lunch. Because he was under strict orders from his cardiologist, the judge would hold court neither on Monday mornings nor Friday afternoons. The lawyers discussed their estimates of the trial's length, and the consensus was four weeks at the least, ten at the most.

A debate ensued over which side would sit at the counsel table closer to the jury. Jamail claimed that courthouse custom gave it to the plaintiff. Bob Brown, one of Miller's partners, said he had never heard of such a rule, but Pennzoil was awarded the choice table anyway. The tables were, however, ultimately configured in a way that permitted each side visual contact with the jury box.

"At least we haven't had the problem of having to decide the shape of the table," the judge said.

On Wednesday arose the critical matter of the order *in limine*— the document that would control what evidence would be withheld from the jury. One by one, Judge Farris began pulling threads from Texaco's case.

Pennzoil moved to block any evidence of "settlement discussions" in the case—a perfectly normal and reasonable request. However, Pennzoil specifically included the meeting at which it claimed Liedtke threatened McKinley with the use of Pennzoil's political influence in Washington. "Mr. Liedtke rattled his sword," one of Miller's partners told the judge. "We intend to prove that." But the judge granted Pennzoil's request. The evidence would be denied.

Judge Farris also agreed to prevent any evidence that Pennzoil was still suing Getty Oil, Gordon and the museum in Delaware, denying Miller the chance to use his "second bite at the apple" argument against awarding Pennzoil damages in this case. By itself, the judge's ruling wasn't remarkable; the existence of separate litigation is often ruled inadmissible. But while denying Texaco the right to say that Pennzoil was suing the Getty group, he permitted Pennzoil to introduce evidence that Getty Oil had sued *Pennzoil* in Delaware, seeking a declaration that it had never reached a contract with Pennzoil—evidence that would help Pennzoil suggest Getty Oil was worried about entering into a deal with Texaco.

Judge Farris ruled that Texaco could introduce no evidence of any "investigations, findings or judgments" pertaining to securities-law violations—namely those that Pennzoil had been accused of in the 1960s—nor any evidence of political fund-raising activities, which would include the $700,000 in Nixon campaign funds that a Pennzoil jet had flown to Washington.

Pennzoil wanted there to be no newspaper or magazine articles admitted, with two exceptions: A *Wall Street Journal* article detailing Pennzoil's "agreement" to acquire Getty Oil—powerful evidence, Pennzoil said, that Texaco knew it was interfering with a contract—and a portion of a *Fortune* magazine article quoting

Sid Petersen to the effect that Pennzoil had a deal. However, an article in which *The American Lawyer* suggested Pennzoil's lawyers were negligent in completing the paperwork on the Getty deal was among those that would *not* be permitted. Texaco's lawyers protested that Pennzoil wanted to admit the articles which seemed to say there *was* a Pennzoil deal and exclude the articles that clearly said there was *no* Pennzoil deal. The judge postponed any decisions until the issue arose in court; his rulings then would favor Pennzoil.

Pennzoil asked the judge to block any evidence involving executives whose employment with Pennzoil had been "terminated"—a request whose significance Texaco immediately grasped; Perry Barber, the general counsel, had left Pennzoil after the Getty Oil deal had fallen through. "Mr. Barber was made the scapegoat for the failure of Pennzoil to obtain a contract from any party in this litigation," argued Bob Brown of Texaco, "for Pennzoil's failure to properly document the contract that they say exists."

"This is just pure speculation and the result of paranoia," Jamail countered, "and an attempt to poison the jury, and it's totally unfair." The evidence of Barber's termination would not be admitted.

Texaco, for its part, also wished a major block of evidence withheld from the jury: facts about the civil war between Gordon's group and Petersen's people. Texaco was particularly anxious to avoid any vivid reconstructions of the meeting at which the Getty Oil board ratified its truce with Gordon while also deciding to sue him—all after Gordon had been excused from the room. Texaco had to make everyone connected with Getty Oil look good in this trial—everyone, that is, but Gordon. "We will be replaying an extraneous dispute before this jury," said Michael Peterson, one of Miller's partners. "The jury does not need to go through two years of internal disputes."

Pennzoil had to fight on this one. Terrell called it "critical evidence for Pennzoil, and I sincerely mean that. . . . It was why Getty Oil Company, which could not rule its own house, became the subject of acquisitions by various oil companies." Jamail added that limiting evidence of the Getty Oil war would cut the "heart" from Pennzoil's case.

Texaco's request was denied. *Pennzoil v. Texaco* was set to become a referendum on *Gordon v. Petersen*—neither of whom was even a party to the case.

*   *   *

The entire top brass of Texaco had shown up in court to attend
what they expected to be the first day of jury selection, but the
length of the pretrial disputes forced the judge to send the jury
pool home until the next day. But the presence of Texaco chairman John McKinley was an opportunity that Pennzoil could not
let slip by. The Pennzoil lawyers were convinced that he would
make a poor witness for Texaco—so poor that they believed
Texaco would never put him on the stand. And Pennzoil could not
compel his appearance, as the subpoena power of state courts in
Texas ranged for only one hundred miles.

But there he was, standing at ground zero.

Susan Roehm, an effervescent young associate from Baker &
Botts, was given the honors.

"Hi, Mr. McKinley!" she said. "I haven't seen you since your
deposition!" And with that she thrust a subpoena into the outstretched hand of the chemist-turned-chairman; McKinley would
be called to the stand after all—as a witness for *Pennzoil*.

· 19 ·

At age four hundred, the "Old Hanging Oak" in downtown Houston is reputed to be the city's oldest living thing, surviving long after the adjacent county courthouse was torn down. According to legend, eleven criminals were hanged from the tree when Texas was still a republic, but years ago a county sheriff hotly disputed the tree's infamy. "We always hung them from a scaffold inside the building," he said.

A few blocks from the oak, on the seedy north edge of downtown, lies the building that replaced the old courthouse, a redstone, turn-of-the-century structure whose front entrance has been bolted shut for years, forcing people with courthouse business to use the easily congested side entrance.

On an insufferably sweltering summer morning, Jim Shannon shuffled into the building with a pair of stereo headphones on his head and the book *At Mother's Request: A Story of Murder, Money and Madness* under his arm. Shannon himself was a bit of an anachronism at age thirty-two, with his beard, longish hair and a record of free-lance political activism that would have done any die-hard McGovernite proud. Shannon recognized Judge Farris as a Republican—a judge whom he had once automatically cast a vote against.

It was a shame that Shannon had dropped out of high school, for there was little doubt that he had a mind bordering on brilliant.

His great-grandfather and his grandfather had been lawyers in Mississippi. His father, a high-ranking aerospace engineer at the Johnson Space Center, had always assumed that Jim would too become a lawyer and thus, as Shannon says, "the legator of the family." Instead, Shannon was a civil servant with the City of Houston, with assignments that had included working with senior citizens and day-care centers, and in the mayor's office, when a liberal politician had occupied that office. He sustained his interest in the law by reading every true-crime story he could lay his hands on.

When he answered his third call for jury duty since 1977, Shannon was hoping not to be picked. Jury service was tedious and time-consuming; the deliberations had continued for so long in the first case Shannon had heard that he had been forced to miss a concert by the Marshall Tucker Band, his favorite rock band at the time. But as Shannon took his place as Jury Candidate No. 52 out of 100, he immediately recognized that he was under consideration for something other than a slip-and-fall case or any other standard courtroom fare. No one could remember so many jury candidates being assembled for a single case outside of a capital murder trial. And he recalled hearing a news report on his stereo radio about a gigantic lawsuit between Pennzoil and Texaco— something about a fight over who really owned Getty Oil.

"Either we're going to a death-penalty case," Shannon said to the candidate at his side, "or it's Pennzoil-Texaco."

Finally Judge Farris climbed the bench and settled the mystery. "This is a civil case," he announced. "But this isn't your average civil case. This is the largest civil case ever filed in anyone's knowledge in Harris County."

A murmur swept the counsel table where a half-dozen lawyers from Texaco were seated. One of them, Richard Keeton, was quickly on his feet.

"Your Honor, may we approach the bench?" Keeton asked.

Speaking sotto voce so the jury candidates could not hear, Keeton explained that the judge's use of such a superlative had only served to legitimize the $15 *billion* dollars of damages that Pennzoil was demanding in the case.

"Your Honor, I want to make a formal motion for a mistrial," Keeton said.

"Denied," the judge answered, and the process of picking the jury was begun.

* * *

The rite of jury selection traces itself to a 13th-century practice in Norman England called the "wager of law," in which each party to a dispute recruited twelve people who would swear to his side of the story. The procedure in 20th-century America bears one striking similarity: each side tries to narrow the field of selection to the jury candidates most demographically suited to its side of the case. The process involves instinct, second-guessing and stereotyping of the highest order. And as both Texaco and Pennzoil would discover by the end of the case, jurors selected on such a basis often behave exactly the opposite of what the lawyers expect.

Texaco, for the most part, wanted professional people—jurors it believed would be smarter and more sophisticated, who would not be romanced by arguments based on handshakes and honor and who would take account of the hard realities of the business world. "We can't have a bunch of rag-pickers deciding this case," Miller explained at the time.

Pennzoil, as its lawyers would later explain, preferred women to men because they would have a "deeper ability to feel outrage," and wanted gentle people of either gender "who wouldn't like rough play." It looked for regular churchgoers who would be offended by indecency, but not fundamentalist Christians who might consider punishment better left to the Lord than the courts. Pennzoil also sought jurors who had not been divorced and people who had spent several years with a single employer—such as Jim Shannon, with eleven years working for the city. "We wanted people who would be shocked at breaking promises," Irv Terrell of Baker & Botts would recall.

The lawyers went row-by-row asking for a show of hands by those who would suffer unduly in a case of six to ten weeks—and the courtroom took on the appearance of a stadium full of thirsty people signaling the only beer vendor walking the aisles. The number of health problems among the 100 was startling: A forty-three-year-old longshoreman was seeing a neurologist for a head injury. Another man had recently fallen off the side of a truck. A fifty-five-year-old chemist at Shell Oil was having chest pains. A young chemical salesman was attending Lamaze classes. One woman wanted to be excused so she could maintain her acne-treatment appointments. Another developed such back pain after hours on the hardwood benches that she asked to leave immediately.

"Let me be sure I'm not the one that's giving you the back pain," the ever-chivalrous Miller inquired.

"You give people a pain in another location," Jamail interjected.

The very people that Texaco most desired—career-motivated executive types—were also the type least anxious to serve. Richard Keeton of Texaco was unmoved by the family problems described by the Indian engineer from Bechtel Incorporated, one of the world's largest corporations. "I think this is a highly intelligent juror," Keeton said. The overworked pipeline manager from Tenneco Incorporated—someone who doubtless appreciated the complexity of contracts—was "the type of juror who ought to be hearing this case," Keeton said.

Neither side wanted ideologues, of which a relatively small number were present. When one woman said she could not participate in any system of law that permitted abortion, Texaco's Miller tried to assure her that abortion was not at issue in the case. "As far as I know," he said, "that's the only thing they're *not* claiming."

But the interrogation of individual jurors was the least important part of the week-long jury-selection process. Its official name, voir dire, for "speak the truth," implies that it is the jury candidates who do most of the talking. The rules in Texas, however, permit lawyers on both sides to lay out their entire case, unbound by the rules of evidence, before they ever give an opening argument. Although this serves the useful purpose of allowing the lawyers to observe the jurors' reaction to various elements of their case, and to ask whether any jurors find their case objectionable, the whole process is tantamount to a jury speech.

Pennzoil's Joe Jamail was acutely aware of the challenge before him. Besides proving that Texaco "tortiously" interfered with Pennzoil's contract, he had to prove that Pennzoil indeed *had* a contract despite the absence of any document saying "contract" at the top—indeed, despite the presence of a memorandum of agreement with a signature page left blank by the board of Getty Oil. Pennzoil's "contract" was a *series* of writings and representations: the Dear Hugh letter in which Gordon agreed to support the Pennzoil deal, the notes reflecting the fifteen-to-one vote by the Getty Oil board, the press release announcing an agreement in principle on the essential terms of the deal, and a whole series of written and oral statements that suggested *intent to be bound*. All told, Pennzoil would argue, they made a contract.

At this early stage of the case Jamail needed a metaphor to make everything plain, and he chose the word *promise*.

"Pennzoil had a promise with Getty Oil and the Getty interests to purchase three-sevenths of the assets of Getty Oil Company. That promise was not just your ordinary promise. It was made by *people*, not companies."

Jamail motioned to the hulking figure sitting uncomfortably on a folding chair near the Pennzoil counsel table. "Hugh Liedtke—who I represent and who is my *friend*, who is seated here today—was the chief executive officer and chairman of the board of Pennzoil. It wasn't a building that made those promises. It was people.

"We are going to be together for a while," Jamail informed the candidates. "I know these seats are hard. But you can appreciate my burden: I think it's the most important case ever brought in the history of America. There isn't any question about that."

Conspicuously absent from the court, however, were the parties at the other end of Pennzoil's promise: Getty Oil, Gordon and the museum, still baking in the oven of Pennzoil's suit in Delaware, a second course in case Pennzoil's soufflé fell in Texas. Jamail, of course, would not tell the jury this, and he had already seen to it that Texaco could not do so either.

Jamail's method of explaining away their absence would turn one of Pennzoil's biggest obstacles in the case into its greatest advantage. The answer was the indemnities granted by Texaco—a rather complex set of agreements that Jamail would succeed in making as plain as vanilla.

He began by outlining how Texaco had "stolen" Getty Oil. "The package was pretty clever," he said, "pretty slick." Texaco offered more money for Getty Oil, but only after forcing the Sarah Getty Trust to break its deal with Pennzoil.

"And does the trustee, Gordon Getty, say that's all right? No.

"Does Mr. Williams or Mr. Lipton for the museum think that's all right? No.

"Does the Getty board say that's all right? No.

"Why? 'Well, the price is right—but what about Pennzoil? What if they sue us?'"

Jamail adopted his most conspiratorial tone.

"I'm going to ask you to listen closely, because this is the fraudulent inducement that Texaco put on Getty to *make Getty crumble*. 'If Pennzoil sues you, then we promise you—and we'll write it in a contract, and it *is* written in a contract—we will

indemnify you, Gordon Getty and the trust. We will indemnify
you, the museum and Mr. Williams. And not only that, but *all of
your representatives*. . . . We're going to indemnify you if
Pennzoil sues you. . . .'

"And that's why Texaco is in here all by themselves. . . . They
are the only necessary party to this action. Whatever would befall
Getty under the indemnity, it's *Texaco's* responsibility."

It was a brilliant sleight-of-legal-hand. The indemnities required only that Texaco cover the cost of legal fees or court
judgments obtained in any lawsuit against the Getty group. They
certainly didn't make Texaco—the lone defendant in the case—
liable for the actions of anybody who wasn't a defendant in the
case. In the course of explaining the absence of the Getty people
from the courtroom, Jamail had made Texaco responsible not only
for its own actions but for those of people who weren't even
around to defend themselves; Texaco was now holding the bag for
everything that Getty Oil had done to Gordon. In fact, Jamail had
made Texaco answerable to every action of every lawyer and
investment banker connected with the Getty group: Marty Lipton
of the museum, Marty Siegel and Tim Cohler of Gordon's group
and Geoff Boisi, Bart Winokur and the rest of Petersen's people.
It was all on Texaco's head.

Texaco would object to Jamail's twisted characterization of the
indemnities; inexplicably, the objection would be overruled.

"This is so vital," Jamail said, returning to the voir dire portion
of his speech. "Is there any member of this panel who could not,
and would not, accept that indemnity . . . as evidence of the fact
that Texaco had knowledge of the Getty-Pennzoil agreement and
binding contract? Is there anyone who would not accept that as
that kind of evidence? If there is, I've got to know that *now*,
obviously. Because that is our proof."

Jamail surveyed the nearly one hundred remaining candidates.
"I see no hands," he said. "I didn't expect any."

The half-dozen lawyers at the Texaco table were dying inside.
Not only had they been denied the right to inform the jurors that
Pennzoil had a second bite at the apple against the Getty Oil group
in Delaware, but now Pennzoil was forcing Texaco to defend
somebody else's breach of contract. They felt like fighters with one
hand tied back—going into the ring with a bigger opponent than
they had expected to meet.

Then Jamail went for the knockout punch.

To the jury he read briefly from a transcript of a news report by the CBS affiliate in Houston, in which his opposing counsel Miller had expressed astonishment that Pennzoil was planning to try the case on the code of the oil patch. "Jesus Christ," Miller had told the reporter, and Jamail was now telling the jury, "they were in New York."

Miller's embarrassment was exceeded only by his outrage at Jamail's stunt. He immediately approached the bench.

"We respectfully move for a mistrial. This case has just commenced and that statement has done us grievous harm. . . . What I object to is reading the curse word that I used."

"There was no curse word," Jamail said. "You said, 'Jesus Christ.'"

"If that's not a curse word in the context in which it was said, I don't know what is."

"Consider it a phrase that I would not have used," Judge Farris said, "but I do not consider it a curse word. Motion for mistrial is denied."

Jamail returned to the Pennzoil counsel table feeling bolder than ever.

"In the statement that I just read to you, Mr. Miller, as spokesman for Texaco, takes the position that the promises and the morality of the marketplace are different in New York than they are here." Under the guise of voir dire, Jamail asked whether anyone agreed with "Texaco's position" that "a handshake in New York is meaningless? I take it you do not."

On the third morning of jury selection, Miller again demanded a mistrial—this time because the *Houston Chronicle* had mentioned the *prior* mistrial demand that Miller had made at the bench. Overruled, the judge said.

Three mistrial motions before the first juror had been chosen. And the fun had only begun.

Texaco began its voir dire in earnest on Monday morning, the fifth day of the trial.

"Now," Miller began, "Mr. Jamail and Mr. Terrell have told you that this is the most important case that has ever been filed in the history of mankind. I take exception to that. I tried a case two months ago where a father was trying to get his children back. . . . This is a suit over money. . . . So much sour grapes is what this case is."

Miller had immediately scored points with the jury, and they would not be his last. Miller talked slowly, in a softer, Oklahoma drawl than the brassy Jamail, and although outside of a jury's presence the expletives ran wild from Miller's mouth, he came off in the courtroom as a true Southern gentleman—sweet and decent, like Gregory Peck as Atticus Finch in *To Kill a Mockingbird*.

One of Miller's strongest points in the case was the extensive evidence that the takeover professionals on the Getty Oil side had invited Texaco's bid. "We did not crash this party," he told the jury candidates. "We didn't butt into anything."

But from his opening shot on, Miller dug himself deeper into the hole that Jamail had started for him. Miller, who never shrank from a fight, decided that he *would* defend Getty Oil, and doing that meant launching a frontal attack on the man who'd caused all the problems to happen: Gordon Getty.

"Why did they ask us to bid? Very simple, straightforward answer: The Pennzoil Company had prevailed upon Gordon Getty—who, you will learn, was not, is not and never will be a businessman." Miller would later add: "Mr. Getty is not a playboy. He is a serious man. He knows next to nothing about business. His interests in life are anthropology, music, poetry, opera—to which I think we would all say, 'Good for him.'

"But there were people on the Getty board of directors who did not feel that this background qualified him to run an oil company. And indeed, as the evidence will show, there were his relatives who felt the same way."

How did Pennzoil fit in? By taking a cheapskate merger deal and getting Gordon to cast his lot with it, "relying first upon his emotional desire to succeed to his father's position, relying also upon his desire to have control of his father's company for once in his life, relying upon not the best motives of human nature, but the worst." Together, Pennzoil and Gordon "put the board of directors of the Getty Oil Company in a vise and squeezed them, to make them commit to sell the stock for less than its worth.

"So there you have it. That's why the Getty Oil Company came to Texaco."

The directors and investment bankers in that Getty Oil boardroom were honorable people, men of their word, Miller said—even though Jamail had derided them as *New Yorkers*. "When you go to New York and you've got a case, you worry about the New Yorkers thinking you're a bunch of gun-toters and whiskey-drinkers and long-neck beer drinkers," Miller said.

"When New Yorkers come to Texas they worry that everybody thinks they're carrying switchblades or worse."

Miller had a special concern on his mind, though he did not state it directly. Sometimes, when Pennzoil's lawyers used the words *New York* in their smearing tones, he thought he heard them saying something else: *Jew*.

"There are some things a lawyer ought not be willing to do to win," Miller said. "And I put at the top of the list the attempt to create prejudice for race or creed or geography or anything like that." Miller demanded to know whether any jury candidates would judge Texaco's *New York* witnesses "by where they're from, or what church they go to."

"Am I dealing with that?"

The candidates sat silently.

Miller had only one thing to prove: that Texaco was an invited guest, that it had not "tortiously interfered" with Pennzoil's deal. But Miller, the combative, all-or-nothing lawyer, would not stop at that. Just in case it came to the conclusion that Pennzoil had a contract—even one that Texaco never interfered with—Miller would not let the jury award so much as a cent of damages to pay Pennzoil for its trouble.

"I want to say something to you and say it so there won't be any misunderstanding about it. We don't owe these people anything. *Not a dime.* That's what we owe them. *We owe them zero.* And I can't take anybody on the jury who's going to say, 'Well, they've sued for fifteen billion. I'm going to give them ten million and call it a day.' . . . Nobody will try to settle this case for us."

By the time it was over, $10 million wouldn't even begin to pay Texaco's legal bills.

Little by little, day by day, the group was whittled down, as candidates raised questions or expressed misgivings. Kay Hartmann, a forty-five-year-old housewife, said she wouldn't believe a word of either side's testimony. "The name of this whole game is greed, and I believe every party is guilty, and I'm really having a problem with the whole thing." She was off the list. Eulalia Cylkowski expressed a philosophical problem with anything dealing in the billions. "It's going to be the consumer who's going to pay for it in the end, in my opinion," she said. She too got the quick hook.

Pennzoil's request for $7.53 billion in punitive damages put some candidates especially on edge. "That just totally offends my sense of morals and ethics,'' said Brad Elliott, a twenty-nine-year-old record store manager. "This kind of money, which basically boils down to what I consider spite money or revenge money, is just an outrage." Elliott was dismissed from the panel to contemplate his outrage. The Reverend Robert Hogan came to the bench to say that he had lost sleep over punitive damages. "I mean that's a lot of money for punishment," he explained. The Reverend Hogan, in fact, was searching his soul over everything he had been told about the case. "There have been times I told my boy that I was going to do something I didn't do," he said. "Would that make me have a contract?" He was excused to resume searching his soul.

Finally the panel was whittled down, and finally the real guesswork began. Each side could eliminate eight candidates for whatever cause it chose, and without giving its reasons. When the "strikes" were over, Pennzoil would claim that Texaco didn't want blacks. Texaco would claim that Pennzoil didn't want Jews. Neither side, however, had a sufficient number of strikes to eliminate either group from the jury altogether.

Fred Daniels . . . Diana Steinman . . . Juanita Suarez . . . Rick Lawler . . . Susan Fleming . . . Lillie Futch . . . Laura Mae Johnson . . . Velinda Allen . . . Israel Jackson . . . Ola Guy . . . Theresa Ladig . . . Shirley Wall.

They were nine women and three men from twenty-five to fifty-six years of age—forty, on average—including four blacks and one Latino. They were eight Protestants, three Catholics and a Jew. They were four clerks, a letter carrier, two professional housekeepers, a registered nurse, a forklift salesman, an accountant, a county-insurance-department employee and a housewife. Not all were native Texans, but of the twelve, all but one had been born south of the Mason-Dixon Line. They would be paid $6 a day; parking their cars in a remotely safe and convenient location would cost $6.25 a day.

The four alternates—Jim Shannon, Doug Sidey, Linda Sonnier, Gilbert Starkweather—were not informed of their status for the good reason that they might pay insufficient attention to the evidence. But Jim Shannon—who by now was well into the copy of the true-crime story he had brought to the courthouse—had counted the order in which the jurors were called and recognized himself as the thirteenth—the first alternate. He went home disappointed that he would have to sit through the entire case knowing he would never help decide the outcome except in the unlikely event that another juror left.

But then again, anything could happen to one of twelve people over six to ten weeks.

The night before Pennzoil officially opened its case, with the jury at last impaneled, there was a knock at the door of Jamail's palatial mansion in the Tanglewood section of Houston. There stood Willie Nelson in a ponytail and a sleeveless black T-shirt with the American flag emblazoned on the back, a Budweiser in his hand. Next to him was Darrell Royal of the University of Texas, retired for more than a decade but still the winningest football coach in Southwest Conference history. Unannounced, they had driven down from Austin in a white stretch limousine to spend the evening knocking back long-necks with their old friend Joe Jamail.

What the hell are you guys doing here? I've got to make an opening argument tomorrow!

But Jamail had them in anyway, and after the week-long voir dire, the distraction was a delight. Jamail went into the attic to find one of his boys' old guitars, on which Nelson played a new song that he would eventually sing at the centennial celebration for the Statue of Liberty. After keeping Jamail up until midnight, two hours past his regular bedtime, Willie and Darrell climbed back aboard the limo for the four-hour drive back to Austin.

Sometime later the limo driver would send Jamail a snapshot he had taken of his passengers while they were standing on Jamail's front porch, waiting to be let inside. After the trial was finally over, Jamail would have the snapshot blown up to the size of a poster to hang in his office, a memento of how a "good-timin' man" celebrated the eve of his biggest case ever.

· 20 ·

Near the huge color photo of Willie and Darrell taken outside of Joe Jamail's home hung another framed picture, but this one was old and faded, in black and white. It showed four grim-looking, mud-splattered Marines hauling a teenage soldier with a gunshot wound through the jungles of Guadalcanal, a photo that over the years had become one of the most widely reproduced images in photographic histories of World War II. The wounded boy on the stretcher was Baine Kerr.

"Your Honor, Pennzoil calls Mr. Baine Kerr as its first witness."

A momentary disappointment filled the courtroom, where many had expected Hugh Liedtke to lead off the case. But Pennzoil's trial team had a good reason for calling Kerr, a friend of Liedtke and Jamail who had spent nearly a decade as the number-two man at Pennzoil.

"Know thy enemy," the warriors of ancient China said, and knowing Dick Miller as well as they did, Pennzoil's lawyers expected Texaco's lead counsel to be wound up like a watchspring by the time the trial began. To subject Liedtke to Miller's first-round punches would constitute a "jury risk": Liedtke would surely fight back, and a show of aggression might alienate the jury. Moreover, Liedtke did not have the patience of Kerr, who as the leadoff witness would have to explain all the key documents through lengthy, dry and often technical testimony. Thus, the cerebral John Jeffers, rather than the combative Jamail or Terrell, would handle Kerr's direct examination.

As he walked through State Court No. 151 to the witness stand, the executive-suite veteran looked like anything but a heroic war veteran. He was small, with a slight build and a soft chin that gave him an appearance bordering on milk toast. Kerr was the classic "inside" executive, a green-eyeshade type who handled the intricate and often grubby legal details required to carry out Liedtke's imaginative business maneuvers.

"I wanted to set the tone," Jamail would recall. "Here's Texaco and its bully lawyer and here's this nice, kindly gentleman simply trying to tell him what happened. But Miller is macho and I knew he'd try to intimidate Kerr." But looks were deceiving. If it came to a fight, Jamail knew that "Baine has got insides of solid steel. He's one tough Marine."

Kerr gave the jury a romantic and somewhat revisionist account of the Pennzoil story—the early partnership with George Bush, the involvement of J. Paul Getty in Liedtke's career—and cast Pennzoil as the kind of company that made life a little better for everyone: the third-largest gold producer, the second-largest producer of molybdenum for stainless steel and the largest sulfur producer. "The clothes you wear, the cars you drive, the fertilizer to raise the crops—they all utilize sulfur," Kerr said.

But Kerr said that none of these achievements were sufficient for the man alongside of whom Kerr had spent his career. Building Pennzoil's oil business into a world-class operation was not merely a corporate objective, Kerr said, but a personal dream, and the dream was Hugh Liedtke's. "He had, I would say, always had an ambition to build a major oil company," Kerr said, and from that moment forward, everything that Texaco had done would be measured against the passion of an old-time oilman trying to attain a lifelong goal.

But even a storyteller as effective as Kerr could not easily overcome a major obstacle facing Pennzoil early in the case.

Pennzoil had to cast itself as the rescuer of Gordon Getty, a Lancelot who leaped into Getty Oil to save Gordon from the fiery breath of Getty Oil's management and to install him in his rightful place as king of the castle. Besides elevating Pennzoil in the jury's eyes, this strategy would belittle the officers and directors of Getty Oil, who had been so ready to break the deal between Liedtke and

Gordon that they would rush into the arms of Texaco. Yet neither
Kerr nor anyone else at Pennzoil had been present for a single
battle in the civil war within Getty Oil. Thus, to lay the necessary
groundwork through its first witness without violating the rules of
hearsay, Pennzoil resorted to using Kerr as a mouthpiece to read
aloud the various legal documents that told the story of the Getty
Oil war—documents that were replete with legalese.

"Whereas . . . aggregate . . . foregoing . . . the undersigned . . ." It was a terrible way to retell such a dramatic story.
It was also the *only* way, but it was boring.

Reconstructing Pennzoil's deal with the Getty interests was an
even greater challenge, but this story, too, had to be told through
documents—the exhibits forming the core of Pennzoil's contract
claim. One by one they went up on an easel—thirty-by-forty-inch
blow-ups of pages from meeting notes, the memorandum of
agreement, the press release, the draft merger documents, all
intended to show that from beginning to end the only essential
term that changed in the Pennzoil deal was the price. And there
was no disputing that that much, at least, had been approved by a
fifteen-to-one vote of the Getty Oil board.

Jeffers orchestrated the paper chase as well as anyone could
expect, summarizing the events through an oversize chronology
and even using a pointer to follow the action on a big map of
midtown Manhattan. But in the three days the testimony consumed, jurors got lost in the maze of offers and counteroffers, of
"stubs" and other technical terms.

And Dick Miller made a point of inflicting all the confusion he
could.

Jeffers asked Kerr whether in his experience as a businessman
an agreement in principle meant that all essential terms had been
reached. "I object," bristled Miller. "That's a leading question."
Which it was.

After Jeffers posed a complex hypothetical to Kerr: "Excuse
me. I'm not sure I understand the question."

In a discussion of what occurred at the Getty Oil directors'
meeting in New York: "May it please the court, that's based on
hearsay."

As Kerr was asked about possible solutions to the problem of
who would buy the museum shares: "The *possible* ways are of no
relevance in this case."

Miller was unrelenting.

"Counsel has asked these leading questions *over* and *over* and

ver and has forced me to get up and object. . . . The *witness* is
n experienced lawyer. The *lawyer* who is asking the questions is
n experienced lawyer. . . ."

Through it all, Judge Farris repeatedly supported Miller, forc-
ng Jeffers to withdraw or rephrase questions, forcing Kerr to eke
ut his narrative, derailing whatever train of thought the jury had
nanaged to climb aboard. "It was like marching to Russia,"
effers says.

Texaco's chief counsel was delighted. *Maybe I've got a fair
hance with this judge after all,* Miller thought. *He's actually
elping me out!*

)ick Miller had known Baine Kerr since the 1950s, when they
vere junior lawyers at Baker & Botts. Although Miller and Kerr
ad gotten along as members of the same firm, Kerr worked in the
orporate department, making him one of those weak-kneed "set-
lers" and "negotiators" for whom Miller held such contempt. As
or Kerr himself, Miller would later have three things to say about
im: "He's a smart guy, an able lawyer, and he's a wimp."

But not all wimp.

Miller spent three days of cross-examination trying to depict
'ennzoil as the heavy and Gordon as its patsy—a man who had
ven permitted his *wife* to recruit members of the Getty Oil board,
omeone who seemed to encounter trouble on every assignment
vith Getty Oil—particularly in the Neutral Zone.

"Well, I've been to Saudi Arabia," Kerr demurred, "and I
hink it's fairly easy to get into trouble in Saudi Arabia."

Miller continued to hammer away: Pennzoil, greedy for oil,
ried to arrange a transfusion of oil-in-the-ground by tapping Getty
)il's "lifeblood" and "taking it back to Houston." Gordon, "a
inger of more than average ability," was greedy for power in his
'dead Daddy's company." Both of them agreed to put the
.queeze on the Getty Oil board—an arrangement that culminated
n the Dear Hugh letter, in which Gordon had promised Liedtke he
vould throw out the board if necessary to get the deal through.

Miller had finally pushed too far.

"How long was Mr. Getty going to put up with this stuff?"
Kerr snapped back. "How long was he going to let them
snooker' him, as they said? I think that at some point he's got to
:xercise his rights." Kerr fidgeted in his seat, and his voice grew
ouder and began reaching into another octave. "The company
going out and finding one of the brothers . . . a *dope addict* living

as a recluse in London, who wouldn't even pay for the medica
expenses of his son who had become a total vegetable! *That's th*
man they got to authorize the lawsuit, and Gordon was a goo
enough person that he, with no legal obligation to do so, paid a
of these medical expenses."

When Kerr had finally left the stand after a total of nearly seve
days, Pennzoil's lawyers had to concede that Miller had scored
few points—although no killer points. Kerr, however, had ac
complished his three most important objectives: He'd gotten th
story on the record; he'd established the foundation necessary t
introduce critical documents that would flow through the jury bo
dozens of times to come; and he had absorbed nearly all of Mi
ler's blows.

"Miller had used up a lot of his venom," Terrell of Baker
Botts would later insist.

Although Joe Jamail was Pennzoil's lead counsel, the job of lay
ing out the company's case fell primarily to Jeffers and Irv Terre
of Baker & Botts. Once the entire case had been "proved up,
only then would Joe Jamail reassume the lead in order to whip th
jury to Pennzoil's side. Jamail would later explain that he felt h
needed to blend into the background for a while because the jur
had seen enough of him during the voir dire phase and would b
seeing plenty more of him—not just at the end of Pennzoil's cas
but during Texaco's case, when he would handle much of th
most hostile cross-examination.

Dick Miller, however, would have another explanation for Jam
ail's willingness to give up the lead. As he watched his old frien
and client delegate the detail work to the younger lawyers who
Miller himself had trained, Miller began to think that Jamail didn
trust himself with the intricacies of Pennzoil's claim. The m
nutiae of the Pennzoil and Texaco deal went over Jamail's head
Miller thought. *He doesn't understand his own case.*

Regardless of the reason, Miller would never have worked tha
way. Although he had assigned more than half of the thirtee
lawyers in his small law firm to assist him, Miller would person
ally oversee nearly every detail. He threw himself so deeply int
the case that he moved out of his house and into the exclusive Fo
Seasons Hotel in downtown Houston, where he lived for four an
one half months in a room adjacent to the suite where two c
Texaco's in-house lawyers, Charles Kazlauskas and David Lu
tinger, set up shop for the duration. *Pennzoil v. Texaco* was th

topic over every breakfast, lunch and dinner, over every cup of coffee that Miller threw down. And although Texaco had its long-time New York law firm, Kaye, Scholer, Fierman, Hays & Handler, on the case, the lone representative of that firm would sit through the entire trial not at the counsel table with Miller, but in the bleachers with the press and public. "One lawsuit, one lawyer."

But Miller may have been selling his adversary short, and indeed many of the lawyers watching the case unfold had to tip their hats to Jamail. It wasn't often that one of the most famous trial lawyers in Texas—indeed, in the whole country—willingly subordinated an entire phase of his case to younger lawyers from a huge, "institutional" law firm. The Baker & Botts lawyers deserved the same sort of credit—for their willingness to handle the nitty-gritty work while watching a personal-injury lawyer stand in the spotlight as lead counsel. Baker & Botts had supervised the entire pretrial phase of the trial, attending some fifty depositions, while Jamail had participated in only a half-dozen or so.

"Joe is a jury genius," Terrell says. "*We* brought organization, manpower and knowledge of corporate deals. We"—Terrell and Jeffers—"were two good trial lawyers, with one extraordinary trial lawyer."

And the fact was, the Baker & Botts lawyers wanted so desperately to win—not only for their client's sake, and to defeat their old boss Dick Miller, but to help restore the luster that their firm's reputation lost when the Pennzoil deal had collapsed in New York. "We were sometimes as much a defendant in this case as Texaco was," Jeffers would explain.

From mid-July to late August, Terrell and Jeffers arduously added brick upon brick to Pennzoil's claim that it had a contract. By reading aloud, by dimming the houselights and playing the video depositions they had taken from Petersen's people, the lawyers added proof of the *intent to be bound:*

—Exhibit #19: The California court affidavit in which Gordon's investment banker Marty Siegel stated that "the Getty Oil board of directors approved a corporate reorganization transaction" and that "Pennzoil, the museum and the trustee have all agreed in principle to the transaction."

—Exhibit #36: The transcript of the California court hearing in which Gordon's lawyer Tim Cohler said, "there is presently a transaction agreed upon." Exhibit #9: The affidavit in which

Cohler, like Siegel, swore the Getty Oil board had *approved* a transaction. Exhibit #10: The same affidavit, resubmitted with handwritten changes, *after* the Texaco deal was in place.

—Exhibit #32: The "fee letter" under which Geoff Boisi of Goldman, Sachs billed Getty Oil $6 million the day after the vote in the Pennzoil deal. Even though the $6 million bill had nothing to do with the Pennzoil deal—Goldman's fee under that transaction would have been $9 million—the timing still looked bad. In addition, the fee letter enabled Pennzoil to show that in the *Texaco* deal, Goldman doubled its fee to more than $18 million. Thus did Boisi, in the mouths of the Pennzoil lawyers, become the "$18 million man."

—Exhibit #60: The marginally legible notes of Stedman Garber, the treasurer of Getty Oil, who upon receiving a report of what happened at the Getty Oil board meeting in New York wrote: "board agreed but for Chauncey that deal should be done." The notes also listed several provisions that Pennzoil considered to be "essential terms."

—Exhibit #80: The handwritten message that Marty Lipton of the museum slid across the conference table to Geoff Boisi, the investment banker, outlining the counterproposal for the $110 in cash and the $2.50 ERC stub. "Museum accepts," Lipton's note read. "Salomon advises fair."

—Exhibit #130: A single paragraph from a *Fortune* article, admitted over Texaco's most strenuous objections, in which Sid Petersen said of the Pennzoil proposal: "We thought there was a better deal out there, but it was a bird-in-the-handish situation. We approved the deal, but we didn't favor it."

Through the worst of Houston's sweltering summer months, the lawyers read aloud from one interminable document after another, trying to establish not only that the Getty people intended to be bound but that the Texaco people had *actual knowledge* of Pennzoil's deal. They introduced a clipping from the January 5, 1984, *Wall Street Journal* describing Pennzoil's deal with Gordon, an article that used the word "agreement" fourteen times. Texaco fought bitterly to keep the article out of evidence as inadmissible hearsay, which, of course, it was. Pennzoil, however, found a rule justifying the exhibit only as a "form of notice" that Pennzoil had a deal of some kind. The article itself was not nearly so damaging to Texaco, however, as an affidavit submitted by Texaco that among the dozens of people who worked on the Getty deal, *not one* could recall seeing the article on the day it appeared.

Thus, January 5, 1984, became known at the trial as the one day in Texaco's history during which no one read *The Wall Street Journal*.

And finally, there were the perfectly legible handwritten notes by Texaco's president, Al DeCrane, and its associate controller, Pat Lynch, reflecting a discussion of the need to *stop the train . . . stop the signing . . . meet with key player Marty Lipton . . . take care of Liedtke . . . leave Gordon with paper*—the smoking gun.

The evidence, however, added tedium to tedium, all the more so because much of it could be introduced only through a witness, and Pennzoil, of course, had no witnesses from Getty Oil or Texaco. As a result, Pennzoil spent days putting on evidence through a television screen.

Of the sixty-three depositions spanning fifteen thousand pages of transcript taken during the "discovery" phase of the case, fifteen had been videotaped by Jeffers and Terrell to assure that the jury got a firsthand look at "the flavor of these people." Night after night Terrell and Jeffers huddled in front of a videocassette recorder and a television set in one of their homes, fast-forwarding and rewinding their way through the hundreds of hours of tape, making notations on the printed transcripts of passages intended for the cutting room floor: the long pauses, the harsh exchanges between the lawyers, the curse words, the unremitting objections announced by lawyers on either side. Then, for hour after hour, sometimes while keeping the jury waiting, they argued with Texaco's lawyers in the judge's chambers over which sections should be restored.

Besides permitting Pennzoil to introduce exhibits, the video depositions also enabled it to retell the story of the Getty Oil war—a story intended to depict Petersen and his people as double-dealers. "All of this evidence about what they did to Gordon Getty," Terrell told the judge at one point, "reflects on whether these are honorable men."

Geoff Boisi, the investment banker who solicited Texaco's bid to defeat Gordon's deal with Pennzoil, fluttered his eyelids, sniffed and chuckled at the lawyers' questions and held forth in the gobbledygook of the takeover game. When asked in the videotape whether he would be testifying as a live witness at the trial, Boisi replied, "If my schedule allowed"—not the kind of answer that impresses jurors who by that time had given up nearly a month of their lives to the case. And when he was asked in the video whether he had come and gone during Gordon's absence from the "back door" board meeting, he swallowed before answering: "I guess that occurred."

Bart Winokur, the outside counsel for Getty Oil who had entered the Getty Oil board meeting in Houston in Gordon's absence, gave intelligent and straightforward responses but appeared a little shifty. He fidgeted in his seat. His eyes darted around the room. His voice at times adopted a somewhat sassy tone. He filled the sound track with the ring of tinkling ice cubes when raising a glass of fruit juice to his lips. There was something just a little unconvincing about Winokur; while he testified that it was not his practice to take notes during meetings, he was holding a pen in one hand.

Sid Petersen, an executive whose career had been scuttled just as it was reaching its peak, came off as smug and defensive, tossing his head, folding his arms across his chest, heaving sighs. When he was asked a question about Liedtke's agreement to honor his golden parachute, Petersen snapped, "I don't give a damn what Mr. Liedtke approved. He had nothing to say about it. *I* had a contract."

Although the Getty Oil battle was a far more colorful story than Pennzoil's "birth of a contract," it suffered from the same drawback; it became *boring*. The hours of video testimony in the darkened courtroom sometimes put jurors to sleep; one, in the second row, once began slumbering with his mouth hanging wide open. One lawyer watching the case began suggesting that a few *Roadrunner* cartoons would help break the monotony. In the morning, as the jurors filed into the jury box, they would experience an immediate letdown if they observed the absence of local newspaper reporters—a sure sign that yet another deposition would be played.

The video witnesses themselves sometimes aroused uneven reactions. Diana Steinman, the only native-born Yankee on the jury, was giving a lift home every evening to alternate Jim Shannon, and although they were prohibited from discussing the case, they sometimes did discuss the personalities. Steinman thought the investment banker Marty Siegel was cute; Shannon thought he was "oily."

Name confusion made the case seem like *The Brothers Karamazov:* Lipton . . . Liman . . . Liedtke . . . Marty Siegel . . . Marty Lipton. . . . Baker & Botts had the sense to design an oversize playbill which it kept on an easel whenever names became a problem; when alternate juror Jim Shannon requested that it remain in sight at all times, Pennzoil did one better by having the exhibit reduced to hand-held size and distributing copies to the jury.

Midway through Pennzoil's case, *The American Lawyer*, a spirited trade magazine, dispatched a reporter to the Houston courthouse for a report on the case of the century. "By burdening the jury with infinitely too much detail," he wrote, "Pennzoil's lawyers seem to be risking defeat in what otherwise might be a strong jury case. It's as if they're trying so hard to be thorough that they're unable to perceive how boring their presentation has become."

On August 14, with Hurricane Danny approaching, Terrell expressed skepticism to Judge Farris over whether court could be held the next day. "Irv, let's don't get into that," Jamail snapped. "We need to *move*."

What moments of excitement that occurred in the first month invariably happened in the jury's absence.

There was, for instance, the "four sinkers" controversy. Oscar Wyatt, one of Houston's most prominent oilmen and surely its most ornery, had a few drinks with James Kinnear, the vice chairman who was Texaco's corporate representative at the trial. Wyatt reported hearing a rumor that Pennzoil had managed not only to obtain four "sinkers" on the jury—jurors known to be sure bets—but that Pennzoil had an inside track with Judge Farris on the preparation of the all-important instructions to the jury. Both sides met to discuss the issue with the judge, and everyone chalked up Wyatt's report to "whiskey talk."

Then, there was the "Getty heiress."

Shortly before the trial had begun, a woman had anonymously called Judge Farris and made extremely threatening remarks about Gordon Getty. The FBI succeeded in tracing the call to Portland, Oregon. Then, midway through the playing of a video deposition, a strange-looking woman wearing several layers of clothing showed up in the courtroom carrying a large piece of luggage. She gave her name, and said she was from Portland.

"I'm the heiress to the Getty will," the old woman said when she was ushered into the judge's chambers. She began sobbing, telling stories of her confinement in jail and mental institutions. When Judge Farris asked why she had threatened Gordon, she replied, "I was mad."

It turned out to be the understatement of the day. The authorities were called in and the woman was hauled to a mental institution in Houston, where she remained until long after the trial had concluded.

Everybody, it seemed, had something against Gordon Getty.

Arthur Liman put the spark back in Pennzoil's case.

It isn't every day that one trial lawyer gets to put another on the witness stand, particularly one as practiced in making jury speeches as Liman. Lawyer Liman had been Liedtke's point man throughout the rough-and-tumble negotiations in New York. He had been at Marty Lipton's New Year's party, where he had first broached the Pennzoil deal to the museum. He had spent hours with Gordon Getty and his advisors. He had spent two days haunting the hallways of the Intercontinental Hotel while the Getty Oil board met. He was a friend of Laurence Tisch of the Getty Oil board. Nobody would be able to summarize all the evidence that Pennzoil had so torturously hauled before the jury in the preceding month quite the way Liman could. Some lawyers lead witnesses; this witness on occasion led his own lawyer.

Liman, although one of New York's best-known securities lawyers, made himself out as a poor man's attorney: the chief investigator of the "bloodbath" caused by the authorities at Attica Prison, head of "the country's oldest poverty law firm," chairman of the commission that Mayor Koch appointed to investigate cover-ups of police brutality by the New York coroner's office.

"Tell the jury what your impression of Gordon Getty is," Terrell asked Liman.

Bob Brown of Miller's firm was immediately on his feet to object, and another lengthy bench conference was under way. Texaco did not want *anybody* saying anything nice about Gordon Getty. "That appears to be totally irrelevant to anything at issue in this case," Brown argued.

"Your Honor, I think Texaco has put Gordon Getty's competence into issue here," Terrell replied.

"Texaco has *not* put Gordon Getty at issue."

Jamail rose to defend the testimony. "The impression that they've tried to give this jury is that Pennzoil took advantage of a *nincompoop*." When Miller jumped into the argument, Jamail spun toward him and pronounced: "You're not rational."

After sending the jury from the courtroom so the argument could continue, Judge Farris proffered his own comment about Gordon. "He has a funny haircut and likes to sing in public," the judge said.

Liman enraptured the courtroom with a spellbinding account of the high-powered dealmaking that culminated with Pennzoil's acceptance of the final counteroffer presented by Getty Oil. Building up toward the momentous handshakes that ended the Getty Oil board meeting, he paused for a glass of water. It was near quitting time for the day, but Terrell begged for just a few more minutes so that the narrative would remain intact.

"Then what happened?"

"Then, within a few minutes, Marty Lipton and Marty Siegel came out, the door opened and they said, 'Congratulations, Arthur, you've got yourself a deal.'"

"And then what did you do?"

". . . I went around the room and I introduced myself and I shook hands with everyone I could. I said, 'Congratulations.' They said, 'Congratulations to you.'"

At last! Pennzoil had presented evidence of a handshake with Getty Oil Company—with *somebody* at Getty Oil, anyway. Liman could not remember exactly whom.

When Texaco's Brown finally got a crack at Liman, he was getting a cold and had just begun losing his voice. Judge Farris recoiled at hearing that Brown was sick: "My cardiologist has told me to stay away from pulmonary diseases," he told Brown.

Judge Farris was beginning to reveal more about himself than his health concerns. This judge, Texaco's lawyers became convinced, suffered from Yankaphobia. He seemed to go out of his way to draw attention to the growing sense of regionalism in the case—and Liman, himself an Easterner, brilliantly cast himself as an outsider in the courtroom, but in a way that only enhanced his credibility with the jury.

He recounted Gordon's hurt feelings at the board's vitriolic reaction to his deal with Pennzoil. "It was very, very hard for him to take," Liman said, "because this was the same board that snuck a lawyer in the side door and voted to sue him."

"Just a minute," the judge said. "Did you say 'snuck'? I thought only Texans used that word."

"Judge," Liman said, "if I stayed down here any longer I think I may become eligible for citizenship." The courtroom roared.

In the midst of a discussion about the reputation that Texaco's investment-banking firm enjoyed in New York, Judge Farris again stopped the testimony.

"Mr. Liman, did you say that the First Boston Company is a *New York* firm?"

"Yes, sir. Even though its name is 'Boston.'"

"If the name is First Boston, it should be in Massachusetts," the judge muttered. "Why didn't they call it First Staten Island Company or something?"

"I *don't know*," Liman replied with wonderment.

But Liman's greatest value to Pennzoil was the eloquence with which he shot down some of Texaco's best arguments—one after the other, like pasteboard ducks in a shooting gallery.

When Texaco's Brown pointed out that Gordon had agreed to support the Pennzoil deal "subject only to my fiduciary obligations," Liman replied: "Fiduciary duties do not permit you to breach a contract. Otherwise you could never have a contract with a trustee or a corporation."

If Pennzoil knew it had a contract, why then were its lawyers working so hard trying to draw up a final set of documents? "Can I use an analogy, sir? If I enter into an agreement to buy a house, I need a deed in order to close it. In order to close *this* transaction, they needed formal documents."

But, Mr. Liman, how can you say "all material terms" had been agreed to when the structure of the merger remained unresolved? "They're lawyers' things," he replied—and what could be less essential than that? "Those things are technical and irrelevant, sir." If that be so, then why were those very issues addressed in *your* memorandum of agreement with Gordon and the museum? "I've often had clients complain that lawyers get paid by the word." More laughter.

Liman's greatest moment occurred when Brown confronted him with the subject of the museum's Marty Lipton, whom Pennzoil's lawyers had already cast as a double-crosser who'd put the strong arm on Gordon to sell to Texaco so the museum could get a higher price. Liman had already testified that he was a dear friend of Lipton, and Pennzoil had made this a case about honor: How could Liman attack Lipton without appearing traitorous?

"My opinion of him and respect for him is undiminished," Liman began. "Martin Lipton was presented by Texaco with an offer he couldn't refuse. What do I mean by that?"

The witness was now asking himself the questions.

"His client was a seller; he was offered a substantial premium over what we offered. Everybody who is party to a contract has the right to either *perform* or to *pay damages* . . . And he was offered in this case an indemnity so that his client wouldn't have to pay damages. . . . And instead of having any kind of risk of

damages for walking out of the contract, the buyer, Texaco, was
going to pick that up. . . .

"It does not in any way reflect on Martin's integrity. I think he
did an extraordinary job for his client in getting them the highest
price and the indemnity. An extraordinary job."

And an extraordinary answer. Liman had not only established
his loyalty to a friend, he had cast Texaco as a Vito Corleone and
administered a further dose of the indemnities to the jury.

Another friendship, however, was continuing to wane in the
courtroom.

At the end of the day, with Liman still on the stand, Jamail
could not resist taking a parting shot at his old friend Miller, who
had been sitting in the audience near the jury box. As Miller
walked through the courtroom gate in one direction, he had
mumbled something while the jurors were exiting from the other
direction. Jamail complained to the judge that Miller was cozying
up to the jurors.

"I don't believe I ever saw a lawyer so goosey," Miller said.

"I am *not* goosey," Jamail shot back.

"Is there an objection or not?"

"I am having mixed emotions about it," Jamail growled. "If
they get close enough to him, they will hate him. I think he
probably ought to sit up here at the counsel table, unless he is
ashamed of representing these people."

Pennzoil had one final building block to lay before putting the
final, emotional flourishes on its case. It had to "prove up" the
greatest demand for damages in courtroom history.

Unfortunately, Joe Jamail was no numbers cruncher. He had
taken 150 credit hours as an undergraduate without ever taking
freshman math, and after he had finally emboldened himself for a
course in home-economics math he got a C. "I couldn't handle
that shit," he says. "People say, 'Jamail, all you know about
math is how to divide by thirds'"—one common formula for
determining a lawyer's contingency fee.

"Well, I do know that."

But Jamail also knew when to rely on expert assistance, and
once Jamail had his number, nobody was more effective at getting
a jury to award it. In all his record-setting damage awards for the
victims of defective products and industrial danger, Jamail had
developed a mastery for making juries understand that a paralyzed
or horribly disfigured man may experience not only the loss of

income for life but also pain and psychological damage. Convincing a jury to put a dollar value on the loss of Hugh Liedtke's once-in-a-lifetime deal would require the identical skills.

How could you put a dollar figure on something like that? "I thought it would be *simpler* than a conscious pain-and-suffering case," Jamail would recall. "You could lay it out for the jury in black and white." Which is exactly what Pennzoil did, on an exhibit some five feet tall emblazoned with bold, black lettering, as two Pennzoil executives took turns explaining the numbers from the witness stand.

It worked like this:

It would have cost Pennzoil just $3.4 billion for those billion barrels, including the share of Getty Oil's debt that Pennzoil would have assumed. One billion barrels was a nice, round number for the jury to ponder; it made the arithmetic simple. It worked out that Pennzoil would have gotten those barrels for $3.40 each.

With the deal destroyed, Liedtke could replace those barrels in only one way: by going out and drilling from scratch. Even if he succeeded in locating that much oil—a highly dubious prospect in the picked-over oil fields of America—it would cost, on average, $10.87 for every barrel found, based on Pennzoil's own experience over the prior five years.

The $10.87 for each of the replacement barrels, minus the $3.40 for each of the Getty Oil barrels, equaled $7.47 in lost opportunity, in *damages*.

Multiply $7.47 a barrel by one billion—1.008 billion barrels, to be exact.

Voilà. Seven billion, five hundred and thirty million dollars. Written another way, $7,530,000,000. Or just $7.53 billion for short. As is common in cases involving sums of fewer zeros, Pennzoil heaped on the identical amount in its demand for punitive damages.

The damages theory was plain and easy to understand; the logic was not only appealing, it seemed unassailable.

And it was utterly outrageous.

It was *three times* the cash outlay that Pennzoil intended to make for its interest in Getty Oil. If Getty Oil's reserves were really worth $10.87 a barrel to Hugh Liedtke, then he could easily have justified paying $10.87 billion for his share—instead of the $2.7 billion he claimed that Getty Oil and the museum had "chiseled" out of him.

If Pennzoil got $7.53 billion to go out and find its billion barrels by putting holes in the ground, it could put the cash in a money-market fund and drill all the holes it required out of only *half* the interest—and still wind up with its $7.53 billion left in the bank when it was all over.

Pennzoil's theory cast tax effects to the wind, did not adjust the dollars spent into the future into the deflated dollars of today and failed to take account of the business risks it would face in owning Getty Oil's one billion barrels. Among other problems, the theory assumed that Getty Oil's "heavy" oil was as valuable as any other kind of oil, which it decidedly was not.

And above all, Pennzoil's deal with Getty Oil had not given it outright ownership of a billion barrels of oil-in-the-ground. Liedtke would get that only *if* after a year, things had not gone well with Gordon and both sides had "split the blanket."

What had Liedtke *really* lost?

Certainly, he was denied the chance to buy 32 million shares of Getty Oil at $112.50 each—shares that turned out to be worth $128 each, according to what Texaco paid. Liedtke had lost the benefit of his bargain: $15.50 a share times 32 million shares, or $496 million.

And yet, for all its imprecision, Pennzoil's damages theory *was* defensible. The widow of an accident victim is entitled to seek damages equal to her husband's entire future earnings, even though he might have been laid off at some point. There was no denying that Getty Oil was a "unique" and irreplaceable asset. Certainly, Liedtke had gotten himself a good bargain.

And there *was* every possibility that Liedtke could have parted company with Gordon after a year and walked away with his billion barrels of oil-in-the-ground—*if* he and Gordon had been unable to come to terms. It was that little word "if" that enabled Jamail to transform a puny $496 million claim for the loss of a bargain into a $7.53 billion demand for the destruction of a dream. If Pennzoil could prove up its claim, that word "if" would be elevated into a $7 billion preposition.

It would be up to Texaco to knock the life out of Jamail's big, big number.

Initially, a jinx seemed to befall Pennzoil's damages case. Well before the trial, Pennzoil had arranged for expert testimony from Pen Thomas, who had served as chairman of Sinclair Oil Company and B. F. Goodrich Incorporated. Shortly thereafter, however, Thomas died.

Jeffers and Terrell made a quick recovery by finding Thomas Barrow, who was nearing retirement as the vice chairman of Standard Oil Company of Ohio, best known as Sohio. Barrow was executive timber through and through: tall and heavyset, with white hair and a powerful, deep voice. And his credentials, ticked off from the witness stand and enumerated on a two-page résumé distributed as Plaintiff's Exhibit #269, were extraordinary: native Houstonian, recipient of the first Ph.D. in geology ever awarded by Stanford University, former senior vice president of Exxon, former chairman and chief executive of Kennecott Copper Corporation and finally vice chairman of Sohio, where he was involved in discovering the Prudhoe Bay oil field of Alaska—North America's richest known deposit of oil-in-the-ground.

Among the thirty-one professional and civic affiliations listed on the résumé, alternate juror Jim Shannon took note that Barrow was a trustee of the Woods Hole Oceanographic Institution—and was impressed to read in the newspaper, during the week that Barrow was on the witness stand, that Woods Hole had found the *Titanic*.

Best of all, Barrow was an unpaid expert.

"What arrangement for compensation did you make with Pennzoil for this testimony you are giving?" asked Jeffers.

"None."

"And why was that?"

"First, I was not interested in being a professional expert witness," Barrow replied. "Second, I felt that the issue of contract was a matter of principle as far as I was concerned and, therefore, I was willing to do it on that basis." The larger-than-life oilman with no connection to the case was doing it out of *principle*. He was a trial lawyer's dream.

Barrow was not, however, entirely unconnected with Pennzoil. Texaco tried to attack his objectivity by noting that he served on the board of Texas Commerce Bancshares with a senior partner of Baker & Botts, and that he had a personal relationship with Liedtke. All of this was true, but Barrow succeeded in deflating the attack by announcing to the jury that he had an even stronger personal acquaintance through his children with Jim Kinnear, the Texaco vice chairman, who was seated right in the courtroom as Texaco's official corporate observer. (Texaco did not point out, because it did not know, that Barrow's next-door neighbor was J.

Hugh Roff, the chairman of Pennzoil's former United Gas subsidiary and a cousin of J. Hugh Liedtke. Roff, however, never
discussed the case with Barrow.)

Barrow wholeheartedly embraced Pennzoil's $7.53 billion "re-
placement model" of damages. "It is not the only model you can
use to measure damages in a situation like this, but, clearly, it is
one of the methods that has validity—if you are sitting in Pennzoil's position." And even that much money might never make
Pennzoil whole, Barrow testified. "In the United States, I would
say it would be very difficult to find a billion barrels. I say that,"
he boasted, "in light of the fact that I'm one of the few people that
ever have. So I know how tough it is."

Barrow testified he was not satisfied merely with endorsing
Pennzoil's theory of damages, so he had undertaken two "alternate theories" to test the $7.53 billion figure. Pennzoil could, for
instance, try to buy a billion barrels in piecemeal transactions;
after analyzing dozens of recent deals, Barrow concluded that this
would cost $8.46 a barrel on average—and $8.46 minus the $3.40
cost of the Getty Oil deal still left a loss of $5.06 a barrel—$5
billion altogether. Barrow's calculations, however, failed to take
account of the biggest recent deals—the takeovers of Gulf Oil and
Superior Oil—which would have knocked a few additional billions off the number.

In another analysis conducted through "the use of modern com-
puters," Barrow had determined that into the next century, the
cash flow from all of Getty Oil's reserves would total $49 billion;
in today's dollars, Pennzoil's share of that amount, minus what it
had agreed to pay for its three-sevenths, was $6.69 billion. This
analysis, however, included the value of Getty Oil's *possible*
reserves, which the other analysis did not.

"Do you feel you appreciate what three-sevenths of Getty Oil
Company would have been worth to Pennzoil?" Jeffers asked.

"It would have made it a completely different kind of oil company."

"Pass the witness."

Miller's partner Richard Keeton, one of Houston's most brilliant lawyers, found it easy to chip away at some of Barrow's
testimony—and Barrow willingly, and honestly, conceded the
points that he had to. The fact that Pennzoil had to spend $10.87
to find a barrel of oil—"that doesn't tell you what the value of that
barrel is, does it?" Keeton asked.

"No."

"You would have to know more facts, wouldn't you?"

"I would have to have more facts."

"Where the crude is located?"

"That would be an important fact."

"Type of crude?"

"That would be important."

"Production costs?"

"That would be important."

"Development costs?"

"Yes."

"Whether it was oil or gas?"

"Yes."

"All those things, you would need to know?"

"You would need to know those things to determine the value of a barrel of oil"—and Pennzoil's theory took account of none of them.

And with that, Keeton entirely dropped the subject. Virtually all of Pennzoil's expert testimony on damages stood unrebutted.

Maybe Texaco had a trick up its sleeve. Or maybe it didn't.

· 21 ·

A thrill swept the jury box when, five weeks into its case, Pennzoil called Gordon Getty as a witness by video deposition. The peripatetic Dick Miller took a seat in the audience, where he could blend in with the public to more discreetly call attention to the laugh lines with snickers and guffaws.

Gordon gave his address as "San Francisco."

"Is that the only home you maintain?" asked Jeffers of Pennzoil.

"It's the only home I maintain," Gordon said, tapping the tips of his fingers together, "but it would be less than fully responsive of me not to add that I have now acquired an apartment in New York and have built a home in Wheatland, California."

Less than fully responsive nearly brought down the house.

"You have been a trustee of the Sarah C. Getty Trust since 1976?"

"Correct."

"Can you give us an approximate date in 1976?"

"That would be June sixth," Gordon answered. "London time."

The jurors had not seen anyone—live or on videotape—quite like Gordon. He was clearly nervous, twitching his mouth, shifting positions from answer to answer, perpetually jerking off his glasses and drilling them into his breast pocket. Judge Farris even got in on the fun. "I notice that when Mr. Getty wants to think, he puts his glasses on, and when he wants to read, he takes them *off*."

During the recording of the deposition, fifteen months earlier, Miller knew that Pennzoil would try to depict Gordon as some sort of tragic hero—and decided he would do his best to advance the impression that instead Gordon was a bumbler. To that end, at one point Miller urged Gordon to hold up an exhibit to the video camera, an exercise that would be bound to come off looking silly on the television screen. Gordon complied, flashing a broad grin as he hoisted the document toward the camera lens. "My right profile," he said.

At that moment, off camera, Jamail raised his middle finger at Miller from across the deposition table.

Few trial witnesses are placed in a position as awkward as Gordon's. Although Pennzoil was still suing him for breach of contract in Delaware—a fact about which the jury continued to remain in the dark—Pennzoil was also anxious to draw out Gordon as a *friendly* witness in this case, as someone whose birthright Pennzoil had tried to restore. Texaco, for its part, remained on Gordon's side in the Delaware case, but was trying to cast him in this trial as an underqualified and inept businessman who had forced Getty Oil into the arms of Texaco by making a selfish deal with Pennzoil.

Gordon, as it turned out, would refuse to perform for either side. He was painfully earnest, yet his recollection of the facts was so sketchy that his testimony could do neither side much value or harm. Partly this was due to the presence during the taping of the deposition of the ferocious Moses Lasky, a lawyer who even in his eighth decade had lost none of his ability to prevent clients from divulging information that could do them the slightest harm. Yet part of Gordon's dearth of memory was just Gordon himself. At times it seemed he had erased his entire file for January 1983 from his computerlike brain; when he did remember, he was overly literal in his interpretation of the questions and maddeningly precise in his responses to them.

"Do you recall signing that agreement on January sixth on behalf of the trust?" Jeffers asked, referring to the Texaco deal.

Gordon examined the document. At length.

"Do I recall signing this?"

Gordon perused the document further.

"Well, on page twelve, sir, I see my signature."

Pause.

"I guess that wasn't your question. Do I 'recall' signing it? Do I 'recall' the event? Do I have an image in my mind?"

Twenty seconds elapsed.

"Sir, I believe so," Gordon answered resolutely. "I believe so. Without betting the farm on it, yes, sir."

But sometimes, when the questions went to the heart of the case, Gordon's answers seemed to border on the deliberately evasive.

"Do you recall what you were celebrating by drinking champagne?" Jeffers asked.

"Yes, sir."

Could this be an answer coming? the courtroom wondered.

"Seemed to me there had been a lot of progress in my ambition to raise the value of the trust," Gordon said.

Jeffers did not wish to settle for the fact that Gordon, Ann and Larry Tisch were hoisting their goblets and saying, *Here's to raising the value of the trust.*

"Can you be more specific about what you were celebrating?"

"I thought the fifteen-to-one vote was progress. That would be specific. I wish I could tell you more clearly what the vote was for, but I can't."

"Okay."

"But I nonetheless thought it was progress."

Pennzoil did manage to extract a modicum of genuinely useful testimony from Gordon. "Would it be fair to say," Jeffers asked, "that when Texaco and Mr. Lipton showed up at this meeting and you found out that the museum was going to sell to Texaco, that you felt *you* had no choice but to do so?"

"I think that's a fair statement."

And later, Gordon said: "I was absolutely convinced that the Pennzoil plan—insofar as at that stage of conception it can be called a plan—was the best possible alternative for the stockholders and the beneficiaries and the museum and all concerned. It was much better, I believe, than any other option available to the company at that time."

Finally, Gordon rose to the occasion on the subject of Liedtke's offer to make him the chairman of Getty Oil. After both sides had already made such an issue of the matter, the jury at last would hear just what Gordon himself had to say about it—sort of.

"You considered yourself to be qualified to serve in that capacity, did you not?" Jeffers asked.

"Don't answer that question," Gordon's lawyer Lasky instructed him. But a stern look crossed Gordon's face. He sat up straight and answered anyway.

"*Certainly* I consider myself qualified to be chairman of the board of Getty Oil."

The deposition's primary value to Pennzoil was its depiction of Gordon as a hapless person, a victim—someone who got chopped down in the one episode in which he had tried to assert himself: the battle for control of Getty Oil.

"I felt sorry for the guy," juror Rick Lawler says.

And although in his deposition Gordon seemed every bit as silly as Miller was making him out to be in the case, the deposition at times also made Gordon seem cute.

On his direct examination by Jeffers, Gordon had mentioned that Arthur Liman, the Pennzoil lawyer, had made an unflattering remark to Gordon about Chauncey Medberry, the Getty Oil director whose Bank of America was being engineered into position as Gordon's co-trustee.

"What did Mr. Liman say?" Miller asked at the conclusion of his cross-examination.

Gordon immediately gave a blushing smile.

"I hate to say it with ladies present," he answered, indicating the stenographer.

"This is nothing other than a search for what happened," Miller assured him. "Can you grit your teeth? She's sure she can handle it."

"Shall I say what he called him?"

"Yes."

Gordon drew his breath.

"He called him 'a duplicitous prick.'"

"That's mild enough," Gordon's lawyer Lasky interjected, and the deposition was over.

"Your Honor," said Jamail, puffing his chest like a peacock, "at this time I call Mr. Hugh Liedtke as a witness for Pennzoil."

Finally, Jamail had his chance to perform with a live witness—Pennzoil's star witness. Putting him on the stand immediately at the conclusion of Gordon's deposition helped to reinforce the idea that Liedtke had been Gordon's rescuer, someone who had attempted to help the misunderstood heir to fulfill his destiny.

The burly Liedtke lumbered alongside the jury box to the witness stand like a linebacker approaching the podium to receive his varsity letter. While Baine Kerr had leaned back in the witness chair, Liedtke leaned forward, in the same position that had caused him to wear grooves in his office desk. A standing-room-only crowd—including Liedtke's wife, Betty Lyn, and some of his children—filled the thirty-five-seat courtroom, and they watched an extraordinarily sophisticated businessman and an equally sophisticated lawyer exude the kind of folksy charm that could only make a jury of Texans glow with warm feeling.

"You have children?"

"Five."

"Would you tell us their names and ages, please?"

Liedtke smirked, fumbling through his wallet for a card listing their birthdates. "I wish you didn't ask me that in front of my wife!"

As a chuckle rippled through the courtroom, Judge Farris chimed in with an endorsement of Liedtke's humor. *"Mea culpa,"* the judge said. "I have a hard time remembering *our* wedding anniversary date." The judge *liked* this witness.

The lawyer and witness had been together three or four evenings a week for the past several weeks preparing for this testimony—not in the fluorescent glare of a law-firm conference room, but over a few drinks in the comfort of each other's homes. "If we have any chance at all," Jamail had told him, "you *have* to come across as the Hugh Liedtke who's sitting in my living room." Only through Liedtke's testimony could Pennzoil clinch honor as the main issue in the case—and only Liedtke could persuade the jury to award damages for the destruction of a man's dream. "Hugh was the moral force of our case," Terrell of Baker & Botts would recall. "He was a rev-up-the-jury witness."

"You and I have known each other a long time, haven't we?" Jamail asked the witness.

"That's true."

"And we are friends?"

"That's right."

Liedtke mesmerized the jury with all of the Liedtke lore—about drilling those 127 wells back in Midland without ever encountering a dry hole, about the attendance of J. Paul Getty at a party to celebrate Liedtke's wedding engagement, about his friendship with George Getty and Jack Roth and the other high-ranking executives of the Old Man's business empire, about his early association with George Bush. Baine Kerr had already told the same stories, but giving it to the jury a second time meant it had to be true.

"Would that be the George Bush that is now vice president?"

"Yes."

There was the account of how Liedtke had saved South Penn Oil Company from extinction and transformed it into the modern-day Pennzoil—a story that gave Liedtke the halo of having worked in a high-level position for J. Paul Getty himself. The unhappy story of the executive thrown out to make room for Liedtke was not included.

"Did Mr. Getty play some role in getting you that position?"

"It wouldn't happen unless he wanted it to happen."

One by one Liedtke ticked off the independent companies that he had formed by spinning off Pennzoil's assets, nearly every one a household name in Houston: Entex Incorporated, which provided natural gas to households throughout the city; United Energy Resources, in whose downtown "plaza" sat the tallest building in Houston; Zapata Drilling Company, one of Houston's biggest hometown operations. The jury, of course, did not have the slightest inkling of the controversy surrounding some of the deals that created those companies.

"Explain, if you will for us, the tender offer that Pennzoil made" for Getty Oil, Jamail asked. "What was it?"

"Well, I'm not a very good one to explain tender offers to you"—this from the man who had *originated* their use in hostile takeovers. "Quite frankly," he finally answered, "anytime you buy anything it's a tender offer. Let's say you buy one hundred shares of stock. You make an offer, a proposal. And when somebody says, 'Okay, I'll sell you my hundred,' you've got an agreement." Liedtke had reduced the most fearsome mode of attack in corporate America—a mechanism governed by thousands of pages of law—into a trade at the flea market.

About an hour into Liedtke's testimony, Jamail turned his witness to the subject of why Pennzoil had leaped into the war for Getty Oil.

"Well," Liedtke said, "one of the principal things that I have been interested in over the years is trying to cause Pennzoil to grow. Different people like different things and for me—and for a long time my brother and Mr. Kerr—our interest has been in growth and not in liquidation." It was an interesting comment from a plaintiff who was justifying his damages claim on his right to liquidate Getty Oil after a year.

Liedtke said he had spent a lot of time keeping tabs on "business opportunities"—as opposed to takeover opportunities—and that before long it had become "increasingly clear that Getty Oil was having great difficulties internally, in that there were obviously three factions and they were all fighting one another.

"And to me, it meant that at some point, something was going to happen. The company might well be sold—or might want to work out some kind of an arrangement with third parties to bring peace and management to the company." Liedtke had made Pennzoil the peacemaker!

Without mentioning the weeks that he had stalked Getty Oil, Liedtke recalled deciding to make his move after reading an account of the Getty Oil war in the December 11, 1983, issue of *The New York Times*.

Miller could no longer restrain himself.

"That document is not in evidence and this dissertation is hearsay and we wish to object." But Judge Farris not only let Liedtke testify about the contents of the article—which was hearsay upon hearsay—he also admitted the article itself into evidence over an apoplectic objection from Miller. Under the rubric of testifying to what he recalled about the article, Liedtke retold the story of the infamous "back-door" meeting of the Getty Oil board—that meeting at which, in reality, people had mostly come and gone through the same door.

"At the meeting," Liedtke said, "Mr. Getty was asked to step out of the room, according to the article, while the rest of the board voted on whether or not to ratify the standstill agreement. While he was *out of the room*, through the *back door*, a Mr. Winokur . . . together with a Mr.—I never know whether it's *Boyzee* or *Bwazee*, who was an investment banker with Goldman, Sachs—came in the back door and suggested to the board of directors while Mr. Getty *was not present* that they join in this lawsuit which Mr. Copley and Mr. Petersen had previously set up, and they voted to do that." Besides suffering from the "back-door" hyperbole—a technicality, to be sure—Liedtke's rendition cited a vote by the board to jump into the family lawsuit, a vote that had never been taken.

"Lawsuit against who?" Jamail asked.

"I mean the suit that the son of his brother who lived in

London—the one reputed to be a dope addict. This is a suit of a young man named—he's a minor—Tara Gabriel Gramaphone Galaxy Getty.'' Close, but not quite. ''This youngster, they got to bring their lawsuit.

''In any event, why, then, the company joined in the lawsuit.''

''The lawsuit was against Gordon Getty?''

''Against Gordon Getty. . . . I will not express an opinion on it.''

And what was in it for Pennzoil to ''break the logjam'' at Getty Oil? Nothing less than the billion barrels of oil-in-the-ground that for three brief days had represented the attainment of Hugh Liedtke's dream.

''George F. Getty Sr.—that's J. Paul Getty's father—worked for fifty years before he died in 1930''—it was actually closer to twenty-five years—''and then J. Paul Getty, who died in seventy-six—they'd taken, you know, eighty to one hundred years to put that position together. It was mostly in the United States and it was something that I didn't believe then nor do I believe now—I know *I* never saw it before and I don't believe in my lifetime I'll ever see it again.''

''See what, sir?''

''A reserve picture like that one.''

Jamail was a lawyer as practiced as anyone in pacing an examination to put time on his side. He walked Liedtke through the events leading to Pennzoil's deal, including Liedtke's stroll through the broken glass of Times Square in the first hours of 1984 and the scheduling of the meeting with Gordon for New Year's Day.

''And did you keep that engagement, that meeting?''

''That's right.''

''Your Honor, if the quitting time is four-fifteen, this is a good place to stop.''

''Very well. We will recess until tomorrow morning at nine o'clock sharp.''

After six weeks of mostly tedious evidence, here was a witness the jury could understand, and Jamail had sent them home hanging on the next word.

Over the next two days, Liedtke bit by bit affirmed every element of Pennzoil's case.

On Gordon's intent to be bound:

''Did you shake hands or say anything?''

"Oh, sure. Yeah. Everybody was, you know, kind of pleased that something had gotten worked out."

On the meaning of Ann Getty's invitation for champagne: "Did she congratulate you, Mr. Liedtke?"

"Oh, sure."

On Gordon's plan to introduce Liedtke as the new boss of Getty Oil: "He suggested that we fly out to California together and walk down the halls of the Getty building there," to assure employees that the company "would continue to grow, and to try to stop the unrest and put some morale back into the organization."

Then, to convince the jury that Pennzoil had reached agreement on all essential terms in the Getty deal, Liedtke one by one ruled out the "open issues" that remained in Pennzoil's effort to draft a final merger agreement.

What about the uncertainty over who would buy the museum shares? "It didn't make any difference to us." What about the supposedly unsettled details of the stock option that Pennzoil had wanted to "lock up" the deal? "The agreement had been reached and the stock option was no longer of any importance." What about the terms of the ERC stub worth $2.50? "Well, Mr. Jamail, you are dealing in minutiae now." Who was going to get Getty Oil's last quarterly dividend? "If you look at it in terms of this deal, it's a fraction of one percent."

And when Jamail asked about the supposedly unresolved issue of continuing Getty Oil's employee benefits programs, Liedtke seized the moment not only to strengthen his own case, but to put a knock on Texaco's. "It was our plan to build Getty Oil. . . . It is quite clear that Pennzoil is a company that builds rather than destroys and liquidates." Pennzoil, he said, had no plans to milk the quarter billion dollars in excess cash in Getty Oil's pension plan—"as Texaco has done. . . . *We* were not going to fire people to finance *our* deal."

Both sides knew that any testimony about laying off people and raiding their pension fund would press hot buttons all over the jury box, so when Miller objected to the line of questioning, the jury was ushered from the room—which succeeded in making Texaco appear to be harboring some deep, dark secret. Jamail claimed that the subject was fair grounds for testimony because Miller had already boasted of how well Texaco had treated the employees of Getty Oil.

"He created this slop and now he's got to swim in it," Jamail said, motioning to Miller. "It's not right for them to tell the jury

that they take such good care of these employees—just like a
kindly old uncle—and then go out and loot their pension fund and
fire eighteen thousand of them.'' (Actually, Texaco had fired
closer to one-tenth that number.)

Miller, however, was prepared with the perfect comeback. ''I
wouldn't think Pennzoil would want to get too far into the record
of its *own* past events, either, and if we're going to open up all
these subjects, then I'd like to have them *all* opened up. I think
we'll come out quite well if we get into *that* kind of contest, and I
think everybody knows what I'm talking about.''

The pension-fund affair was quickly disposed of, and Jamail
returned the attention of his witness to the main issue: the theft of
Hugh Liedtke's deal.

Should Texaco have known that Pennzoil already had an agreement? Jamail asked. ''People don't get that place in a corporation
without doing their homework,'' Liedtke answered. Would you,
Mr. Liedtke, have interfered with someone else's deal? the lawyer
asked. ''Mr. Jamail, I would consider that about as arrogant and
unethical a thing to do as I can possibly imagine.''

And explain to the ladies and the gentlemen of the jury how
Texaco succeeded in breaking your deal?

It was the question that cast Marty Lipton of the museum, one
of the nation's most famous lawyers, into the Svengali of *Pennzoil
v. Texaco.* Testifying purely to his speculation, Liedtke told the
jurors that it was Lipton who had forced Gordon into breaking his
deal with Pennzoil. To amplify on the point, Jamail had distributed copies of the fifty-year-old document that had established
the Sarah C. Getty Trust—and made it clear that the jury understood Gordon was not permitted to sell the family's shares, except
to avoid economic loss. ''Under the Texaco proposal,'' Jamail
asked Liedtke, ''what happened to the trust stock?''

''In my view, the trust was *forced* to sell and *frightened* into
selling''—and it was Lipton who put on the squeeze. It happened
that night at the Pierre, when Gordon and McKinley of Texaco
were fencing over what Texaco would pay. Lipton, Liedtke suggested, was called to the scene to function as the enforcer.

''He wasn't up there just to pass the time with good old Gordo
in the middle of the night,'' Liedtke barked. Once Lipton had
made a deal to get top dollar from Texaco for the museum shares,
''He goes up there and tells him, 'Gordon, if you don't sell, we're
going to put the pants on you. . . .'

''I think that they *forced* Gordon Getty to sell.''

Was Texaco so desperately short of oil that it would even grant indemnities to protect Gordon and Lipton from a lawsuit by Pennzoil?

"I think so, Mr. Jamail," Liedtke said in his most self-assured tone. "In fact, I would be positive in my own mind. That's a *desperate* situation."

"Pass the witness, Your Honor."

Miller's cross-examination of Liedtke would throw yet two more longtime acquaintances into mortal combat. Although Miller and Liedtke had never forged a real friendship, Miller and Liedtke's brother Bill had, and Miller had once briefly represented Liedtke as a lawyer. None of that, of course, now mattered in the slightest.

"Miller's going to try to bait you," Jamail had warned Liedtke, "but I don't want you getting into any fishmonger fights with him, and if you do, it will make me mad."

"How will I know if you're mad?" Liedtke had asked.

"Because you'll see the back of my ass going right out that courtroom door, that's how!"

"You'll *leave*?"

"You're goddamn right."

Miller had estimated he would spend only a few hours cross-examining Liedtke; instead he spent nearly six days. But in an entire week of cross-examination—from Tuesday, September 3, to the following Tuesday—Miller would only make Liedtke stronger. The more forceful the questioning grew, the more forceful the answers became.

Miller demanded to know why Liedtke never called anyone at Texaco to complain about the theft of $7.53 billion worth of oil-in-the-ground. "Suppose I'm in the banking business and somebody's stolen the Brink's truck," Liedtke answered. "Do I call the robbers or do I call the police?"

When Miller demanded to know whether the Getty Oil board had a fiduciary duty to consider the higher offer from Texaco, Liedtke turned the tables and asked about the board's "fiduciary duty" to the employees of Getty Oil. Pennzoil would have kept every last one of those 19,400 Getty employees—but they got "slaughtered" under the Texaco deal. "You are *dead* if you get fired," Liedtke snapped.

Liedtke suggested that the original handwritten notes of the Getty Oil board meeting in New York would have resolved any controversy about whether Pennzoil had a contract. But the notes were "flushed down the toilet," or perhaps "put in the shredder."

Or maybe set fire to? Miller asked facetiously.

"They *were* pretty hot," Liedtke replied.

Liedtke, however, left himself wide open at one point in the discussion of the discarded notes. When he declared, "I am not in the business of destroying evidence," Miller immediately opened the deposition of Pennzoil's corporate secretary, Sally Hazen. She had testified before the lawyers that she had destroyed her own handwritten notes of a Pennzoil board meeting—two days *after* the Getty Oil notes had been destroyed.

"Did you know that?" Miller asked.

"No." Score one round for Texaco.

On occasion the audience put on a better show than the witness. Miller complained that a battery of Liedtke friends and family had overtaken the ringside seats. "They're all sitting on top of the jury," he told the judge. "They respond to everything that Mr. Liedtke says."

Jamail would have none of it.

"I haven't heard a cheering section except for those chickenshits sitting back there from Texaco," he said. "Get that on the record as I said it," he instructed the court stenographer.

Later Bob Brown of Texaco complained that Terrell of Baker & Botts was waving his arms, tossing his head, groaning, laughing and whispering audibly during Liedtke's testimony. "Entirely unprofessional and distracting," he called Terrell's behavior.

Judge Farris held up a pad of contempt-of-court citations. "I will, *in fact,* enter the attorney's name, write the violation, the date, the time of day, the amount of the fine, *the days in jail.* . . . I will just say, 'Mr. Jamail, that's two for you.'"

"Why do you pick *me* out?" Jamail protested.

Although Liedtke was performing brilliantly all by himself, he got some help from the bench. For it was during Liedtke's cross-examination that Farris made one of his most perplexing rulings on the case.

Miller was convinced not only that Pennzoil never had a contract, but that Pennzoil *knew* it didn't have a contract. Getting Liedtke to admit to that would be impossible, of course. But he had a piece of evidence that he thought would go a long way toward convincing the jury, and he wanted to thrust it under Liedtke's nose. It was a sworn statement by Richard Howe, who had recently succeeded Baine Kerr as president of Pennzoil. The statement, submitted to the Delaware court only two days after Pennzoil's deal had fallen apart, showed that Pennzoil knew its final merger documents had to go back to the Getty Oil board for approval.

Although Judge Farris had refused to let Texaco divulge the existence of Pennzoil's Delaware suit against the Getty group, this document had been filed in connection with *Getty Oil's* suit against *Pennzoil*—the one in which Getty sought a court judgment that it never had a contract with Pennzoil. And Judge Farris had freely permitted Pennzoil to discuss *that* lawsuit.

He would not, however, let Miller use the document.

"I've already ruled that you cannot tell the jury about the lawsuit," the judge said.

"I'm not doing that!" the exasperated Miller replied. This sworn statement had to do with the other lawsuit—the one the jury had already heard about time and again.

"If you give them that statement, you'll be telling them part of it," the judge said. "I won't allow that."

Miller had to resort to a lame line of questioning. Did Mr. Howe ever tell you that your deal required further approval by the Getty Oil board? he asked Liedtke.

"Not that I recall."

For two days, the dauntless Miller tried everything he could to suggest that Liedtke knew he had no contract. There was, for instance, Miller's "gun at the head" theory.

After the Getty Oil board had made its fifteen-to-one vote, Pennzoil never withdrew its $100-a-share tender offer. There was a threat implicit in keeping the tender offer out there after the deal supposedly had been reached—a kind of Damocles sword hanging over the heads of Petersen's people. If they had tried to back away from an important provision of the deal, or had insisted on some new term that was costly to Pennzoil, then Liedtke already had a mechanism in place to buy as little as 11% of Getty Oil in the public markets, combine with Gordon's 40% and oust the entire board and management of Getty Oil. Wall Street has an expression for such a weapon: *a gun at the head*.

Earlier in the case, Miller had gotten Baine Kerr to admit that he had heard of that expression, although Miller failed to develop the issue very effectively. Now, with Liedtke on the stand, he had another chance. Keeping the tender offer on the table was purely a "ministerial" thing, a decision by some lawyer that had no effect

on the deal, Liedtke asserted. Moreover, Liedtke had never even
heard the expression "gun at the head"—and told Miller he
couldn't believe that Baine Kerr had heard of it either. Then
Miller struck.

"Now, Mr. Liedtke, would you say that having a gun at somebody's head while you are negotiating the terms of a definitive
merger agreement is an advantage or a disadvantage?"

Liedtke stammered. "Well—"

"That's not too hard, is it?"

Liedtke called it a "stop beating my wife" question, adding, "I
have testified under oath that I have never heard that before you
brought it up today." Miller brandished a copy of some testimony
in which Baine Kerr had acknowledged some vague acquaintance
with the expression.

"Well, it's very clear, isn't it, that Mr. Kerr had heard of it?"

"After you badgered him and badgered him, he said he *may*
have heard of it."

"Badgered him and badgered him?" Miller asked. "I asked
him two questions about it!"

"Mr. Miller, I don't want to argue with you."

"Would you like to look at it?"

"I don't need to look at it," Liedtke answered angrily. "It's all
been presented to the jury. They can make up their own minds
about your tactics."

"I'm sorry. I didn't hear you."

Liedtke looked at Jamail, who was beginning to rise from his
chair—the signal to Liedtke to back off. *The hell with diplomacy,*
Liedtke thought.

"I said the jury can make up its own mind about your tactics."

"Mr. Liedtke," Miller replied, "you have me at a disadvantage, sir. I cannot answer you."

Farris spoke up. "We have reached the end of the day, so we
will stop on that bittersweet note." Liedtke looked back at Jamail—who instead of walking out the door was now smiling
broadly. Liedtke had made Miller out to be the bully, and
Liedtke's argumentativeness had put the testimony of someone
else altogether at issue. One of Miller's strongest facts had been
buried, lost in the argument.

Before Liedtke had left the stand for good, the undaunted Miller
took a further shot at proving Pennzoil knew it never had a contract. Why, Miller asked, did Pennzoil find it necessary to keep
three PR people stationed in New York during the three days that

it was trying to complete the final paperwork? After all, the deal had already been publicly announced. Didn't Pennzoil *already* have a contract?

Liedtke misstepped. "This was a big event in the history of our company," he answered, "and I'm sure they wanted to be around."

"Well," Miller shot back sarcastically, "the 'big event' had already taken place, Mr. Liedtke."

No answer.

"*Hadn't* the 'big event' already taken place?"

Initially Liedtke answered lamely. "The formalization of a big event is another big event," he said. Immediately, he recognized that the response was insufficient; he had to do better. For a second or two he stammered—and then he hit upon an analogy.

"Let's say it's like a hard-fought bill in Congress," Liedtke finally added. "The thing gets done. Everybody knows it's done. That's a big event.

"But eventually, why, they type it up and they take it over there to the president. He signs it. He gives his pen away—you know, half a dozen pens. It's the formalization of the thing that's another big event for the news people." It was a stunning, brilliant answer.

Like a basketball coach calling time out when the opponent scores several unanswered points, Miller turned to Judge Farris.

"Does Your Honor plan to take a recess this afternoon?"

"Last-Word Liedtke" had struck again.

On September 4, midway through Liedtke's testimony, a state appeals-court judge in Louisiana handed down a little-noticed opinion in the lingering litigation over Pennzoil's natural-gas-supply problems in that state. As a Pennzoil subsidiary, the opinion stated, United Gas Pipeline had "lied" on pumping reports filed with the Federal Power Commission, had "falsified its reported reserves" and had exploited its influence with the federal regulators to help it "defeat the entire structure and jurisdiction of the Louisiana Public Service Commission and *breach its contracts.*"

The twelve jurors and four alternates were running out of small talk by the third month of the case. The Houston Astros were already out of the pennant race. The weather was the same sweltering stuff day after day, and hurricane season had passed. The mayoral race was intensifying over a gay-rights issue, but the jurors had already debated it ad infinitum. The one remaining subject they had in common—the case—was also the one subject they were prohibited from discussing.

So after a while, some of the jurors began talking about Diana Steinman.

She was an outsider, the only native-born Yankee and the only Jew in the group, a twenty-nine-year-old who had been in Houston only three years working as an office temporary. Some of the jurors began drawing Diana out about her personal life over lunch, or during the long stretches of time when the lawyers and the judge were arguing with the jury out of the courtroom. All the jurors were stir-crazy by this time, but the confinement had a more pronounced effect on Steinman. Somehow she became convinced that she had become the main topic of conversation even when she was *not* in on the conversation.

On the morning of September 10, Steinman walked into the judge's chambers all by herself after telling Bailiff Carl Shaw that she desperately needed an audience. The lawyers were present as well.

"I find there's being created an uncomfortable situation with all the other jurors," she said meekly. "And yesterday afternoon all I did was to try to stop from bursting into tears. I find they're talking about me often when I'm not present. And this was confirmed to me yesterday by a few minor incidents and I find it very upsetting."

"Tell me how you know they were talking about you," Judge Farris asked.

Steinman insisted that the jurors had begun keeping track of her whereabouts so they knew when it was safe to gossip about her. In the middle of her story, the judge temporarily excused her from the courtroom. It was one of the few occasions in the trial where Jamail and Miller had no cause to disagree. Steinman had to go.

"She may contaminate the rest of the jury," Jamail told the judge. "I don't know what the hell she's going to do."

Steinman was called back and told she should speak to no one about the case. "When you go out, don't stop," the judge said.

"I'm taking the stairs," Steinman answered. "The fastest way out."

When the remaining jurors were led into the jury box, Judge Farris informed them that one of their number "was unable to continue" and had been excused. And while all the jurors wondered what had happened, Jim Shannon wondered how he would get home that night.

But that was a small inconvenience. For the man who had figured out that he was the thirteenth juror—and thus the first alternate—immediately recognized that he had become the twelfth juror.

Miller ended his cross-examination of Liedtke with a potshot at his old friend and opposing counsel Jamail. When Liedtke suggested that Geoff Boisi of Goldman, Sachs wanted a higher price for the Getty Oil shares only to increase his fee, Miller turned to the Pennzoil counsel table and glowered. "I bet you know lawyers who charge as much as thirty or forty percent of what they collect, don't you?"

Miller turned to face the jurors. "I bet we could even find one here in the courtroom if we look, couldn't we?"

Jamail had a specific reason to keep his redirect of Liedtke brief: There was another witness Jamail wanted to get on the stand that day. Jamail did take the time to explain away the absence of the Getty interests from the lawsuit by giving the jurors one more dose of the indemnities.

"It was because of the indemnity, the complete and total indemnities, that Texaco had given to every Getty entity and every Getty lawyer and every Getty representative," Jamail asked, "that it would be useless in our opinion to bring them in this lawsuit in Texas?"

"Yes, sir."

Miller leaped to his feet to object—Getty was not in the suit, of course, because unbeknownst to the jury, it was still being sued in Delaware. But Miller's objection was too late; Liedtke already had answered.

Jamail continued his redirect long enough to take Liedtke past the lunch break, excused him from the stand and summoned Pennzoil's final witness to the stand.

"I call Mr. McKinley as an adverse party, Your Honor."

The jurors would see the contrast in the two chairmen without even the interruption of the afternoon break.

Dick Miller liked John McKinley a great deal, but he had no pretensions about the kind of witness he would make. McKinley was an engineer who acted like an engineer, with "zero sense of humor." He had never given trial testimony in his life. Although one of the world's most powerful industrialists, he was bound to make a weak witness.

As he watched McKinley approach the witness stand and Liedtke walk away from it, Jamail could already taste the blood. Calling someone on the other side as *your* witness, least of all the chairman and chief executive of the other side, bordered on suicidal. But Jamail had felt certain that Texaco itself would never put McKinley on the stand. Texaco had three other high-ranking executives through which it could give its side of the story; two of them—President Al DeCrane and General Counsel William Weitzel—were lawyers, and the third, Vice Chairman Jim Kinnear, was sitting through every day of the trial. The only way Jamail could assure getting McKinley on the stand was to call him himself.

"I knew he was arrogant," Jamail would later say. "I wanted the jury to see him in his true light."

In fact, McKinley was not an arrogant man—distant and wooden, yes, but seldom condescending or superior, and never mean. Yet by the time Jamail was through with him, McKinley would appear to some in the courtroom as the Godzilla of Big Oil.

Although he would keep McKinley squirming on the stand for a week, Jamail had a relatively narrow objective: to demonstrate that Texaco had *actual knowledge* of the Pennzoil deal, which was a requirement in proving Pennzoil's tortious interference claim, and to suggest that Texaco had acted with *malice*, which was necessary to collect punitive damages.

Jamail immediately established that McKinley had a duty as chief executive officer to make sure that Texaco's lawyers and investment bankers abided by company policy.

"Is it Texaco's policy, sir, to interfere with other people's contracts?"

"No," McKinley replied.

"All right, sir. We'll come back to that."

Jamail then reminded the jurors of a key element in Liedtke's testimony: that a company's top management—not its lawyers—are the ones who negotiate contracts. It was Jamail's way of convincing the jury that Pennzoil had a done deal, even though Pennzoil's lawyers had failed to get a signed contract.

"The lawyers don't make the definitive agreement, do they?" Jamail asked McKinley. "It's the top executives, such as yourself, that bind the company. Isn't that correct?"

McKinley pondered the question for a few seconds, looking for trap doors. But he had to answer Jamail honestly. "Yes," he replied.

Jamail made sure the answer had made the proper impression in the jury box. "I think we're on the same wavelength," he said.

One by one, day by day, Jamail phrased his questions in a way that gave McKinley no choice but to answer yes or no.

You knew that Gordon and the museum had signed a memorandum of agreement? *Yes.* You could have stopped your deal when you found that out? *Yes.*

"You didn't do that, did you?"

"I did not halt anything. . . ."

"And that was *intentional*? You did it and you knew what you were doing?"

McKinley hemmed and hawed, but finally answered, "Yes. No question. . . ."

"Texaco's success, very simply, defeated this arrangement that Pennzoil had with Getty." It was a statement rather than a question—and although the answer was obvious, it was not one that McKinley was anxious to give.

"Is your answer, 'Yes, it did'?" Jamail finally demanded.

"Well, you just keep using the word 'defeat,'" the exasperated McKinley finally replied. "We weren't out to defeat Pennzoil and we were out—"

At this point even the judge began helping Jamail. ". . . answer it yes or no," the judge said, and *then* explain.

". . . The answer to your question was, I suppose, yes."

At every opportunity, McKinley tried to extricate himself from incrimination by tailgating his answers, slipping in the words that everyone he talked to at Getty Oil assured him they were free to deal. But the jury quickly grew weary of the refrain.

When he was permitted the opportunity to explain himself, McKinley made an extremely persuasive case for the diligence of his efforts to rule out that Pennzoil had a contract. He ticked off practically a dozen people connected with Getty Oil who had personally assured him that Pennzoil had no contract. If Pennzoil really did have a deal, then either McKinley was perjuring himself over and over—a possibility that no one entertained—or else McKinley had been *lied to*.

He said that numerous investment bankers had assured him Getty Oil was free to deal. "We've been told about the financial interests of investment bankers in these kinds of things," Jamail countered. Sid Petersen had told Texaco that Getty Oil was free to deal. "Were you aware of Mr. Petersen's prior efforts to remove Gordon Getty as sole trustee?'' Jamail asked. And after McKinley said that Gordon's investment banker had told Texaco that Pennzoil had no deal, Jamail thrust at McKinley a copy of Marty Siegel's California court affidavit, forcing McKinley to read aloud for the jury the section referring to ''the transaction approved by the Getty Oil board of directors.'' That sworn statement, Jamail pointed out, had been filed with a court only two hours before McKinley had his meeting with Gordon Getty at the Pierre.

Jamail did not deflate the credibility of every person who had repeated the ''free to deal'' refrain to McKinley, but he cast sufficient doubt on some of them that the jury could easily think the worst of all of them. And besides sending plenty of verbal signals to the jury, Jamail sneaked through a few visual ones from his table right next to the jury box. Dick Miller complained that Jamail had strategically positioned an article detailing complaints against Texaco by Getty Oil employees right on top of the Pennzoil counsel table. And Jim Shannon, by this time a full-blown juror, would recall having a clear view of the big, bold headline on a *Fortune* article which the judge had refused Jamail permission to introduce. The headline read: HOW TEXACO OUTFOXED GORDON GETTY.

Jamail was not about to waste another opportunity to draw his smoking gun: the handwritten notes produced from Texaco's own filings, notes that McKinley had not seen since his deposition fourteen months earlier.

''Nobody at Texaco was trying to 'stop the signing.' Is that your testimony?''

''No one that *I* know of was attempting to stop the signing,'' McKinley firmly answered.

''Would you please turn to page eight, sir?'' Jamail asked. McKinley complied. ''What's the first thing you see?''

McKinley peered through the bottom of his glasses.

''The words 'stop signing,''' he replied.

Jamail paused to let the point sink in.

When the subject matter was technical, McKinley often got the better of Jamail, whose acquaintance with the intricacies of tender-offer regulations couldn't compare with that of the Texaco chairman. On another occasion, McKinley teased Jamail over the variety of colors that Jamail had used to highlight a document he had asked McKinley to read. ''The highlights are in blue on this page,'' McKinley cracked.

Jamail narrowed his eyes. "Mr. McKinley, you are very astute—and had you used some of that astuteness" before bidding for Getty Oil, "perhaps you wouldn't be here today."

The testimony that left the most lasting impression on the jury occurred on the morning of Jamail's third day of cross-examination. Jamail wanted to know whether anything besides the bid by Texaco had prevented the completion of Pennzoil's paperwork on the deal. Yes, McKinley said. He knew that Getty Oil was also soliciting offers from other oil companies.

But do you *know* of anything?

"Yes, I know of possible things. . . ."

Jamail would not quit without the answer he wanted.

"I'm not looking for *possible* things, Mr. McKinley. I'm looking for *actual knowledge,* Mr. McKinley. And the truth is, is it not, sir, if Texaco had not stepped in and made its bid at the time, you know of *no reason* why this definitive merger agreement would not have been completed, do you?"

"I don't know of any specific reason why it might not have been completed," he finally admitted.

In the eyes of the jurors, the chairman of Texaco might just as well have pleaded "no contest."

Dick Miller's examination of his client was mercifully brief. When Miller tried to depict Texaco as a hometown company since 1902, McKinley quickly volunteered that the company had almost immediately left Houston. McKinley tried to suggest that Texaco had plenty of oil-in-the-ground before the Getty deal, thanks partly to its operations in Saudi Arabia—a fact that only established Texaco's close ties with the notorious OPEC. McKinley also detailed how kindly Texaco had treated Getty Oil's employees, avoiding any wholesale firings and awarding generous severance to those who were dismissed. But the discussion of severance gave Jamail a beautiful opening for his re-cross examination.

He asked whether McKinley had gone into the Getty deal expecting to fire people.

"Yes," McKinley blithely answered. "It was my hope that there would be some redundancy of personnel and that they could be terminated. Yes."

Jamail filled his face with the greatest disbelief he could muster.

"That was your *hope*?"

Jamail again rubbed McKinley's nose in Texaco's sorry exploration record. He asked about Texaco's "Mukluk misadventure"—the most costly dry hole in history—and got McKinley to admit that the event had probably played a "subconscious" role in the decision to go after Getty Oil. He tried to make Texaco somehow responsible for the skyrocketing price for Middle Eastern oil. And before finally letting McKinley leave the stand he asked, "Why was Texaco trying to take care of Liedtke?"

Repeatedly, McKinley ducked the question—and repeatedly Jamail asked it again.

"Why would Texaco at all be discussing 'taking care of Liedtke'?"

Miller, a type-A personality if there ever was one, could stand it no more.

"One reason this examination is taking so long is that counsel just asks the same question over and over and over and over and over and over and over and over."

Jamail never did get an answer, but he was satisfied just the same. Moments later, McKinley at last was free of Jamail's torment.

"Your Honor, at this time I rest Pennzoil's case." It was September 18, 1985—nine weeks after the jurors had first assembled. That evening, when stenographer Jacquelyn Miles was transcribing the day's proceedings, the record reached its 13,993rd page.

The jurors were sent home so that the judge and the lawyers could spend the remainder of the day arguing Texaco's "motion for instructed verdict"—a routine filing midway through a case in which the defendant claims that the plaintiff has not met its burden of proof.

Pennzoil had not proved it had a contract, Texaco's lawyers asserted—and with no contract there could be no interference. Pennzoil's deal with Getty Oil was "subject to" the signature of Getty Oil Company on its memorandum of agreement; the signature had never been produced. It was "subject to" a definitive agreement, which had not been produced. It was "subject to" shareholder approval, and Pennzoil had not proved that Gordon *and* the museum, as 52% shareholders, would have voted for it.

And Pennzoil had not proved a "meeting of the minds" because it had not disproved the existence of "open issues."

"The terms of the contract claimed by Pennzoil are too indefinite, too vague, too uncertain, to result in an enforceable contract," said Bob Brown of Texaco.

To no one's surprise, the judge denied Texaco's demand to call off the trial right then; such motions are granted rarely.

What *had* Pennzoil proved?

That hands had been shaken. That a host of documents *had* been signed, even if none was a definitive agreement. That the Getty Oil board approved *something* by a fifteen-to-one vote. That the same board of Getty Oil wanted desperately to avoid giving Gordon Getty control. That Gordon felt pressured by Texaco to break his deal with Pennzoil. That Gordon and the museum had asked for indemnities against a lawsuit by Pennzoil—and refused to warrant their clear title to the shares they owned in Getty Oil. That Texaco was seriously short of oil and that its investment bankers had at least discussed a way to *stop the signing* and *take care of Liedtke*. That Pennzoil *felt* it had been damaged by $7.53 billion.

Many of the details had escaped the notice of some jurors; a few long stretches of the case had been genuine sleepers. It was a good case, yes, but hardly an overpowering one. As Jeffers would recall: "We were a little ahead when we closed out our case." But only *a little ahead*.

Jamail, however, knew that a plaintiff in a personal-injury case—or any other tort claim—often makes its best points beating up on whatever "wrongdoers" the other side dares to put on the witness stand. And Jamail felt cockier than ever as Texaco launched its defense.

At one point Jamail slipped a note across the Pennzoil counsel table to John Jeffers.

> JJ,
> I'm beginning to like this fucking case.
> —JJ

· 22 ·

"**I**t's obvious this man has come here to do a hatchet job!" Texaco had just begun its case through Bart Winokur, the Philadelphia lawyer who had succeeded Lansing Hays as the top outside attorney representing Getty Oil. In barely an hour on the stand, Winokur had begun making mincemeat of Pennzoil's case, and Joe Jamail couldn't stomach it.

The issue was whether Winokur could state a legal opinion from the witness stand, but there was much more at stake. Winokur *was* adversarial, although in an articulate and unemotional way. In long, cleverly phrased answers, he had already attributed the downfall of Getty Oil to the whimsy of Gordon Getty, had cast Pennzoil as the King Kong of the takeover game and worst of all had begun chipping away at the credibility of the evidence that formed the foundation of Pennzoil's contract claim. Winokur's very appearance as Texaco's leadoff witness had stunned Pennzoil, whose lawyers had spent nights preparing for an executive from Texaco to take the stand.

"This real volunteering witness is obviously exuding hostility," Jamail cried. "I can look at him and see it."

Miller's hair-trigger temper exploded.

"We bring a witness in here, the *first witness* who has appeared in our case, and he has to take this kind of abuse from a lawyer who knows little or nothing about this case." Miller paced angrily. "They can't stand for us to put on our evidence. They sit there wounded and bleeding. . . . It doesn't exactly deserve a medal for bravery.''

Pennzoil's lawyers, in fact, had overreacted to Winokur's testimony, for they did not realize that a greater hostility was beginning to build on the bench and in the jury box—with Winokur as the target. For despite his effectiveness in depicting the fear and pressure inflicted by Gordon and Liedtke, he did so in a way that made him seem sarcastic, distant and cold. Winokur was almost too smart, too glib, for his own good—and for Texaco's.

"We're not talking about a five-cent candy bar,'' Winokur said, explaining the company's wish to hold a "controlled auction" to escape from the Liedtke-Gordon deal. "We're talking about a ten-billion-dollar company.''

Winokur was every bit as colorful when he questioned the truthfulness of Pennzoil's official tender-offer documents, in which the company stated its intention to "participate in a constructive way" in the affairs of Getty Oil. "If you want to be constructive, you don't hold a gun to somebody's head,'' Winokur said. "I mean, if somebody came to your front door and had a paintbrush in one hand and a gun in the other and said, 'I heard your house needs painting,' you might suspect he's there for something other than painting your house.''

And the tender offer itself put such pressure on Getty Oil that it could be described only as a *bear hug*. "The term bear hug is meant to imply, unlike a normal hug from a human being you can disengage from, the bear is a lot stronger and once he hugs you, you're stuck.''

Pennzoil was drawn to Getty Oil in the first place only because Gordon had spilled so much *blood in the water*—"a situation where a target company has been wounded, in effect has bled. And as we all know in that situation, when sharks are around, as soon as there is blood in the water, they immediately close in on the victim.'' As Winokur turned the pages of his takeover lexicon, a Texaco lawyer listed the terms one by one on a big, white marking board.

There was something about Winokur that Judge Farris apparently didn't like. At one point during the day, after addressing his bailiff Carl Shaw as "Citizen Shaw,'' the judge explained, "We're trying to sound like Philadelphia.'' At another point, the judge angrily accused him in the jury's presence of trying to "slip in" a legal opinion about when a deal is a done deal—which, in fact, Winokur had. Then, at the end of the day, a fight ensued over whether Winokur was speculating about what was in the minds of

the Getty Oil board—which, in this case, he was not. The judge
sent the jury out and gave Winokur another harsh lecture.

Miller rose to his defense. "He may be at a disadvantage because he talks a little funny, if I can say so," Miller said, with
Winokur still seated in the witness chair. "But for there to be a
suggestion that the witness has slipped something in—"

"Mr. Miller." Judge Farris scowled. "Are you *quite through*
lecturing me?"

"I didn't mean to be lecturing you."

"That's what you *have* been doing, and I will have no more of
it. Understood, sir?"

Instead of backing down, Miller unwisely asked Farris to take
back his accusation that Winokur had slipped in improper testimony. Everyone in the courtroom looked on in thunderstruck
silence.

"I will *not* do it!" Farris pounded the desk. "Counsel with this
witness," he told Miller, "because I will *unload* tomorrow if it
happens again. . . . We are recessed for today"—and he stormed
from the bench.

A wonderful way for Texaco to end the first day of its case.

Despite all the advantages of going second in a trial, there is one
overriding disadvantage in a particularly long case: The jury begins losing its goodwill and patience. That message came through
loud and clear when Texaco asked the court to accommodate Bart
Winokur's wish to celebrate the Jewish High Holidays back home
in Philadelphia.

Judge Farris had been giving the jurors Friday afternoons off,
but Texaco asked for an exception so that Winokur could complete his appearance and remain in Philadelphia the following
week for Yom Kippur. "This has caused a great deal of con-
sternation among the jurors," Judge Farris answered. "In view of
the great unrest amongst the jurors, now in their twelfth week at
six dollars a day, I feel that it is unfair to ask them to give up their
Friday afternoon, which they had counted on, to allow this witness to have a better schedule."

The judge launched into a discussion of the High Holidays that
sounded like he was reading from the script of *Fiddler on the Roof*.
Winokur would be excused from testifying the following week in
time to reach Philadelphia by "sundown," when the holiday began, and then would have to return. The judge also took advantage
of the occasion to tell a Jewish joke, according to Winokur and
Miller. Pennzoil's lawyers do not recall hearing the joke, which

was never recorded in the court transcript. Miller and Winokur, however, would later recall the joke in virtually identical words. In any case, as an ethnic slur it was relatively mild—a shaggy dog story, really, with the punch line, "Funny, you don't look Jewish."

In addition to religion, pregnancy was also an issue in the scheduling of witnesses.

Pat Vlahakis, the lawyer who had helped Marty Lipton represent the museum, was newly pregnant when the trial began, and Texaco wanted desperately to call her as a live witness. But Pennzoil's case had already dragged on far longer than anyone had expected; unless Texaco could have interrupted Pennzoil's case for a Texaco witness, Vlahakis' pregnancy would have been too advanced to permit her safe travel by the time the defense had begun.

"*I* didn't get the woman pregnant," Miller had told the judge.

"He would have if he could have," Jamail replied gratuitously.

Expecting Pennzoil to interrupt its case for a Texaco witness had been asking a lot. And John Jeffers of Pennzoil hadn't been at all interested in having a pregnant witness testify for the opposition before a jury with eight women. Texaco thus had had to settle for flying to New York on a weekend to take Vlahakis' testimony in a video deposition, which Texaco played for the jury during the two days that Winokur was unavailable to testify.

Vlahakis made her message clear: As far as the museum was concerned, Pennzoil had no deal. She gave a colorful account of the chaos that reigned in the offices of Pennzoil's New York law firm the night of January 3, 1984, when dozens of lawyers and PR people struggled to agree on a two-page press release announcing Pennzoil's agreement in principle. This press release—now floating through the jury box as Plaintiff's Exhibit #5—was one of Pennzoil's strongest pieces of evidence, because it showed that all the Getty interests had signed off on everything that Pennzoil considered an "essential term" in its contract.

Vlahakis, however, discredited the document as the product of battling lawyers who had no idea what the deal was all about. She testified that Pennzoil's lawyers refused formally to approve the news release because they believed there was too much "squabbling" occurring over the terms of the deal—and because no one wished to interrupt Hugh Liedtke's celebration at "21" to obtain his approval of the press release. She recounted how she stood up to the lawyers for Pennzoil and Getty Oil when a complication arose over the museum's desire to sell its shares ahead of other shareholders, recalling her stern announcement that there was no deal as far as the museum was concerned. "I said it emphatically and I said it several times," she recalled.

It was some of the strongest testimony Texaco had yet elicited. During the taping in New York, Jamail, a light smoker, bummed a number of cigarettes from Charlie Kazlauskas, a chain-smoking Texaco lawyer, giving Kazlauskas the distinct impression that Jamail was worried.

Yet in the end, Vlahakis wound up inflicting as much damage on Texaco as support. The jury heard her recount how Marty Lipton had ordered her to stay away from a drafting session with Pennzoil's lawyers so that the museum could negotiate a higher price from Texaco.

"We were dealing," she explained matter-of-factly.

And the definition of agreement in principle that she gave the jury seemed tailored to the facts: The phrase means "that the parties are *attempting* to negotiate a transaction, and *have not* reached an agreement, and in fact *are not* bound by anything—and can *walk away* from those negotiations at any time."

In keeping with its success in turning adversity into advantage, Pennzoil also managed to extract one of its most valuable pieces of testimony from Vlahakis. When Harold Williams of the museum had signed Pennzoil's memorandum of agreement for submission to the Getty Oil board, his lawyer Lipton had handwritten some words next to the signature line, declaring that if the plan were not approved by the Getty Oil board, then ". . . the museum will not be bound in any way by this plan and will have no liability or obligation to anyone hereunder."

Those handwritten words were powerful evidence for Texaco that the museum was not part of the Pennzoil plan—until John Jeffers of Pennzoil got his chance to interrogate Vlahakis during the deposition.

"You wanted to be doubly sure" that the museum was not bound? he asked her.

"We wanted it to be absolutely, positively clear."

"No board approval, no one is bound?"

"If this is rejected, no one is bound. That's correct."

After getting Vlahakis to agree with everything he was asking, Jeffers quickly switched the issue around.

"And what's the converse of that?" he asked.

"Sorry?" she asked.

"What is the converse of that?"

Vlahakis could see that Jeffers was playing a linguistic game and tried to extricate herself. But it was too late.

"If it's not rejected, then people are bound," she acknowledged.

"If it's approved, people *are* bound?"

"That's correct."

Jeffers had elicited not one new fact, but his questioning had left a deep impression on the jury: If that fifteen-to-one vote *was* intended to approve all essential terms of the Pennzoil deal by the board of Getty Oil, then the museum—as well as Gordon and Getty Oil itself—*were* bound. All that mattered now was what the Getty Oil board had really voted for on that night in New York.

Bart Winokur spent his final hours of friendly interrogation accusing Baker & Botts of permitting Pennzoil's deal to fall through. He laid out the strange story of the messenger who supposedly got lost delivering Pennzoil's draft of the final merger documents. In Winokur's mind there was no doubt where the fault rested for the absence of a final merger document: with the corporate lawyers from the esteemed firm of Baker & Botts.

Winokur also laid out all the open points that Pennzoil's lawyers had failed to resolve. On the matter of the ERC "stub"—the kicker intended to add $2.50 in today's dollars to Pennzoil's $110 cash price—Winokur identified six details potentially worth hundreds of millions of dollars that had remained unclear. Treatment of employee benefits, the method of purchasing the museum shares, the terms of Pennzoil's "lockup" option—all were left open in the documents drafted by the lawyers from Baker & Botts, Winokur said.

Nowhere in the courtroom did the testimony leave a stronger impression than at the Pennzoil counsel table, where Irv Terrell of Baker & Botts had fire in his eyes.

Before releasing the witness to Terrell, one of Miller's partners, J. C. Nickens, launched into a flourishing finish for his direct examination.

"Mr. Winokur, in your role as a negotiator and advisor to the company, did the Getty Oil Company have an agreement with Pennzoil on January third?"

"No."

"Did it have an agreement with Pennzoil on January fourth?"

"Absolutely not."

"Did it at *any time* have an agreement with Pennzoil?"

"No."

Nickens moved to undo some testimony by Liedtke that under

the "custom" of the oil patch, there was no doubt Pennzoil had a done deal.

"How many mergers or acquisition kinds of transactions have you been involved in in your career?"

"Hundreds."

"Are you familiar with *any* custom and practice, on the basis of that experience, with regard to whether such transactions are reflected in writing?"

"Yes."

"What is that custom and practice?"

"They are *always* reflected in writing."

It was an exceedingly strong finish.

"We pass the witness, Your Honor," Nickens said.

Irv Terrell rose from the Pennzoil table knowing that he had plenty of specific facts to attack, but there was one overriding issue in his mind. As he would later explain it, "We had to make the jury dislike him." Only a few minutes remained until the end of the day, so Terrell engaged in only some light jousting before trying to end the day on a high note.

"Mr. Winokur, we're going to break in a moment, and I want you to know what we're going to talk about first thing in the morning so that there won't be any surprises. We're going to talk about the November 11, 1983, Getty Oil board meeting. You remember that, don't you?"

Winokur, a corporate lawyer, committed the unpardonable sin of throwing a hanging curveball to a trial lawyer.

"I remember the part that *I* was at, yes."

Terrell swung for the home-run fence.

"Yes!" he cried, spinning on his heels. "*You* were only there for part of it, weren't you?"

"That's correct."

"*You* were only there for the part that Gordon Getty was not there for, isn't that right?"

Nickens leaped from the Texaco counsel table to object.

"It's immaterial and irrelevant to any issue that will be before this jury," he asserted.

Terrell shot back: "This man came in and did something I think this jury needs to know about. . . . This man's credibility—he's sworn here, he's the lead witness for Texaco. *They* are vouching for his credibility, and I think the jury needs to know the facts."

Judge Farris dismissed the jury for the day and instructed Bailiff Carl Shaw to close the door behind them.

"Mr. Terrell?"

"Yes, sir?"

"I don't want you—or indeed any lawyer on either side—to tell the jury what a bad guy somebody is. Just go ahead and question him." The judge turned to Winokur. "If they think you're a bad guy after they hear the answers, so be it." And then to Terrell: "But don't again start slicing on him."

"I won't, Your Honor."

Terrell returned to Baker & Botts and began sharpening his knives.

"Did anybody communicate with Tara Getty, the fifteen-year-old that filed this lawsuit?"

"Not that I'm aware of."

Inexorably, almost every trail down which Terrell pushed Winokur led back to the same place: the boardroom at the Getty Oil research center in Houston, where in the midst of ratifying a truce with Gordon, three of Petersen's people—Winokur, special counsel Herb Galant and investment banker Geoff Boisi—had entered to discuss the company's joining in the lawsuit that the company had helped to arouse in the name of Tara G. G. G. Getty. In the shorthand that Pennzoil had adopted for the trial it continued to be known as the back-door board meeting, even though there was really no back-door maneuver at all.

Certainly, the back-door meeting was relevant to Pennzoil's contract claim. It helped establish why Getty Oil was "in play," it bolstered the theory that Hugh Liedtke had ridden to the rescue of Gordon and it depicted some of the people who "reneged" on Pennzoil's contract as two-timing double-crossers.

But what was the relevance of the back-door board meeting to Pennzoil's case against *Texaco*? On the surface, none. Below the surface, plenty. Jamail and his co-counsel had talked about indemnities at every opportunity. If somehow the jury were convinced that Texaco was accountable for everything that Getty Oil had done, perhaps the jury just might take that into account in awarding punitive damages.

Terrell showed Winokur a copy of Plaintiff's Exhibit #60—the handwritten notes recorded by Stedman Garber, the Getty Oil treasurer, while he was receiving a report by telephone on what happened at the Getty Oil board meeting.

"How do you account, Mr. Winokur, for this note that says, 'Board agreed, but for Chauncey, that deal should be done'?"

Winokur explained that after approving Pennzoil's $112.50 price, the board only wanted to *try* to do a deal.

"In good faith?" Terrell asked.

Winokur was silent for a second.

"*In good faith*, Mr. Winokur. . . . Were you trying to do that in good faith?"

"I always try to do everything I try to do in good faith."

"You try to do everything you do openly and honestly, Mr. Winokur?"

"As openly and honestly as I can."

"That's what you did on November eleventh with Gordon Getty, isn't it? Openly and honestly?"

"I think we acted in that situation perfectly properly."

Terrell zeroed in on the departure of the Getty Oil lawyers from their drafting session with Pennzoil the night that Texaco began trying to cut a deal of its own. Did Getty Oil's people deal openly and honestly with Pennzoil? Terrell demanded.

"Depends on what you mean by 'openly and honestly. . . .'"

"Did you do that, Mr. Winokur, on January fourth and fifth? Did you? Yes or no?"

"Did I do what?"

Neither Winokur nor Texaco realized that behind the expressionless faces in the jury box, the annoyance was building into rage. Rick Lawler, for one, wished he could reach into the witness stand and shake a straight answer from Winokur. *Quit playing these semantical games!* Lawler kept screaming inside. *Tell it like it was!* "I wasn't there to waste my time watching a joust between two lawyers," Lawler would later say.

Unwilling to concede the slightest ground to Terrell, Winokur grew evasive even on the most direct questions. When Terrell asked whether Gordon was adequately represented by investment banker Marty Siegel, Winokur replied, "It depends on what you mean by 'adequately represented.'" When Terrell asked whether all the parties had agreed to the press release announcing the Pennzoil deal, Winokur replied, "Well, it depends on what you mean by 'agreed to.'"

"That arrogant ass sat there and stonewalled like H. R. Haldeman," Juror Jim Shannon would recall.

But the questions intended to leave the deepest impression on the jury weren't really questions at all, but little jury speeches that Terrell wove into his interrogation. One such speech occurred after Winokur asserted that he and Sid Petersen were concerned that Gordon and Liedtke could somehow rip off the public shareholders in their plan to take control of Getty Oil for themselves.

"Mr. Winokur," Terrell shot back, "Mr. Petersen didn't care

about the minority shareholders. He just wanted to make sure Gordon Getty didn't come out on top! This was just like the November eleventh board meeting all over again. People were running around and doing things behind Mr. Getty's back.

"You wanted to *obstruct* the deal, didn't you?"

"I object to the question as being argumentative, Your Honor!" Nickens of Texaco interjected.

"That's sustained."

But the objection didn't matter, for Terrell had planted in the jury box a suggestion that Getty Oil walked away from the Pennzoil deal solely to spite Gordon Getty. Pennzoil's lawyer would see to nurturing that idea into a fully grown belief that would bear fruit once the jury began deliberations.

Joe Jamail, an incurable note-taker, had been scribbling like mad through much of Winokur's cross-examination. At one point a cute little moniker occurred to him, which he jotted down for future use.

The note read, "Back-Door Bart."

Every trial lawyer has three audiences: the jury, the judge and the record. Merely to convince the first two is insufficient; each side must see to it that its case is properly "preserved" in the official transcript. Announcing "I object" won't do; the lawyer must state specific reasons for the objection, and often cite a specific rule or case to defend it. If the judge refuses to admit certain evidence, the lawyers must see that it somehow reaches the record anyway, even if outside the jury's presence. Appeals courts do not look generously on lawyers who cry, "Here's something we forgot to get in the record."

Among the half-dozen trial lawyers on Texaco's side, no one concentrated solely on such matters of procedure and law. Pennzoil, however, had W. James Kronzer.

Among the diverse legal personalities gathered around the Pennzoil counsel table, Kronzer was surely as unusual as any other. He was a rather ungainly figure—a tall, portly and slightly rumpled man of middle age who spoke in a perpetual mumble. He was a loner in the legal community, operating a one-man law firm in an out-of-the-way town called Bay City. A painfully proper air hung around Kronzer, who was the one fellow most apt to wear a white shirt and tie when he went out drinking with his lawyer buddies on the weekend.

But for all his unlikely qualities, Kronzer had over the years become one of the true giants of the Texas bar, an expert in

appellate procedure of such reputation that Judge Farris practically
genuflected when Kronzer walked into the courtroom for a pretrial
hearing and took a neutral seat in the audience. "I have taken note
that a very prominent member of the bar, particularly in the appellate area, is present," the judge had said. "Mr. Kronzer, are you
for or against any of these lawyers?"

"We're sort of in a lull down in Bay City, Your Honor,"
Kronzer demurred. "I thought I would come in and observe an
interesting evening and occasionally hold Mr. Jamail's hand."

Kronzer was the high priest of Texas civil procedure, a walking
encyclopedia who could step into a heated debate before the bench
and cite as many cases from memory as it took for Pennzoil to win
the point—and when that proved insufficient he could cite the
legislative history of a Texas law and the views of some professor
who might at one time have published a comment on the subject.
He was also the most erudite member of the Pennzoil team, mus-
tering just the right pun, joke or literary allusion to make a complex point seem simple and clear.

Kronzer also liked a friendly little bet.

One of the great mysteries shrouding Texaco's case was
whether Dick Miller would put the vaunted Marty Lipton on the
witness stand. Some of the Pennzoil lawyers were convinced that
he would: Lipton was too important a player; the jury would think
Texaco was hiding him. Yet Pennzoil's lawyers were also convinced that Lipton would be as poor a witness as Texaco could
present. For one, in Pennzoil's view, Lipton had a lot of double-
dealing to account for. For another, the lawyers thought, Lipton
would have to undergo a wholesale character transformation in
order to get this jury to like him. Lipton was too arrogant and
hifalutin'—he was too *New York*.

Pennzoil's lawyers wanted so desperately to see Lipton on the
stand that they decided against taking his deposition; that way,
Texaco could not get away with playing a video or reading a
transcript of his testimony aloud. Pennzoil wanted Lipton to appear out of his element.

Kronzer offered to wager $5000 that Miller would *not* put Lip-
ton on the stand. Terrell, although convinced that Lipton would
appear, did not share Kronzer's taste for such high stakes and
agreed to take the bet for $500. Jamail sided with Terrell and
demanded $250 of the action.

"They *have* to bring Lipton," Terrell said.

* * *

One day, shortly into Texaco's case, Judge Farris wearily climbed the bench and horrified the courtroom. His lips were blue.

Almost from the beginning of Pennzoil's case, the jurors and lawyers had watched the judge's health deteriorate a little each day. But throughout October, after Texaco's case was finally under way, he was regularly absenting himself from the bench to take calls from his physician—and sometimes knocking off for entire days.

Dick Miller swore the judge had lost twenty-five pounds. On occasion, he needed assistance getting into his chair. Every heart in the courtroom went out to him.

When the testimony had begun, Miller was greatly encouraged by the judge's readiness to sustain his objections. But as the case wore on and the judge's condition declined, the rulings from the bench seemed to fall more and more often in Pennzoil's favor, just as they had during the critical pretrial phase of the case.

"At the beginning of the trial I think he resolved to try to do the case in an evenhanded way," Miller would later say. "As the case wore on, the judge was simply unable to maintain what for him was an artificial position."

For all of his anger at the judge's rulings, Miller would also recall, "I admired his bravery." And Miller continued to feel strongly that whatever adverse rulings were thrown down from the bench, he could bring that jury to his side. *I can win this case,* Miller thought.

"What in the world are you doin' in Houston, Texas?"

It was the first witness that Miller himself was handling in Texaco's case, and he wanted a friendly, free and easygoing examination. But Geoff Boisi of Goldman, Sachs, who had played a critical role as Getty Oil's investment banker, did not pick up his cue.

"I'm testifying before this court," he replied humorlessly.

Miller continued trying to make his witness appear folksy— with the unintended effect of making him seem curious. Miller introduced the personal lawyer who had accompanied Boisi to Houston as the man sitting in the back who was "looking real serious." Miller also pointed out that Boisi was the first investment banker to testify in the case. "You are the only investment banker in this room," Miller told him. "Don't use any code words on us."

The trial had entered its 100th day, and Miller, sensing the jury's

growing impatience, was determined to keep the examination brief
and to the point, which it was. In parts, it was also extremely
effective. Boisi recalled his unwillingness to declare the Pennzoil
deal "fair" to the public, an opinion based on nearly a year of
experience in analyzing the value of Getty Oil. He recounted the
intensity of the pressure imposed on the Getty Oil board by the
combination between Pennzoil and Gordon, and how their proposal
to take Getty Oil private had been structured in a way that was
almost impossible to defeat. He depicted a board gripped with
anxiety, wishing solely to get a deal that would be fair to the public.
And it was to that end, Boisi said, that he began placing a series of
phone calls to other oil companies, including Texaco.

And at no time did Pennzoil have a deal, he said. The Getty Oil
board voted fifteen-to-one solely to deal with Pennzoil at a price of
$112.50—every other major aspect of the deal remained unre-
solved. An extensive discussion of these "open issues" did, in
fact, occur during the board meeting, Boisi said, even though the
discussion wasn't reflected in the sixty-eight page notes of the
meeting; the corporate secretary, Boisi explained, happened to be
seated at the opposite end of the long conference table where the
open issues were being discussed.

Miller concluded his three-hour examination by asking how the
directors felt when they learned that Gordon had entered into the
"secret agreement" with Pennzoil known as the Dear Hugh letter.

"They felt under even more pressure," Boisi said. "This was
the first time true acknowledgment was made that they would be
thrown out of their positions if they did not knuckle under to the
proposal.

"There was a combination of outrage and a feeling of terrific
pressure."

"No further questions," Miller said.

Again, it was Irv Terrell handling the cross for Pennzoil, and
the Dear Hugh letter was as good an issue as any with which to
begin attacking Boisi.

"Are you telling the jury," Terrell asked, "that fifteen members of the board of directors—when they voted fifteen-to-one on
the evening of January third—did not have the courage to stand up
and say, 'No, you bullies, we won't do this.' That only Chauncey
Medberry could muster the courage to ward off Pennzoil?"

"I am not saying that."

"Okay."

Terrell had just squeezed some of the life from Texaco's claim that Pennzoil and Gordon had pressured Getty Oil into seeking another buyer for itself.

Boisi had experience testifying in court cases, but not before a jury that had come to relish hostile exchanges as a source of excitement in a long, tedious trial. Jurors would later recall feeling resentment that Boisi, seated only a few feet from the jury box, refused to make eye contact with any of them. His height forced him to hunch over the microphone, which made him appear all the more uncomfortable, and Dick Miller had to admonish him to keep his voice up. Boisi's habit of fluttering his eyelids deepened the perception that he was arrogant. And he quickly fell into the same kind of evasiveness that killed Winokur with the jury.

Boisi had also never faced a trial lawyer quite like Terrell, who would turn his back to Boisi, heave a sigh, stare at the ceiling, put his head on the table and jangle his pocket change whenever he felt it necessary to cast doubt on Boisi's responses.

"You remember the November 11, 1983, board meeting in Houston, don't you?" Boisi had been part of the trio that had entered and left in Gordon's absence.

"I remember the November meeting; I don't remember the day," Boisi coolly replied.

"And you remember it was in Houston?"

"I remember that there was a Houston board meeting."

These were not the kind of answers a jury in its 100th day of trial wanted to hear. By now some of the jurors were openly rolling their eyes and muttering to one another. *Why do we have to put up with this?*

"Did you feel that was an honest thing to do?"

"I guess I do."

At another point Boisi testified that he had only "skimmed" the Pennzoil memorandum of agreement when it was distributed to the Getty Oil board.

"When was the last time you read this memorandum of agreement, Mr. Boisi?" Terrell then asked.

"I would say when I skimmed it at the board meeting." Texaco's second live witness had never once read through a document that the laymen of the jury had already been through a dozen times or more.

Terrell did attempt to extract facts from Boisi—some of them among the most important that Pennzoil had yet received. Boisi acknowledged that under the ill-fated Pennzoil deal, Goldman's fee would have been about $9 million; under the Texaco deal, he said, Goldman received more than $18 million—ample motive in the jury's eyes for Boisi's involvement in killing the Pennzoil deal. When Terrell asked Boisi what the Getty Oil board approved fifteen-to-one, Boisi responded "primarily the price"—the first evidence from anyone on Texaco's side that it was not *solely* the price that the board had approved.

And then there was the matter of the "courtesy call"—the call that added a new and unexpected brick to the foundation of Pennzoil's "tortious interference" case.

Everything presented from Texaco's side of the courtroom to that point suggested that Getty Oil continually *invited* Texaco into the bidding for Getty Oil. But Boisi said that when he spoke to Texaco president Al DeCrane the morning after Pennzoil had secured its agreement in principle, he, Boisi, did *not* solicit a bid. Boisi's call the morning of January 4, 1984—the day after the Getty Oil board meeting in New York— was made "out of courtesy," in thanks for Texaco's earlier expression of interest. Even Terrell was taken by surprise—but he immediately grasped the value of the testimony.

"So your evidence is that when you talked to Mr. DeCrane on the fourth, you did *not* solicit an offer? It was merely a *courtesy call*?"

Boisi gave a long-winded explanation.

"Was the answer to my question yes or no?" Terrell demanded. "Did you solicit him on the morning of the fourth?"

"Well . . ."

"Yes or no?"

"I would say no," Boisi finally answered.

The grumpy Judge Farris, although continually scolding Terrell for asking too many long and complex questions, also did Terrell the favor of unloading on Boisi just as he had on Winokur. "Mr. Witness," the judge barked, with the jury absent, "I want you to always, always, always, always, always, always, always, in a yes-or-no question, answer yes or no if you understand it. *Then* go ahead and give an explanation."

This lecture proved valuable to Terrell shortly after testimony resumed in front of the jury.

"Did the board vote at any time on January second and third to *authorize* you to go out and sell the company?"

It was a question that Boisi could easily have spent all afternoon answering—describing the months over which the board had resigned itself to selling the company, the obsession of the board with obtaining the best price possible, the commitment of the board chairman himself in the effort to a bidder at a better price—but there was no use in even trying to explain.

"No," Boisi replied. "Not in a formal sense." Suddenly it appeared to the jury that Sid Petersen's investment banker had cut himself loose from a board that had approved a deal with Pennzoil. Through "courtesy calls" or otherwise, Geoff Boisi had succeeded in doubling his $9 million fee by paving the way for an interloper like Texaco.

After his day and a half of cross-examination, Boisi returned to New York and had a conversation with his pal Marty Lipton, who was one of sixty-six people identified by Texaco as a potential witness. Boisi asked Lipton what arrangements he had made for his personal security if he were called down to Texas to testify.

"He had never felt so threatened," Lipton would recall.

On October 15 the jury got its first look at someone who had actually voted at the Getty Oil board meeting in New York. He was Henry Wendt, the president and chief executive officer of SmithKline Beckman.

Wendt created an enormously powerful presence resolutely striding to the witness stand. He was tall, trim and patrician—impeccably dressed, with a strong jaw and just the right wave in his hair. "If you called central casting and said, 'Send me a CEO,' they would not send Gregory Peck," Charlie Kazlauskas, the Texaco in-house lawyer attending the trial, said. "They would send Henry Wendt."

"Does SmithKline Beckman manufacture any products that the folks on this jury might recognize?" asked Bob Brown of Texaco, who handled the direct examination. Wendt ticked them off: Contac, Sine-Off, Allergan and others. Giving rapid-fire answers, Wendt looked the jurors straight in the eyes and spoke with such power that Brown asked him to back away from the microphone.

Brown came right to the point, showing Wendt a copy of Pennzoil's memorandum of agreement. "I explicitly voted *against* the approval of that document," he said. At Pennzoil's initial price of $110 a share, "the whole thing, I felt, was highly suspect."

Question: How about the Dear Hugh letter?

"I saw it as an act [by Gordon Getty] to sell the other shareholders down the river."

Question: When you learned midway through the board meeting that Goldman was shopping the company, what did you think?

"I was all for it."

Question: The meeting notes say the board voted fifteen-to-one after Harold Williams "moved that the board accept the Pennzoil proposal"; what exactly *were* you voting on then?

"One hundred and ten dollars a share" and the ERC stub worth $2.50 in present value. "That was it."

Question: You mean you were not voting on *other* terms, as they were enumerated in the original memorandum of agreement?

"Definitely not. The memorandum of agreement had been voted down and was never raised again."

Question: What did you think when the meeting was over?

That the management and advisors of Getty Oil and Pennzoil would meet and "pursue this price proposal and other elements that would lead to a definitive agreement."

Question: Why did you urge the board to award Sid Petersen a new and improved golden parachute?

"It was quite clear that the company was going to be sold," Wendt said, quickly adding: "To Pennzoil or someone else."

Question: After that meeting, did you shake hands or offer congratulations to Arthur Liman of Pennzoil?

"No."

Were you in a congratulatory mood?

"No, I was not."

Question: What did you think when you were informed of the Texaco proposal?

"It's the very result I hoped to achieve."

"We pass the witness," Brown said.

It had lasted about two hours. It was thoroughly devastating testimony to Pennzoil—and all of it had registered in the jury box. *This ballgame might as well be over,* thought James Kinnear, the Texaco executive who was observing the trial.

Judge Farris knocked off early to keep a doctor's appointment.

"The most flagrant grab of court authority I ever witnessed!"

Jamail was furious. Henry Wendt had flown back to Philadelphia before Pennzoil ever had a chance to cross-examine him. Moreover, Texaco was saying that another obligation would prevent his return for two weeks. "I move and insist that we recess until tomorrow morning and they *bring him back* and put him back on the stand so we can cross-examine this man!"

Miller tried to do some explaining to Jamail.

"I don't respond to you," Jamail snapped at Miller.

A communication breakdown had occurred; Texaco thought

Pennzoil had approved Wendt's lengthy absence, and Pennzoil thought it was only consenting to a brief departure.

"Every single witness we have got is from fifteen hundred miles away," Miller said. "For us to have to put up with this kind of baloney from Mr. Jamail is ridiculous. . . . Every time they get in a tight spot it's the same thing."

Judge Farris, feeling grouchier than ever, did Jamail one better. "Have him here Monday morning at nine o'clock or I will strike his testimony," he commanded Miller. "We will continue this morning with whatever witnesses you have."

The lawyers filed out of judge's chambers and into the courtroom. And suddenly, Jim Kronzer, Pennzoil's appellate expert, owed Jamail and Terrell each $250.

Richard Keeton of the Texaco counsel table was standing before the bench. "This is Mr. Martin Lipton of New York, Your Honor."

Seldom do lawyers experience such anguish over whether to put an important participant on the stand. The decision usually is cut and dried: If he will play poorly, don't; otherwise, yes.

The debate over whether to put Lipton on the stand reached into the highest executive levels of Texaco headquarters in White Plains, New York. "Everyone was torn," in-house counsel Charlie Kazlauskas says today.

On the one hand, every single witness, testifying either live or by deposition, had emphasized Marty Lipton's role as a go-between—first in bringing the Pennzoil deal before the Getty Oil board and then in helping the Texaco deal to fruition. Lipton was in the best position to rebut the testimony of Arthur Liman, who had given such convincing accounts of all the offers and counteroffers, the supposed handshakes and congratulations, involved in the Pennzoil deal. Moreover, there was no disputing that Lipton was a lawyer of singular brilliance who might be just the man to clinch Texaco's case. Finally, to those who knew him and had seen him in a relaxed atmosphere, Lipton, in fact, came off as a pretty regular guy. He had the large figure and—without his heavy-duty eyeglass lenses—the friendly face of the fatherly television actor Tom Bosley; his slightly nasal voice even made him sound like Bosley.

Among those arguing in favor of Lipton's testimony was Miller's highest-ranking partner, Richard Keeton, who before going into business with Miller had handled a number of cases on referral from Lipton.

Dick Miller was never absolutely committed against calling Lipton, according to other Texaco lawyers, but he recognized the risks of doing so more than the rest. As he would later explain, "He's not exactly from Fort Worth. . . . Marty is a boy who grew up in Jersey City." The long-play duration of the trial was itself presumption against calling *any* witness who wasn't absolutely necessary. Moreover, Lipton was a lawyer whose time no one was anxious to waste; at the moment he was up to his ears in representing his friend Laurence Tisch, who was undertaking a major investment in CBS, and was representing Beatrice Foods in another major takeover story playing out on the business pages.

But above all, the question facing Texaco was whether "in the ambience of that trial, was Marty going to play bad?" recalls William Weitzel, the general counsel of Texaco. "East Coast" was going over poorly in this Gulf Coast courtroom, and if anyone personified the East, it was the Jersey-born Lipton.

The pro-Lipton forces had carried the day, and Marty Lipton took the bench wearing a double-breasted suit coat with a bright red handkerchief flaming from the pocket.

"Mr. Lipton, this jury has heard your name quite a bit so far in these many weeks," said Keeton. "I would like you to please give the jury a little bit of background concerning yourself."

Everything about Lipton's credentials bespoke either takeovers or *New York:* personal experience in more than one thousand mergers and acquisitions . . . author of the definitive treatise on mergers (He did not give the title: *Takeovers and Freezeouts*) . . . special counsel for the City of New York in the effort to avert the city's bankruptcy in the mid-1970s . . . trustee of New York University . . . trustee of Manhattan Eye, Ear and Throat Hospital.

"Are you a stranger to Houston, Texas?"

"No," Lipton replied. "I have spent quite a bit of time in Houston." But ticking off the variety of Houston energy companies he had represented only served to strengthen the impression that the M&A professionals in New York were the ones responsible for the adverse local impact of takeovers.

Keeton's examination was short, to the point and underscored each of the main items in Texaco's case. Lipton told the jury how Gordon's alliance with Liedtke had defeated the plan to put Getty Oil up for a "controlled auction," how Pennzoil had importuned the museum to join with Gordon and throw out the board if it didn't go along with the joint-ownership plan, how the museum joined in the memorandum of agreement only to have it "considered" by the Getty Oil board. He talked about the slow speed of Pennzoil's lawyers in drafting the final documents, the ambiguity surrounding who would purchase the museum shares, the understanding that the final deal with Pennzoil required further approval by the Getty Oil board.

But Lipton's testimony was wholly dispassionate and colorless. He had to be repeatedly admonished to "speak up," and his recollection of the critical events seemed so superficial that he sometimes appeared to surprise even the lawyer who had put him on the stand. Keeton showed Lipton the critical sentence in the notes of the Getty Oil board meeting which stated that Lipton's client, Harold Williams, had "moved acceptance of the Pennzoil proposal." Lipton testified it was his understanding that Williams was only making a motion to *negotiate* with Pennzoil—not to complete the deal.

"Is that what Mr. Williams stated to the board?" Keeton asked expectantly.

"I don't remember what Mr. Williams stated to the board," Lipton replied matter-of-factly.

On occasion Lipton even appeared argumentative with the Texaco lawyer. At one point, for instance, Keeton admonished him against making any legal comments. "I don't think I was about to do that," Lipton retorted.

In a few cases, Lipton's testimony against Pennzoil only backfired on Texaco. For nearly four months Texaco had been trying to convince the jury that Pennzoil and Gordon were trying to rip off the public shareholders with their proposed buyout price of $112.50 a share. Lipton, however, recalled a meeting with Getty Oil's investment banker at which Lipton "argued very strongly" for Geoff Boisi of Goldman, Sachs to bless Pennzoil's $112.50 price with a fairness opinion—testimony that only helped to convince the jury that Pennzoil's price was a fair one after all.

Lipton also recounted how he had assured Texaco's executives that Pennzoil never had a deal. But in doing so he also recalled how the Texaco people "questioned me closely as to whether I believed that the museum was in a legal position to contract with Texaco—whether they were breaching, or inducing the breach of contract." Although Lipton was trying to dramatize Texaco's diligence in establishing that the museum was free to deal, he had placed the words "inducing the breach of contract" in the mouths of Texaco's executives.

Jamail had been sitting back silently throughout the direct testimony. But he perked up when Keeton raised the sensitive subject of Lipton's instruction that his associate Pat Vlahakis should not attend a drafting session with Pennzoil so that she could assist in the talks with Texaco.

"Did you prevent her from going to the meeting, or ask her not to go to the meeting in order to in some way delay or frustrate the negotiations with Pennzoil?"

"No."

Jamail rose to object. "He told the lady not to go to the meeting," Jamail said. "To explain some alibi or to make some self-serving statement—I object!"

Keeton's line of questioning had been entirely proper, and Judge Farris refused to sustain the objection. "You will have the opportunity on cross-examination to correct that, if you feel it needs correcting," the judge said.

"And I will!" Jamail answered. With that little interruption Jamail had skillfully sent a message to the jury: *Don't believe what Lipton tells you until I'm through with him.*

Like Texaco's other witnesses, Lipton defined agreement in principle in the negative: "It means an actual agreement has not been reached . . ." he said. And in his experience, he said, "there's *always* been a definitive, executed agreement in every transaction I've been involved in that was consummated."

Hugh Liedtke's ire toward Marty Lipton had been festering ever since Pennzoil's ill-fated days in New York. In his deposition Liedtke derisively referred to the New Year's Day gathering at Lipton's apartment as Lipton's "floating cocktail party." Liedtke had also accused Lipton and his "flacks" of planting news articles suggesting that Baker & Botts had botched the Pennzoil deal.

Liedtke's full fury was unleashed through the person of his friend Joe Jamail, who rose from the Pennzoil table for the first cross-examination he would conduct in the case.

"Mr. Lipton," Jamail said, pacing slowly along the jury box, "you seem to be a man who uses his words precisely. Do you?"

"I try to."

"Did I understand you earlier to testify that not *one* of your clients ever intended to be bound [prior to] signing a definitive merger agreement?"

"That's correct."

"And you meant it when you said it?"

"Yes."

Jamail hunched over the Pennzoil table and stood up with a three-page document in his hand.

"Were you involved in the acquisition by Esmark of Norton Simon?" Jamail asked, producing a preliminary acquisition agreement valued at nearly a billion dollars between the companies.

"Although this was not a definitive merger agreement, this *was* a binding agreement. Isn't that correct?"

"Yes, sir."

"Sir?"

"Yes, sir."

"All right, sir. Now, let's talk for a minute about Mr. Siegel," Jamail said.

Jamail was now striking even closer to home. Gordon's investment banker Siegel was "the other Marty," a member of Lipton's "Royal Order of the Golden Joystick," practically a co-inventor of the "poison pill"—a friend of Lipton.

"Would you say Mr. Siegel is a man who uses words precisely, correctly?"

"Yes."

"And Mr. Siegel is a knowledgeable man in the mergers-and-acquisitions trade?"

"Yes."

"And a man who would not use words foolishly—if under oath, especially?"

"That's correct."

Jamail handed Lipton a copy of Plaintiff's Exhibit #19—the California court affidavit that Siegel filed after the Getty Oil board had voted on the Pennzoil deal.

"Would you read it aloud, please?" And the jury heard once again about how the Getty Oil board had approved the Pennzoil deal—only this time from the mouth of Marty Lipton.

"Mr. Lipton," Jamail said, "he used the words, 'transaction now agreed upon,' didn't he?"

"Yes, he did."

"Well, do you believe he knew what he was saying?"

"No."

"You just disagree, is that it?"

"I disagree."

"You've never been shown this, have you?"

"I have not."

Jamail had had only a few minutes to begin his examination, and already it was the end of the day.

"Tell him to get rid of that fucking handkerchief," Dick Miller told his co-counsel, Richard Keeton.

It went on and on: the King of Torts against the Big Deal Lawyer, a man in boots against a man in pin-stripes, the UT lawyer against the NYU lawyer—and the jury hung on absolutely every word. Jamail sought first to wipe out Texaco's theory that the Getty Oil board had been pressured by Pennzoil and Gordon into its fifteen-to-one vote.

"You're paid large amounts of money to go and advise these people?"

"That's correct."

"To make sure they don't act unreasonably under pressure, right?"

"That's correct."

"To make sure they don't succumb to pressure?"

"That's correct."

"You felt terribly pressured?"

"Yes, I did."

"*You* didn't crater, did you?"

"I hope not. . . ."

"Are you here telling this jury that you, who pride yourself as being the leading takeover lawyer in the country, could not prevent that board from just *cratering* because of this 'assault' by Pennzoil? Are you telling us that these sixteen board members with their lawyers, their advisors—all sophisticated men, educated people who have pressures every day—succumbed to Pennzoil's pressure and voted fifteen-to-one on the counterproposal that they themselves made?"

"No, I don't think the pressure by Pennzoil motivated them. I think what—"

"That answered the question."

When Lipton tried to explain his answer, the explanation was lost in a tirade of objections and harsh admonitions from Judge Farris to "be more succinct."

"Mr. Lipton, I want to be real sure that I understand this."

Any trial lawyer worth his fee is spontaneous, extemporaneous—willing to abandon even the most carefully planned line of questioning to seize a small piece of testimony and enlarge it into something the jury will not forget. Just such an opportunity arose when Lipton testified that in his opinion Gordon could never have completed the Pennzoil deal without hiring lawyers who specialized in complex oil-and-gas transactions.

"Are you saying that two people cannot agree unless they hire a bunch of lawyers to *tell* them they've agreed?" Jamail demanded.

"I'm not saying that at all, Mr. Jamail. I'm saying that two people who are contemplating an agreement with respect to a ten-billion-dollar transaction would be awfully foolish to do it on the basis of an outline and the absence of an expert's advice. . . ."

Jamail knew what he wanted.

"Mr. Lipton," he said, glaring, "are you saying that you have some distinction between just us ordinary people making contracts with each other, and whether or not it's a ten-billion-dollar deal? Is there a different standard in your mind?"

"Yes, indeed."

"At that point," juror Jim Shannon would recall, "my jaw just dropped."

Jamail waited a full five seconds to let the response sink in.

"Oh," Jamail said. "I see."

"So if it wasn't a bunch of money involved in this Getty-Pennzoil thing, it could be an agreement?"

"Well, if there was five or ten dollars involved, I guess you might say that."

In fact, Lipton was articulating a legal principle codified in 17th-century England and known as the Statute of Frauds, which held that complex transactions of great size or complexity *do* impose a higher requirement of agreement. But as far as the jury was concerned, Marty Lipton had just made honor in business contingent on the number of dollars involved.

As for the memorandum of agreement:

"You meant it when you had Mr. Williams sign it, didn't you?"

"He certainly meant to sign it, yes."

"Well, did he mean to honor his signature?"

"Well, I can't answer that question because I don't know what you're referring to. I'm sure Mr. Williams always honors his signature."

"That's *exactly* what I'm referring to."

Jamail was questioning Lipton about the role of Pat Vlahakis in the controversy over the purchase of the museum shares.

"This escrow agreement prepared by your partner would insure that you had some advantage over the other shareholders, financial advantage?''

''You said Ms. Vlahakis was my partner. She's my *associate*.''

Lipton, who would later explain that he was only trying to leave an accurate record, immediately saw Jamail's eyes light up, and recognized that he had committed a terrible blunder. Lipton had slighted not only someone he worked with, but a young woman— a pregnant woman at that.

In the jury box, Lipton's cavalier reply put everything in perspective. ''When I heard that, I realized that Marty Lipton had just reinstated the caste system,'' Jim Shannon would say.

There was even a shudder at the Texaco counsel table. ''There was no excuse for that,'' one of the lawyers would recall.

Back to Harold Williams' motion to ''accept the Pennzoil proposal,'' as Dave Copley, the secretary of the meeting, reported in his notes.

''I told them I wanted it clearly understood that the museum would be a signatory to such a plan only for the purpose of having it approved—I mean—I'm sorry. I said 'approved.' Only for the purpose of having it *considered*. . . . This whole plan was not something that the museum wanted at all.''

Jamail was incredulous at the suggestion that Lipton's client would move adoption of a plan he did not support.

''Well, then why—who—did somebody have a gun at Mr. Williams' head?''

''No.''

''Did somebody *force* him to make the motion to adopt the Pennzoil proposal?''

''No.''

''He did it, didn't he?''

''No.''

''He *didn't* 'move to adopt the Pennzoil proposal'?''

''What Mr. Williams did was move to approve going forward with discussions and negotiations with Pennzoil. . . .''

''Was Mr. Copley *sober* when he was taking these notes?''

''I don't know, sir. I don't know.''

Mercifully for Texaco, the end of the day finally rescued Marty Lipton from the stand—at least until tomorrow.

To juror Rick Lawler, Lipton had already come across as a brilliant lawyer who fell apart when he was forced to deal with the real world—''a genius who can't tie his shoes.''

"Texaco," Lawler would say later, "should have left Marty Lipton in New York."

Throughout the morning that Marty Lipton was on the stand, Judge Farris was gripped with such pain in his head and neck that he could barely function. He visited an orthopedist, who immediately referred him to the M.D. Anderson Hospital, a nationally prominent tumor clinic in Houston. Judge Farris was dying of cancer.

An immediate hiatus was called in the case.

Four days later, Judge Farris dictated a letter from his hospital bed to the two administrative judges with jurisdiction over Harris County:

Dear colleagues:
   As you know by now, my physical disabilities make it impossible for me to continue to preside over the above-styled cause, which has now gone 15 weeks.
   It is my understanding that Defendant has three more weeks of testimony left, to which must be added the time involved in putting together the charge—a tedious chore—as well as time for arguments and jury deliberations.
   Certainly, none of us would ask the jurors—$6-a-day jurors— to wait five, six or seven weeks to resume the trial, and certainly I will not ask them. I am, therefore, taking this step to formally request a replacement district judge be appointed to preside over the case.

The duty to choose a new judge fell to Thomas J. Stovall, a former Harris County judge who now supervised all the courts in a wide area of South Texas from a crackerbox bank-building office in the tiny community of Seabrook. Judge Stovall had ruled out more than a dozen candidates for the case when suddenly he heard an old friend from San Antonio was riding the circuit as an itinerant judge.

Maybe Sol Casseb would step into the case.

· 23 ·

When Joe Jamail was a little boy and his family traveled to San Antonio, they would often drop in to visit the Casseb family. Both the Jamail and Casseb clans were involved in the grocery business, and both were Lebanese.

It was during these visits that Joe Jamail became a friend of Solomon Casseb Jr., then a teenager bound for St. Mary's College in San Antonio. Sol would later attend the University of Texas Law School, eventually to become a state-district judge in San Antonio, where occasionally in the 1960s Jamail would present a case in his court. After Casseb left the San Antonio bench in 1968, he and Jamail would make a point of getting together every now and then for a few drinks and a few laughs about old times.

"Shit," Jamail would recall affectionately. "I've known Sol Casseb since I was a child almost."

Texaco, however, had a link of its own. Judge Casseb was an acquaintance of W. Page Keeton, a distinguished former University of Texas Law School dean, whose son, Richard, was now working with Dick Miller on the Texaco case. And before joining Miller's firm, Keeton had referred a number of divorce cases to Judge Casseb, who was then in private practice.

Would the associations stand for anything? If so, which would prove stronger?

* * *

When he applied to become a roving state judge in May 1985, Sol Casseb had not tried a case in nearly seventeen years. He had not, however, been forgotten in the Texas halls of justice. As an administrative judge in San Antonio, he had adopted the revolutionary concept of requiring that judges wear black robes, and had outlawed such venerated courtroom customs as smoking, eating and newspaper reading—standards of protocol that were eventually adopted statewide. In his obsession with judicial efficiency he created a docket system that gave San Antonio the distinction of having the fastest justice in the state.

But outside the Texas bar, it was Casseb's clothes that made the deepest impression.

In the 1960s, when he accompanied Judge Stovall to lobby a state legislator for a judicial pay raise, Casseb chose to wear a silk shirt, a silk suit and a $300 pair of alligator shoes. The legislator got one look at Casseb and asked, ''You're here for a raise?'' When he hung up his robe in 1968, it became apparent all over town that Sol Casseb was San Antonio's best-dressed man; on occasion he carried a hardwood cane with a gold knob. In 1977 the local chapter of Women in Communications honored him with its annual Fashion Plate Award.

After leaving the bench, Casseb adopted the legal specialty of handling divorces for the very rich. But in May 1985, just before the Pennzoil trial had begun, he had set aside his private practice and obtained an appointment as a roving judge, traveling to anywhere in the state that a temporary judge was needed to help relieve a crowded docket or fill in for an absent jurist.

Judge Casseb was divorced, and thoroughly enjoyed his appointment as a traveling judge—so much so that he supplemented his paltry expense allowance from his own pocket. ''They allowed me twenty-three dollars a day for my meals,'' he would later say. ''I drink more than twenty-three dollars' worth of Chivas and soda before I even sit down to eat.'' Despite his advancing years, Judge Casseb was a hardy, vibrant man in top physical condition—and quite a handsome one, with a full head of hair and a rigid physique.

''He is a vigorous gentleman with the ladies,'' Judge Stovall says.

Casseb was presiding over a case in Corpus Christi when Judge Stovall called, requesting that he come to Houston. On October

28, 1985, Sol Casseb succeeded the ailing Anthony Farris as the trial judge in *Pennzoil v. Texaco*.

Texaco moved for a mistrial. There was no way, the company insisted, that a brand-new judge could acquaint himself with 5000 pages of motions and pleadings, 150 trial exhibits and the 17,901 pages of testimony that had preceded him. Confusion over the rulings of the prior judge would force the trial to stop and start "like a milkman's truck," Texaco said. "The jury would be sent out and brought back into the courtroom again and again like a yo-yo." Moreover, only a mistrial could cure the prejudice that Texaco had already suffered as a result of Judge Farris' on-again, off-again trial schedule. While Pennzoil was presenting its case, testimony had been heard during 74% of the available courtroom hours, but during Texaco's case, only 38% of the available time, Texaco claimed.

"While Pennzoil enjoyed the opportunity to present its case in an orderly and uninterrupted fashion for nine weeks, Texaco has been forced to put on its case in bits and pieces, torn apart by repeated interruptions and hiatuses," Miller declared in the motion for mistrial.

Never one to waste time, Judge Casseb read the motion for mistrial without ever leaving the bench, overruled it on the spot and gave each side ninety minutes to summarize everything that had already occurred in the case.

"If y'all would like to do it now, I'm ready to hear from you."

"*We* would," Jamail responded. And after three hours, the new judge was ready to go.

While expressing sympathy for the medical problems of Judge Farris, the no-nonsense visiting judge could hardly believe the unbridled fashion in which his predecessor had conducted the case. Instantly Judge Casseb infused the courtroom with etiquette, elevating the court stenographer, Jacquelyn Miles, to the title of "Miss Reporter" and bidding good morning to the "ladies and gentlemen of the jury," something that Judge Farris had rarely done. After discovering that every wall clock in the courthouse was set to a different time, he instructed the lawyers and jurors to synchronize watches. ("I don't have a watch," juror Jim Shannon announced.) Judge Casseb cut the lunch break from ninety minutes to an hour. He extended the quitting time to 5 P.M. from about 4:15. And he decreed that the jurors and lawyers would no longer have Friday afternoons and Monday mornings free.

Judge Casseb vowed to himself that the case would be concluded in three weeks.

But if the judge intended to favor his childhood friend Jamail, he did not show it when Marty Lipton returned to the stand.

"Your Honor, may I approach the bench on this matter?" Jamail asked when the judge had sustained an objection by Texaco.

"No, sir."

Judge Casseb also curbed Jamail's talent for inciting the jury. "Just ask the questions now, counsel," the judge instructed him. "Save the jury speeches until later."

But Jamail needed little assistance from the bench in disposing of Lipton. For when Lipton returned to Houston after the ten-day delay for the substitution of judges, he had left whatever fight remained in him back in New York. "The jury thinks you're fencing with them," Texaco's Keeton had told him. "Just answer yes or no." And yes-or-no answers were the kind of responses that delighted an incriminating interrogator like Jamail.

Jamail noted that the indemnity contained in Texaco's contract protected the museum against any lawsuits arising from the Pennzoil agreement.

"It's referred to as the Pennzoil *agreement,* isn't it? You are a precise man. You could have stricken that language, couldn't you?"

"Yes," Lipton replied. He had, in a sense, just conceded that Pennzoil had a deal after all.

"You at least had Pennzoil on your mind enough to ask for and get an indemnity against any claim brought on behalf of Pennzoil, didn't you?"

"Yes."

"Very well."

And to suggest that the Getty Oil board had approved a whole lot more than Pennzoil's $112.50-a-share price, Jamail drew the attention of the jury to page 58 of the Getty Oil board notes, which read in part: "Mr. Stuart inquired whether *other provisions* of the Pennzoil offer presently before the board would be applicable, to which Mr. Lipton responded, 'Who cares?'"

Lipton initially denied ever making so cavalier a response—a comment which suggested that the Getty Oil board had dealt with all the *essential terms* of the Pennzoil deal. Lipton's recollection

was that he had said, "Who knows?" But later he added, "I think
it's a mistake, but I will defer to his notes."

When Lipton finally left the stand for good to return to the CBS
takeover and his other pending deals, he had left few admirers
behind in the courtroom. Oddly, however, one of them was Joe
Jamail.

"What did they want Lipton to do? Lie for them?" Jamail
would later ask. "Lipton wouldn't do that. I guess he just figured,
'Well, I'm not going to lose my dignity; I'm going to answer him
truthfully,' which is what he did."

Despite his good intention of completing the case quickly—and
thus freeing the jurors from their confinement—Judge Casseb's
fast-track schedule had an unintended effect on the jury. According to some jurors and lawyers in the case, it aroused racial
resentment.

One of the black members who could commute to downtown
only by bus felt the court was insensitive to her bus schedule.
Another black juror complained that she had already arranged an
important activity with her child on a Friday afternoon; Judge
Casseb refused to relent. And yet, when a white woman on the
jury complained of a scheduling problem for the same day, Judge
Casseb finally agreed to make an exception and give the jurors the
afternoon off after all.

Black members of the jury were furious. *He's prejudiced. . . .
He's discriminating*, they said. White jurors tried to suggest that
the judge had consented to the day off only because a second juror
had requested it, not because the second juror happened to be
white. But in a situation where all jurors were being deprived of
their rights—to talk about the case, to read the business page of
the newspaper, to travel out of town during the week and otherwise to maintain normal lives—it was easy for anyone to feel put
upon. Although many of the jurors had become fast friends—
black and white alike—some began to worry that any polarization
of the group would make it more difficult to reach a verdict.

Their personal strains mounted every day. Jim Shannon, who
would sometimes spend lunch breaks in his office at city hall,
discovered that he would be returning to a new job and a new
boss. Rick Lawler, who was making a forty-five-minute drive
each way instead of the usual commute of only a few minutes, was
going into the office in the evenings, trying as best he could to
keep the customers in his sales territory from defecting to the

competition. Another juror walked one day into the warehouse where he worked as a clerk and found a stranger sitting at his desk. "May I help you?" the stranger asked. One of the jurors contracted phlebitis and had to keep her foot elevated in the jury box.

Dick Miller, who in twenty-five years had tried only one case that *wasn't* before a jury, sensed this group's growing impatience and resolved to get his case out as fast as possible. He had started the case with a list of sixty-six potential witnesses; he had called but four live ones: Winokur, Boisi, Lipton and Wendt.

Miller began ruling out witnesses. After keeping Sid Petersen on tenterhooks for weeks, Miller decided against calling him. Not only was time short, but Petersen would make an "uptight" witness, Miller concluded. And although Miller considered Petersen to be smart and shrewd as a businessman, "he wasn't enough of a son of a bitch to deal with that hostile environment," Miller would recall. He ruled out Bob Miller, the former president of Getty Oil, and Harold Berg, the former chairman, even though they could have conveyed down-home images as strong as Hugh Liedtke's. He had considered calling Morris Kramer, the Skadden, Arps lawyer who had rendered an opinion to the Texaco board that Pennzoil had no deal; after Lipton, however, Miller decided that Texaco would have to get by without live testimony from any additional New York lawyers. He might have called Moulton Goodrum, Joe Cialone or any of the other Baker & Botts lawyers whom he believed had botched the Pennzoil deal—but calling a lawyer who represented the opposition was always risky, and the percentages on risky witnesses were not favoring Texaco so far. Finally, Miller also had three experts ready to puncture Pennzoil's $7.53 billion damages figure; he remained undecided about exactly what to do with them.

But Miller was convinced of one thing: He would not let that jury go into its deliberations thinking that Pennzoil had a contract.

To that end, on October 31, Miller called Chauncey Medberry to the stand. In his former capacity as chairman of Bank of America, Medberry could brandish the credentials of a man who had been the nation's number-one banker—a real banker, not an investment banker. He was the lone dissenter in the fifteen-to-one vote by the Getty Oil board, which made him seem a man of principle and resolve. And he was such a tough yet sincere old codger that his manner bordered on cute.

"Well, first off, there was no binding agreement on anybody, as I look at it," Medberry told the jurors. Even though Gordon and the museum had signed Pennzoil's memorandum of agreement, "the *corporation* had never agreed. There was no meeting of the minds on the thing and it had not been executed. . . .

"You would not sell a ten-billion-dollar corporation based on a discussion that takes place in the middle of the night. There would have to be a document with all the terms agreed upon, hammered out, negotiated. *That's* the way business is conducted in this country."

It sounded a whole lot better than Winokur, Boisi or Lipton seeming to tell the jury how business was conducted in New York.

"Here is a buyer," Medberry said, "sensing an opportunity for a quick kill, hovering around, pressing. They had encamped two of the major shareholders, who, for different reasons, would have favored a quick deal with Pennzoil, and the other directors who were sort of handcuffed. . . .

"Now, that's really the whole story."

Jeffers could not undo Medberry's testimony about the board's lack of *intent to be bound*, but he annihilated Medberry's claim that Pennzoil had played unfairly with the Getty Oil board.

"Bank of America makes money financing unsolicited takeover offers from time to time, doesn't it?" Jeffers inquired.

"It has in the past," Medberry said, going out of his way to add, "It reviews them very carefully before it does it."

"You don't have any ethical or moral objections to the bank making some money on unsolicited tender offers?"

Medberry immediately got his dander up.

"Well, now, let's not be quite so cavalier about it," he snapped. The bank always considered the qualifications of the raider and whether the target was inadequately managed, as well as "the ethics of the situation."

"Did you know that the Bank of America was one of the banks that was providing the financing for Pennzoil's tender offer for Getty?"

Medberry tried to extricate himself by pointing out that he had retired from the bank by that time, but the damage had been done. "Jeffers played out the rope," Jim Shannon would recall, "and left Medberry swinging."

Laurence Tisch took a day from his takeover waltz with CBS to climb into the witness stand for Texaco, and was nearly as effective as Wendt had been: articulate, self-assured and unwavering in his assertion that the Getty Oil board never intended to bind itself to an agreement with Pennzoil. "I know what I voted on," Tisch declared.

Jamail had resolved to put on his kid gloves for Tisch. "I had just bled Lipton," he would later explain. Jamail's only cheap shot was going out of his way to establish that Tisch's Loews Corporation had been in the casino business.

Tisch generally gave straightforward and unconditional answers, which made it easier for Jamail to extract peacefully the testimony he wanted. Tisch—held out to the jurors as one of Wall Street's most experienced investors—said that investment bankers usually do not submit their bills until a deal is done. Was it not unusual then, Jamail asked, that Goldman, Sachs submitted its initial $6 million bill to Getty Oil the day after the fifteen-to-one board vote? Yes, Tisch said. That was unusual.

Tisch also acknowledged that his friend Marty Lipton had an interest in seeing Gordon abandon his arrangement with Pennzoil in favor of the Texaco deal. "In order for the museum to get the one hundred twenty-five dollars, Gordon Getty had to agree to sell the trust shares for one hundred twenty-five dollars?"

"Put it that way," Tisch said, "yes."

Only once did Tisch adopt the evasiveness of Texaco's earlier witnesses. The much-discussed toast of champagne that Tisch had lifted with Gordon and Ann was intended to celebrate not the completion of a deal, but only "the acceptance of a *price* that could lead to the agreement on a contract."

"You're not friends with Gordon Getty?" Jamail inquired at one point.

"No, sir."

"Did Gordon Getty think you were *his* friend?"

"Define 'friend' and I'll answer the question," Tisch commanded.

"Sir," Jamail replied, "I can't define the New York friendship."

Like Pennzoil, Texaco suffered the problem of having to put on much of its case through depositions. Unlike Pennzoil, however, Texaco had videotaped none of its. The jury quickly learned that the only thing worse than a video deposition was a non-video deposition. Texaco did play some of the video depositions that Pennzoil had taken, but for the most part two Texaco lawyers would have to read aloud by the hour, turning pages by the hundreds and struggling vainly to inject some life into the dry, dry words with a little inflection. If Pennzoil had needed *Roadrunner* cartoons to maintain the jury's interest, Texaco needed the Rockettes. Even Judge Casseb took note in the official bench record of "tired, bored and somewhat sleeping jurors."

Moreover, the depositions presented by Texaco were hardly uniformly favorable *to* Texaco, just as those played by Pennzoil were a mixed bag. Texaco's investment banker, Bruce Wasserstein of First Boston, testified that he felt Getty Oil was free to deal, but also recounted advising his client of the need to "stop the train . . . to stop the museum" from signing away its shares to Pennzoil. Stuart Katz, a Getty Oil lawyer, recalled how Pennzoil's lawyers had bungled the drafting of documents but also acknowledged leaving the Pennzoil drafting session for a meeting with Texaco and never disclosing where he was going. Harold Williams, president of the museum, said when he moved "acceptance of the Pennzoil proposal," he was referring only to the $112.50 price—not an entire contract. But Williams also testified that the museum had a tacit understanding to sell its shares to Texaco before John McKinley ever paid his visit to Gordon Getty—which only strengthened Pennzoil's claim that Texaco and the museum had forced Gordon to sell by threatening to relegate the trust to minority ownership in a company majority-owned by Texaco.

Each deposition heaped minutiae upon minutiae. The jurors heard about Bart Winokur's taste for hot chocolate, about Marty Siegel having to use the lobby bathroom at the Pierre because Ann Getty was sleeping in the bedroom, about the eating habits of the Pennzoil PR people who were present in New York for the deal.

One disadvantage of going second in a trial—that the other side has already begun using up the jury's patience and goodwill—was doubly true in this case. "We were told the trial would be six to eight weeks," Shirley Wall, one of the jurors, would later say. "Anything over that seemed to be a lot of time."

But Pennzoil, too, was arousing great restlessness in the jury box. Although the attacks on Winokur, Boisi and Lipton had been stimulating, the jurors wearied of the relentless parade of Pennzoil exhibits under the nose of every witness called by Texaco. Over and over, the jurors had to sift through their piles of documents to fetch the Cohler and Siegel California court affidavits, the Getty Oil meeting notes, the Dear Hugh letter, the memorandum of

agreement, the press release announcing an agreement in principle, the "deal should be done" notes by the Getty Oil treasurer,
the *take-care-of-Liedtke* and *stop signing* notes of Texaco's executives—so much so that some of them already had entire sections memorized. Pennzoil also confronted nearly every witness
with the "back-door" board meeting. Invariably, the cross-exam-
inations by Pennzoil seemed to last longer than the direct exam-
inations by Texaco.

"It reached the point where it was almost ludicrous to listen
anymore," Rick Lawler would recall. The jurors began mumbling
the answers to questions before they were given—particularly
when a question by a Texaco lawyer called for the answer, "free
to deal." And when Pennzoil would ask yet another witness about
the back-door board meeting, a murmur would rise from the jury
box: *God, Pennzoil, why am I hearing this again?*

Even though Texaco was the only defendant in the courtroom,
it had long since become clear to the jurors that Getty Oil was
every bit as much on trial. Hopefully, Judge Casseb's instructions
would sort everything out.

Once the $500 in winnings on the Lipton bet had been distributed
at the Pennzoil table, only one element of Miller's strategy remained the subject of intense speculation among the Pennzoil
lawyers: Would Texaco contest Pennzoil's $7.53 billion damages
figure?

The answer was not as predictable as it might seem. Many of
the nation's finest trial lawyers consider it strategically unwise for
the defendant to so much as *mention* damages, let alone put on an
extensive rebuttal against whatever lofty sum the plaintiff is demanding. The theory is that questioning the damages only "digni-
fies" the request. Suggesting an alternate figure, or merely
chipping away at the plaintiff's demand, might prompt the jury to
award a compromise amount, where otherwise it might award
nothing at all.

Terrell and Jeffers of Baker & Botts knew as well as any
members of the Houston bar that Dick Miller was just such a
lawyer. Miller took pride in his distinction as a "bet-the-company" lawyer, and in years of risk-taking he had accumulated an
unassailable record of zero-dollar verdicts for clients that might
otherwise have faced at least nominal dollar judgments.

But this was not a $10,000 slip-and-fall case, or a $100,000
intersection collision, or even a $1 million death claim by the

widow of an offshore oil worker swept away at sea. This was the greatest damages demand in history. Would Dick Miller bet Texaco for $15 billion?

At the moment, with only a few days remaining in his case, Dick Miller himself was not sure.

He had been sending mixed signals all along. When Pennzoil had called one of its executives to help lay out the framework for its damages theory, Miller's partner Richard Keeton had told the court he would probably spend at least three hours conducting the cross-examination; yet the next day, when Pennzoil had passed the witness to Texaco, Keeton spent only a few minutes attacking the testimony, and then only lightly. Similarly, Keeton had made virtually no effort to undermine the damages testimony of Tom Barrow, Pennzoil's expert.

On the other hand, Pennzoil knew that Miller had already lined up his own damages experts—a move that was decidedly out of character for him. Like many lawyers, Miller often refused to put on damages experts not only to avoid dignifying the other side's number, but also to prevent any cross-examination. It was fine for the defense to call Dr. Brokenbone to the witness stand for expert testimony that the plaintiff's medical bills were unreasonably high. But that only gave the plaintiff's attorney the opportunity to ask, "Well, Doctor, how do you suppose it felt when the engine mount ripped through the dashboard and cut off my client's legs? Did he experience pain, Doctor? Did he suffer, Doctor?''

Miller had lined up three damages experts. But would he actually put them on the stand?

The Baker & Botts lawyers began placing anonymous calls to the switchboard of the Four Seasons Hotel, where the out-of-town Texaco people holed up, inquiring if any of the experts had showed up. One day, as Texaco appeared to be bringing its case to a close, they were disappointed to find out that the experts had checked in.

It looked like Bet-the-Company Miller was going to lower his stakes after all.

Dick Miller had presented the museum story through Lipton—with little success. He had presented the Getty Oil story through Winokur, Boisi, Wendt, Medberry and Tisch—with mixed success. Now, in the final week of his case, it was time to tell the Texaco story.

First to the stand was James Kinnear, the vice chairman who

had sat through the entire trial. The jurors liked Kinnear, a friendly-faced man with a wry smile. On the stand he told a brief, articulate account of Texaco's involvement, emphasizing how at every step Texaco was assured by the Getty interests that they were completely *free to deal*.

Kinnear was followed to the stand by Bill Weitzel, the Harvard-educated general counsel of Texaco who had been at the side of Chairman McKinley throughout the deal. Weitzel gave an extremely convincing rendition of Texaco's refrain that the Getty interests were "free to deal." Weitzel, himself a former trial lawyer, also avoided falling into the trap doors that Jeffers opened for him. But in addition to being colorful and shrewd, Weitzel also proved somewhat excitable. On the cross-examination Jeffers drew out one of the most vitriolic responses of the trial while trying to depict Texaco as a company desperate for oil-in-the-ground—and Weitzel as an executive obsessed with fulfilling a mission.

"When you went to New York to see Mr. Lipton, at that point there was no way that you, Bill Weitzel, were going to turn back and say, 'Sorry, we've run into some contract principles here. Sorry we rushed you up here, board.' I'm saying your state of mind, Mr. Weitzel, when you left White Plains that night of the fifth to go see Mr. Lipton, was that you were *not going to turn back*."

Weitzel practically leaped from the witness stand.

"That's *really wrong* because if I ever had the *slightest idea* that there was any contract—a binding contract between Pennzoil and Getty—Texaco would not have moved one inch. Because I know Mr. McKinley would not have moved against my legal advice. So that's just *really wrong*." Despite his harsh exchanges with Jeffers, Weitzel was convinced he had left a positive impression on the jury. "I thought there was no way they didn't believe me," he would recall. *"I was telling the truth."*

Texaco's last live witness was the one that Pennzoil had expected to see first—Al DeCrane, a smoothly polished and youthful-looking lawyer of fifty-four years who had delivered Pennzoil its smoking gun: some of the handwritten notes of Texaco's internal-strategy sessions. The Pennzoil lawyers knew that despite his authorship of Pennzoil's best "interference" evidence, DeCrane would be a tough match. "We hoped to reach a draw with him," Irv Terrell would later recall.

In a case where Texaco had accused its adversary of arousing religious and regional bias, Miller appeared to go out of his way to establish DeCrane as a father of six who attended Notre Dame—biographical details that ran little risk of alienating the three Catholics on the jury. Miller's objective with DeCrane was simple: explain away the handwritten notes.

"They were strictly for my purposes," DeCrane said.

"Did you ever have any idea they would be an exhibit in a lawsuit?"

"No."

"Or that they would be the subject of court testimony?"

"No. That never entered my mind at the time I was making these notes."

"Or that questions might be asked—great, detailed questions concerning the notes themselves, and what they meant?"

"No. Not at all."

Thanks to Miller's skillful—and unabashedly leading—questioning, he and his client offered credible explanations that made the notes seem less insidious, without appearing to cover up their true meaning. Miller asked questions not only about DeCrane's own notes, but those taken by Pat Lynch, Texaco's associate controller, during the same series of meetings with Texaco's advisors from First Boston.

Only have an oral agreement: Miller emphasized the word *only;* what more evidence could exist that Texaco didn't think Pennzoil had a done deal? "That was simply a shorthand reflection that there's nothing final," DeCrane affirmed.

Stop train: "Well, if they already had a contract on January the third, the train was in the station, wasn't it? How about that?" Miller asked.

Option to Pennzoil. . . . Could be a way to take care of Liedtke. Liedtke could tender his option shares under Texaco's offer and walk away from his broken deal with a profit of $120 million. "He wouldn't go away empty-handed?" Miller asked. "That's right."

DeCrane also passed the buck, attributing some of the notations purely to the candor of the takeover strategists from First Boston.

Trust: 1930. . . . Might have to go the tender offer route to merge Gordon out or create concern that he will take a loss: DeCrane had three stars alongside this reference to the problem of getting Gordon to sell. "At this point, I didn't discuss our philosophy with them," DeCrane said—the philosophy that Texaco only conducts mergers the "friendly" way. "We had retained them as advisors. . . . They were making some suggestions as to alternatives which could be considered.''

If there is a tender offer and Gordon doesn't tender then he could wind up with paper. "That was never discussed in any detail," DeCrane insisted. "It was not a plan of Texaco's, but that was what people were talking about. . . . It's legal and it's legitimate and it has been done.''

DeCrane withstood Terrell's cross-examination exceptionally well—but as he had with all of Texaco's witnesses, the Pennzoil lawyer managed to extract a state of mind, if not any new facts, that played nicely into Pennzoil's case. "We had to make him out as someone who was very cold in his decision-making processes,'' Terrell would recall.

He seized on DeCrane's scribbling that Pennzoil *only has an oral agreement.*

"Do you live up to your oral agreements, Mr. DeCrane?''

"I live up to my contracts.''

"Do you live up to your oral agreements?''

"I live up to my contracts and commitments.''

"I want you to tell this jury *right now* whether you live up to your oral contracts. *I* want to know.''

"I live up to all of my contracts. . . .''

The live testimony of the trial ended on an interminable episode of one-upmanship. In his cross-examination of DeCrane, Terrell tried to impeach Texaco's witnesses from Getty Oil, claiming that they were obligated to "cooperate" with Texaco's defense under the indemnities they had received. When Miller conducted his redirect examination of DeCrane, he, too, took up the issue of credibility.

"Oh, one other thing,'' Miller said to DeCrane. "'Cooperation' doesn't mean lying under oath, does it?''

"It certainly doesn't,'' DeCrane answered.

Terrell then stood up for a thirty-second re-cross examination.

"And, Mr. DeCrane, if this jury believed that people had come in here and lied to it under oath, you'd want the full power of the court to redress that, wouldn't you?''

DeCrane sat silently.

"Wouldn't you?''

"I don't think anyone has lied that I'm aware of.''

"If they have, and the jury believes they have, you want the full power of the court and the jury to put an end to that, wouldn't you?''

"I believe that justice and truth should be what we seek in this whole proceeding," DeCrane answered.

Terrell concluded his re-cross. Miller rose for *further* redirect examination.

"Would that apply to Liedtke?"

"That applies to everyone."

Terrell again.

"Would it apply to Mr. Lipton?"

"It applies to everyone."

"Mr. Boisi?"

"To everyone."

"Counsel for Texaco rests," Miller announced. It was November 12, 1985—eighteen weeks after the jury had first assembled for voir dire.

Miller was as convinced of victory as he could be. Sure, Pennzoil had scored plenty of points, but not in the areas that counted. Pennzoil had failed to overcome the fact that its deal with Getty Oil had remained "subject to a definitive agreement"—and there was no definitive agreement among the reams of exhibits that would be going into the jury deliberation room. Moreover, the existence of the "open points" in the Pennzoil deal was unrebutted; only their significance had been contested. Finally, every single witness for Texaco had established that Getty Oil, Gordon and the museum were all *free to deal*. Not one had seen the Pennzoil people shaking hands with anyone—except, perhaps, with the people on their own side.

There was no point in showing any weakness now by disputing the $7.53 billion figure, Miller figured. Why remind the jury of the figure at all? They hadn't even heard it for a month.

Then, Miller saw Judge Casseb's jury instructions.

· 24 ·

Ladies and gentlemen of the jury, the court is now going to read you what is known as the Court's Charge.''

At last, after testimony by thirty-five witnesses spanning sixty-seven days and twenty-three thousand pages of transcript, after weeks of doubt over who was really on trial, the jurors finally would be told just what they were supposed to do.

For the most part, it would be Pennzoil telling them.

In Texas, as in many other states, juries in civil cases reach their verdicts by answering questions called "special issues." The lawyers on both sides submit their own suggestions to the judge, who has final authority over the number and order of the questions, and most importantly over the manner in which they are posed. The judge must take account of the applicable law—in this case, the law of New York—as well as all the evidence in the case. In some cases the outcome of the case can easily ride on the phraseology of the special issues and the instructions accompanying them.

By his own admission, Judge Casseb was utterly unfamiliar with New York contract law. Pennzoil and Texaco might have given him a crash course by filing legal briefs summarizing the many case precedents involved—except that Judge Casseb didn't believe in briefs. "Just give me your cases," he told the lawyers. When both sides heaped a stack of prior judicial decisions on the bench, he noticed that only three had been cited by both sides. It was those three on which Judge Casseb decided to rely.

As for the evidence, Judge Casseb had heard but two weeks of it. He had missed the first twenty-five witnesses. "I would like to go on record in stating that I am at a disadvantage because I did not read the full parts of this evidence," he told the lawyers as they conferred over the charge. "And I guess I'm going to have to assume the consequences of it because I guess it will be my error if any error is committed." But when he called the lawyers together for a conference on the charge, the no-nonsense judge wasn't about to waste time. "We'll be here all day until we finish it," he said.

And when it was done, on the most important issues, the charge tracked almost verbatim with the language suggested by Pennzoil. "We took a real fucking on this charge," Miller would say later.

The judge began reading aloud to the jurors. "Special Issue Number One," he said. "Do you find from a preponderance of the evidence that at the end of the Getty Oil board meeting of January the third, 1984, Pennzoil and each of the Getty entities— to wit, the Getty Oil Company, the Sarah C. Getty Trust and the J. Paul Getty Museum—intended to bind themselves to an agreement that included the following terms. . . ." And he enumerated three: a price of $112.50 to the public shareholders, the intended ownership by Pennzoil of three-sevenths of the company, and a plan to divide the company's assets within a year.

Nowhere did the question ask whether Pennzoil had a "contract." In fact, the word agreement appeared in the charge thirty-two times and the word contract not once. Joe Jamail had whipped out his Webster's New Collegiate and read to Judge Casseb the third-listed definition of agreement: "a contract duly executed and legally binding." Pennzoil had also pointed to the fact that the words contract and agreement are often used interchangeably in jury instructions—as indeed they are. But in this case "agreement" was eerily familiar of Pennzoil's memorandum of agreement and the Getty Oil press release announcing an agreement in principle—two documents whose authenticity could not be challenged. Worse still, the instructions accompanying Special Issue Number One said that "An agreement may be oral, it may be written or it may be partly written and partly oral"—an instruction that was defensible under the law, but one that seemed to echo the handwritten notes in which Texaco's president wrote: *Only have an oral agreement.*

"Special Issue Number Two," Judge Casseb continued. "Do you find from a preponderance of the evidence that Texaco knowingly interfered with the agreement between Pennzoil and the Getty entities, if you have so found? . . . A party may interfere with an agreement by persuasion alone, by offering better terms, by giving an indemnity against damage claims to the party or parties induced to breach, or by any act. . . ."

On the Pennzoil side of the courtroom, right next to the jury box, the view was *tough luck for Texaco*. Six days earlier, as Texaco neared completion of its case, Judge Casseb had surprised both sides by asking them to submit written suggestions for the charge that same day. Pennzoil, however, already had a draft ready. In fact, its lawyers had been working on a proposed charge for months. It had expanded its back-office legal team to include Mark Yudoff, dean of the University of Texas Law School and a contracts professor for fifteen years. Dean Yudoff had gone through at least ten drafts of a proposed charge working with Randall Hopkins, a Baker & Botts partner who had spent months making himself the resident expert on New York law.

Dick Miller's small law firm, the Pennzoil lawyers reckoned, had stretched itself too thin.

Nonsense, Texaco would later say. Its lawyers had been hard at work on the charge for weeks as well—and had submitted an extraordinarily detailed proposal to the judge as a way of *trying* to convey some of the favorable evidence that Judge Casseb had never been exposed to. "We were trying to educate Casseb on the case," Bill Weitzel, Texaco's general counsel, would later say. The judge may have appreciated Texaco's lesson, but he did not adopt Texaco's proposed charge. And Weitzel says that when a Texaco lawyer delivered one draft of the proposed charge to Judge Casseb's room at the Hyatt Hotel, he was startled to find three Pennzoil lawyers—Jamail, Jeffers and Kronzer—comfortably seated in the judge's room. (The Pennzoil lawyers, however, had gone there to deliver some materials of their own, as the judge, in open court, had invited them to do.)

"Special Issue Number Three," Judge Casseb continued. "What sum of money, if any, do you find from a preponderance of the evidence would compensate Pennzoil for its actual damages, if any, suffered as a direct and natural result of Texaco's knowingly interfering with the agreement between Pennzoil and the Getty entities, if any?"

There was that word "agreement" again, the people on Texaco's side groaned, one of thirteen formal, written objections that Texaco had filed to Special Issue Number Three alone. In all, Texaco had filed seventy-three separate objections to the charge— all of which Judge Casseb had overruled on the spot.

"Special Issue Number Four," the judge continued. "Do you find from a preponderance of the evidence that Texaco's actions, if any, were intentional, willful and in wanton disregard of the rights of Pennzoil?

"Special Issue Number Five. What sum of money, if any, is Pennzoil entitled to receive as punitive damages?"

A cadre of Texaco's top executives had just arrived in Houston for the closing arguments. "Damn it, Dick!" one of them said. "Can't you do something about this charge?"

But Dick Miller had only one last defense to prevent the charge from exploding under Texaco: his closing argument. *I've got to piss on the gunpowder,* he thought.

Judge Casseb gave each side four and a half hours for summation, allowing the plaintiff, as is customary, the first word as well as the last. Right to the final phase of the trial, Pennzoil maintained the team approach: Jeffers and Terrell would open the argument for Pennzoil, dividing three hours between them, leaving the final ninety minutes for Jamail.

Miller thought it was typical that his old friends from Baker & Botts wanted to add their two cents' worth to the closing argument. That was just how his old law firm had always worked: Everybody wants all the credit and recognition they can get. And yet, throughout the trial, the division of duties between Jamail and Baker & Botts had worked beautifully, just as it would now in the final nine hours of trial. Miller, on the other hand, had personally supervised the completion of his case while working feverishly on his closing argument. When his time came to speak, some people in the courtroom sensed he was exhausted. Early in his argument, he confused Sherlock Holmes and Oliver Wendell Holmes.

But even on Miller's worst days, his flashes of brilliance could not help shining through. On the theory that a good offense is the best defense, he talked about what kind of company Pennzoil was.

"It's a company run by lawyers who know a lot about takeovers and who understand power and the force of money—and who, as they looked at the saga of the Getty Oil Company unfolding, could sense the weakness of that company.

"What they sensed even more was that what Gordon Getty

wanted above everything else was to be the head of his daddy's company. They knew that. They could feel it. And they were smart enough and shrewd enough and clever enough to figure out a way to use that.''

Miller glared at the jury and recounted Pennzoil's method of approach: a hostile tender offer between Christmas and New Year's.

''And why did they do it? Because they knew that's when the company would be least able to defend itself—and they were right about that. People on vacation. People gone. People not working. People taking holidays. People are scattered all over everywhere—except the Pennzoil team.

''If they wanted to buy some Getty Oil Company oil, why didn't they just go to them and say, 'Look, we'd like to buy some of your oil. We want to be a big company. We need some oil. You've got lots of oil. You guys don't know how to run your business anyhow. Your chairman's a boob. You dumb klutzes can't run this company. Sell some of this oil to us, but we only want to give you three dollars a barrel.'

''You say, 'Well, gee. That's not fair.'

'' 'Well, by God, we'll take you over then.' ''

Miller slowly began casting his glance toward Hugh Liedtke, who was leaning forward in a folding chair near the counsel tables.

''So what would y'all say is the most prevalent human emotion? It's fear, isn't it? And a skillful negotiator—and nobody denies that this fits Mr. Liedtke—understands fear, understands how to use it. And what he did by that tender offer was tighten up Gordon Getty's gut so tight with fear that he was willing to deal.

''What they had in mind was making this deal with Gordon Getty and putting this 'divorce clause' in. You tell me: Do you really think that a man who has been the chairman of a major oil company for twenty years is going to participate in a five-billion or six-billion-dollar deal and let *somebody else* run it? Do you really think that J. Hugh Liedtke was ever going to let Gordon Getty run the Getty Oil Company—even if he owned fifty-seven percent of it?

''Not quite.

''But what they did understand was his intense desire to be the head of his father's company. So Mr. Liedtke said, 'You are going to get a big office and it's going to say on the door that you are the chairman, and you are going to decide such things as what we do about the war in Bangladesh and how we stand on international monetary policy and we'll just run the rest of it.'

"And so they had his number. You know how I know that? Because they made a deal with him in two hours. In *two hours* they had talked Gordon Getty into that memorandum of agreement and he'd never seen them before in his life. Fantastic!"

So much for Pennzoil's "negotiations" with Gordon. What about its "negotiations" with the board of Getty Oil?

"Nobody with any authority at Pennzoil ever talked to anybody with any authority with the Getty Oil Company—except one time, and then all they said was, 'We don't want to talk to you. We don't want to buy any oil from you. We have got a deal with a forty-percent stockholder. Get lost.'

"And they sat over there in the Waldorf Towers and talked to each other.

"Don't tell me that Texaco beat Pennzoil. You beat yourself, Pennzoil.

"So arrogant, proud—so full of yourselves that you will not talk to anybody."

And where was the famous handshake?

"They never shook hands with anybody except themselves, as if it makes any difference whether you shake hands or not. And it doesn't. . . ."

And what about Texaco?

"I guess to my mind the most persuasive piece of evidence we have about the issue of tortious interference is the *overwhelming pervasiveness* of the fact that *we did not crash this party*. We were invited to the party. We didn't look them up. They looked us up. 'Mr. Texaco, we want you to bid. We're in a squeeze and we think the offer we're getting is shameful.' And the largest stockholder has joined with an outsider to try to buy the public shares at what they think is an outrageous price.

"This has got to be the first case in the history of mankind where the white knight got sued by the dragon!

"It's got to be one of the dullest cases tried in the history of mankind—when the most exciting thing in the case was when I came into the courtroom with my fly unzipped."

Hour after hour it went on, and finally Miller addressed the subject of Pennzoil's $7.53 billion damages demand—something he had not contested through the trial itself.

"My little grandchild has got a little book that says, 'How much is a million?' And there is a little thing in the book; it says it would take you ninety-five years to count to a billion.

"God, I don't know whether you *could* do it in ninety-five years. But these people sit out there and file this. Why did they do that? Why did they sue for this monstrous sum of money?

"I think it's basically to get publicity. To satisfy the desire for revenge for getting beat out. These people can't bring themselves to admit they got beat. They want to think they got cheated. So that's the reason for this case. . . . If you give them *one thin dime*, they are going to be encouraged. . . .

"*They didn't have a contract*. Never had a contract. What they have got is a company run by three lawyers who have got this lawsuit and they are trying to hit a big gusher here in court.

"That's what this lawsuit is about."

Unfortunately, Miller slipped up at the very end, when he asked his co-counsel whether he had overlooked anything and was told he had forgotten to mention that Pennzoil really had no interest in taking care of Getty Oil's employees. "The fact that I did not mention it when I should have does not mean it's not important," Miller said apologetically, and he sat down, spent.

Although his closing argument was over, his most costly oversight was yet to occur.

Joe Jamail had become the King of Torts convincing juries to send messages to corporate America—messages to build safer cars, safer guns, more reliable pharmaceuticals, better drilling rigs. And as he always told juries, the way to send that message was to render the kind of punishment that companies easily understand: big-money judgments.

He and his co-counsel were confident, but they did not believe they had the case in the bag. Already, an in-house Pennzoil lawyer was making plans to draw up press releases announcing a loss as well as a victory. The outcome seemed sufficiently up in the air that Jim Kronzer, Pennzoil's procedures expert, was concerned that Jamail might try *too* hard in his closing argument. Just before Jamail rose to speak, Kronzer slipped him a note that simply said, "Cool." Jamail grinned quizzically at Kronzer from across the table. "I don't want to put any heat on you," Kronzer whispered, "but this thing is up for grabs."

Still, Jamail knew there was only one way to bring in the "gusher" that Texaco feared—to arouse the jury's passion against Texaco and to embolden it to award billions of dollars. Jamail had to throw a thunderbolt into the jury box.

For only the second time in Jamail's career, his wife, Lee, came to the courtroom. The first time had been nearly thirty years earlier, in the case where Jamail had sued the city of Houston for putting a tree in the path of his client's car. And on that occasion, Lee Jamail, with a needle and thread in her purse, had come to the courthouse only to repair the broken fly on her husband's pants.

"There were so many misstatements in what Mr. Miller has told you that I'm not going to try to answer all of them," Jamail began.

"He starts out by giving you four hours of excuses for Texaco's conduct and he ends it up by saying to you, 'Don't take (away) our company because if something went wrong, it was Getty that did it.'"

Jamail stood silent for a second.

"They bought and paid for this lawsuit when they gave the indemnities!"

Actually, Texaco had bought only a lawsuit that might be filed *against* the Getty people, and that lawsuit was sitting in Delaware. But Jamail was getting the last word; no one could now refute him.

In his summation Miller had also called the jury's attention to the absence of the word contract from the judge's instructions, urging them to impose the highest standard possible in deliberating over whether Pennzoil had a done deal.

"Judge Casseb is one of the most honored and brilliant judges in America!" Jamail responded. "If he had wanted to say to you, 'contract,' he would have said it. I have no quarrel with 'contract,' but it's his charge, not Mr. Miller's."

Miller had also reminded the jury of contradictions in the testimony of Pennzoil's witnesses. Indeed, when asked what they considered to be the "essential terms" of their deal, Pennzoil's own people could not seem to agree.

Jamail admitted that some of Pennzoil's witnesses may not have told the same "pat story." "*He* didn't have that trouble," Jamail said, motioning to his opponent. "Every witness *they* brought reminds me of one of those squeeze dolls. You squeeze it, it says, 'Free to deal, free to deal, free to deal. . . .'

"They spent a lot of time by their own admission with these witnesses. They had no choice but to come and give you some suggestive after-the-fact opinion. That's what they did. I suppose we have to start with Mr. Bart Winokur and Mr. Boisi back in Houston. You remember him, 'Back-Door Bart'? Snuck in the back door when they got Gordon Getty out of the room? Kicked him off, or tried to instigate a lawsuit to kick him off the board of his own family trust?"

Then there was Lipton.

"To him, money is the object. That's what it is. That's what he said. He says people can agree without lawyers, but in his mind the amount of money makes the difference. And that's that specialized group calling themselves mergers and acquisition lawyers, carving up corporate America to their liking. They get together. They're all within a couple miles of each other. One wins one day and another wins another day—and the investment bankers win *all* the days and *they win with these indemnities.*"

Except occasionally to quote some passages of testimony, Jamail never looked at his notes. He rambled from one issue to the next and then back to the first issue. His argument was disjointed, undisciplined—but it was working.

"Again, if Judge Casseb wanted to say 'written contract,' he knows how to do that. He didn't. The law requires no such harsh burden of us.

"Stop the train. They didn't stop it. They derailed it, *blew it up!*

". . . I thought you were going to throw up listening to all the depositions you had to hear. . . .

"There has been no dispute about damages in this case. None. Texaco, through its lawyers, does not think you will assess those kinds of damages. They think you are not big enough to do that. I think you are."

One by one, Jamail looked the jurors in the eyes. *I can see it, smell it, taste it,* he thought. *They're throwing away all the bullshit after four and one half months.*

"Hugh Liedtke is my friend," he said, turning to face the oversize figure sitting in the courtroom. "Baine Kerr is my friend. And I can't divorce myself from that. But I can't divorce myself from *these* facts, either: They honestly endeavored to build their company.

"Hugh Liedtke, who got started with the help of J. Paul Getty. You heard the testimony. *The man who came to his wedding!* Liedtke a stranger, interloper, going to rape that company? He got his start from that family! Believed he could work with them—and he could have, if it had not been sabotaged.

"You people here, you jury, are the conscience, not only of this community now in this hour, but of this country. What you decide is going to set the standard of morality in business in America for years to come.

"Now, you can turn your back on Pennzoil and say, 'Okay, that's fine, we like that kind of deal. That's slick stuff. Go on out and do this kind of thing. Take the company, fire the employees, loot the pension fund. You can do a deal that's already been done.'

"That's not going to happen."

Jamail paused.

"'I have got a chance. Me. Juror.

"'I can stop this. And I am going to stop it. And you might pull this on somebody else, but you are not going to run it through me and tell me to wash it for you.

"'I am not going to clean up that dirty mess for you.'

"It's you. Nobody else but you. Not me; I am not big enough. Not Liedtke, not Kerr, not anybody. Not the judge. Only you in our system can do that. Don't let this opportunity pass you. Do not.

"You can send a message to corporate America, the business world. Because it's just people who make up those things. It isn't as though we are numbers and robots. We are people. And you can tell them that 'you are not going to get away with this.'

"I ask you to remember that you are in a once-in-a-lifetime situation. It won't happen again. It just won't happen. You have a chance to right a wrong, a grievous wrong, a serious wrong."

Jamail lowered his voice to a whisper.

"It's going to take some courage. You got that. . . . You are people of morality and conscience and strength.

"Don't let this opportunity pass you."

It was Friday at 2:30 P.M. when Jamail sat down. Judge Casseb ordered the twelve jurors into the deliberation room and assumed the unhappy duty of dismissing the three jurors who were told for the first time that they had been alternates all along. It was a sad moment, and for the first time in his career the judge requested an ovation from the audience.

One of the alternate jurors, a machine-shop manager named Gilbert Starkweather, approached Miller following his dismissal to speak and shake hands for the first time after their four and one half months in the courtroom together.

"You really did a great job," Starkweather said.

It was exactly what Miller did not want to hear. "It troubled the hell out of me," Miller says. "It told me he thought I had the hard side of the case."

Inside the jury room, the dozen remaining jurors took their seats around an old metal conference table, exchanged glances, drew their breath—and opened the floodgates of discourse. Although some had spent the last four and a half months gossiping about the witnesses and the lawyers, they had done their best to avoid discussing the issues in the case. Now, almost *everyone* was talking—at once. It was pandemonium.

Eight years earlier, a Houston jury had spent hours deadlocked over what had appeared to be an open-and-shut child-molestation case. Several little girls had taken the witness stand and told the same story about the neighborhood flake. In the minds of eleven jurors there was no doubt that the children were telling the truth. But the twelfth juror thought the testimony seemed rehearsed and harbored grave doubts about the prosecution's case.

The twelfth juror in that case was Jim Shannon.

Shannon had had more than "reasonable doubt" about the guilt of the man—the standard applied in criminal cases—he had grave doubt. He decided he could not live with himself if he let the jury send a man to jail when Shannon thought he might very well be innocent. But Shannon also realized that if he refused to relent, the jury would be hung and the defendant would be subject to another trial. *There might not be someone like me on the next jury,* Shannon thought.

So Shannon cut a deal in the jury room. He would vote for conviction, but only if all the other jurors promised to impose nothing harsher than probation when the time came to deliberate over the defendant's sentencing. The other jurors bought it, and the man was found guilty but set free.

"In my experience, there are shepherds and there are sheep on juries," he would later say. And Shannon, the oldest of eight children, had no doubt about which direction the sheep on the Texaco jury should turn.

"It wasn't even a close call," he would explain later.

Shannon considered the conduct of Texaco and Getty Oil reprehensible. "I was pissed off at the temerity of these fuckers," he would recall, referring to some of Texaco's witnesses. As for Texaco itself, Shannon was "morally outraged," among other things, at the claim that no one at Texaco had read *The Wall Street Journal*'s account of the Pennzoil "agreement."

And all of the so-called open issues that Texaco had raised, and the hypertechnical argument that Pennzoil knew it didn't have a contract because it never withdrew its tender offer—that stuff could not begin to outweigh the outward manifestation by everyone connected with Getty Oil that they knew they had a deal with Pennzoil, he believed.

It might have been different, Shannon would also recall, if Texaco had truly tried to heap the blame on the Getty people—"if they had only said, 'They lied to us.'" But Texaco had tried to defend Petersen, Winokur, Boisi, Lipton and everyone else on the other side of the deal—everyone, that is, but Gordon.

Shannon was on a mission.

When he followed the jurors into the deliberation room on Friday afternoon, Shannon immediately sensed that a voting bloc had been formed in this case as well. The black jurors, who had continued feeling resentful toward Judge Casseb, were sitting together at one end of the table. Thus, Shannon nominated one of the black jurors, Fred Daniels, to serve as the foreman. Everyone liked Daniels; he was quiet and dignified—a letter carrier who wore a sports coat to court. Daniels, however, declined to serve. A white woman, Shirley Wall, was then nominated and agreed to preside. But a moment later, after recognizing that it was the foreman who stood up in court, she too withdrew.

"I got nominated by default," Rick Lawler would later recall, and he agreed to serve.

It was already clear that the jury was leaning to Pennzoil's side. But the group's thoughts were ill-formed; they were still digesting the closing arguments, trying to see through the smokescreens that the lawyers on both sides had developed, feeling each other out without wishing to overcommit themselves. Some pawed through the pile of exhibits left in the jury room; Shannon got a small thrill examining the original yellow legal pad on which the president of Texaco had written *only have an oral agreement . . . take care of Liedtke.*

Lawler tried to focus the group on Special Issue One; there was no point in discussing anything else until the group had decided whether Pennzoil had a contract. One by one, the jurors spoke—some strongly in Pennzoil's favor, citing the California court affidavits, the Getty Oil meeting notes and other exhibits. A few held back, saying virtually nothing. A few others expressed strong misgivings about Pennzoil's case, citing the "open issues" that suggested Pennzoil did not have a contract, as well as the testimony of the three Getty Oil directors who said that the board had voted only to enter into negotiations with Pennzoil.

Finally, Lawler presided over a straw poll. Seven said Pennzoil had a contract; the remaining five—accountant Susan Fleming and all four blacks—said it did not, although one of them later changed his vote. Everyone felt discouraged; a verdict in either direction required ten votes, and at the end of the day, about ninety minutes after they had received the case, ten votes seemed as distant as ever.

Jim Shannon went home for the weekend with a splitting headache, wishing he were still an alternate.

On Sunday morning, Rick Lawler got his first chance of the weekend to get away from the family and collect his thoughts. He drove to a quiet inlet along Galveston Bay and went crabbing, using chicken wings as bait.

"I ran the case through my mind, looking at the intent of each guy," he would recall. And next to each guy was a piece of evidence screaming *intent to be bound*. Gordon's lawyer had stood up in a California court referring to "the transaction agreed upon." Gordon's investment banker had filed a court affidavit to the same effect. Gordon himself drank champagne with Larry Tisch. As for the museum, Lipton specifically refused to warrant clear title to the museum's stock to Pennzoil and had demanded Texaco's protection in any breach-of-contract suit by Pennzoil. And for Getty Oil itself, Sid Petersen had told *Fortune*, "We approved the deal. . . ." The company's investment banker, Geoff Boisi, had submitted his bill the day after the Pennzoil deal and the day before the Texaco deal—even though the bill was scheduled for presentation no matter what else happened. The company's lawyer, Bart Winokur, had signed off on a press release announcing the terms of the Pennzoil deal.

"The action surrounding their intent was overwhelming," Lawler recalls concluding.

Then Lawler thought about the legion of takeover professionals on the Texaco-Getty side: midnight meetings in hotel rooms, lawyers and investment bankers passing information about someone else's deal to one another, playing both sides against the middle, one group playing other groups against each other. "The people in my company would never consider dealing in the way these people dealt," he recalls thinking.

Guilty, Lawler concluded, picking up his crab pail and returning home.

The weekend was therapeutic for all the jurors. Each had gone through the same exercise as Lawler, in some cases over and over.

Some jurors, such as Shirley Wall, had taken notes throughout the testimony that they could now use to refresh their recollections.

Monday morning, when the deliberations resumed in a more businesslike and less stressful atmosphere, the jury was nine-to-three on Special Issue Number One. The bloc of opposition was quickly cracking, and no longer were the jurors polarized along racial lines.

Everyone in the room harbored deep resentment toward Texaco's first three witnesses: Winokur, Boisi and Lipton. But Texaco's next three witnesses were different. Wendt, Medberry and Tisch had actually voted in the Getty Oil board meeting, and the holdout jurors continued to be gripped by their testimony that they never intended to be bound to Pennzoil.

Then Shannon remembered the notes taken by Stedman Garber, the treasurer of Getty Oil, while he was receiving a telephone report on what had happened at the board meeting. Shannon slipped the document to the center of the jury's conference table: *Board agreed but for Chauncey that deal should be done.*

Shannon was coming on too strong. *Okay, Jim, whatever you say,* Susan Fleming said sarcastically. "Quite a few other people at the table told him to be quiet and sit down," Lawler would recall. But with the Garber note and other evidence, Shannon finally helped turn around one of the three remaining holdouts, leaving just Fleming, and Velinda Allen, a twenty-five-year-old hospital clerk, by themselves on Special Issue Number One.

The jury had mustered the ten votes necessary to find that Pennzoil had a contract.

They moved to Special Issue Number Two: Did Texaco knowingly interfere? The *stop the signing* notes remained fresh in the jurors' mind because their author, Al DeCrane, had been Miller's last witness. DeCrane and Miller had done a good job trying to suggest that Texaco felt Pennzoil did *not* yet have a deal, and that Texaco was only recognizing the need to move with dispatch. But staring at such words as "stop train" could not help creating an impression of *interference.* Again it was ten-to-two.

Special Issue Number Three: Actual damages. The $7.53 billion question.

The jurors discussed what a horrendous amount of money it was. Rick Lawler would recall having an empty feeling in the pit of his stomach. A few expressed concern about how well they would sleep at night knowing they were responsible for inflicting such an assessment. Some expressed less concern for the economic well-being of Texaco than for the personal feelings of Dick Miller—a man whose courtroom temperament was jocular, sometimes even sweet, a lawyer who struck many of the jurors as a man of greater sincerity than the often acerbic Jamail.

Yet little choice seemed to exist: Judge Casseb's instructions said the law entitled Pennzoil to all the damages necessary to make itself whole, and Pennzoil had documented the number backward and forward. Pennzoil's damages expert—an impressive witness, an unpaid witness!—had called the number "conservative," and indeed Pennzoil had tossed out some even higher numbers based on more liberal economic assumptions.

What had Texaco given the jury to work with? Zip. Shannon read aloud from Judge Casseb's instructions: "You will not consider or discuss anything that is not represented by the evidence in this case."

"It isn't our place to fabricate a case for Texaco," Shannon declared. Seven billion five hundred and thirty million it was, by the same vote of ten-to-two.

Next, the jurors had to take up the business of punitive damages.

The jury had been deliberating for barely six hours, including its abbreviated session on Friday afternoon. But out in the courtroom, where the Pennzoil and Texaco lawyers stood their vigil, it seemed like an eternity.

Then Judge Casseb took the bench. "I have received two notes from the jury," he announced. "The first one requests the definition of 'wanton disregard.'" The second question asked whether the Texaco annual report was in evidence.

At the Texaco table, every heart sank. It was obvious that the jury had already reached Special Issue Number Four: whether Pennzoil was entitled to punitive damages. There was no telling for sure, but it appeared they had concluded that Pennzoil had a contract. And the only reason the jurors could be interested in Texaco's annual report—which wasn't in evidence—would be to satisfy themselves of how much Texaco could afford to pay.

Judge Casseb returned a note saying, "Wanton is used in its ordinary sense and is synonymous with reckless or heedless disregard of the rights of others." As for figuring out what Texaco could afford to pay, the jurors were on their own.

*   *   *

Punitive damages were quickly proving to be the most contentious issue in the jury room. The jurors had had little flexibility on the issue of actual damages, but they had all the latitude in the world on punitives. "Now it is really at our discretion," Shannon would recall thinking.

Six of the jurors wanted to sock Texaco with the full $7.53 billion—a few had even entertained the notion of *increasing* the amount. The two women still holding out on Special Issue Number One would not go along with anything but zero. The remaining four jurors were in favor of awarding punitives, though not in the amount requested by Pennzoil.

Shannon put himself in the third group by design. "I wanted to be able to argue from a neutral position," he would explain.

They began trying to trade. *Should we cut it by one half? One third? One quarter?* Lawler tried to play Solomon, obtaining one group's agreement to reduce the assessment if another group increased its figure by the same amount. Some jurors refused to go below $3.8 billion; some refused to go higher than $2.5 billion. "It was crazy," Lawler would recall. "People were throwing numbers all over the table."

Yet no matter how they cut it, the outcome involved billions. *How can we just pick a figure involving billions out of the air, on some caprice?* some asked.

One thing troubled Lawler throughout the discussion. It was apparent that most jurors—particularly those arguing for the big figures—wished to punish the Getty people as much as the Texaco people, especially the three takeover professionals whose credibility Texaco had vouched for on the witness stand. There was Lipton, who in the jury's view had played both sides of the street in the Getty deal, who had insulted his "associate" Pat Vlahakis and who had told the court that the amount of money may control when an agreement has been reached. There was "Back-Door Bart" Winokur, who had walked away from a drafting session with the Pennzoil lawyers to cut a deal with Texaco and who had done nothing but fence on the witness stand. And there was Geoff Boisi, the "$18-Million Man," who had doubled his investment-banking fee by delivering Getty Oil from Pennzoil to Texaco—and who had insulted some jurors with his evasiveness. Juror Theresa Ladig was offended that some of Texaco's witnesses would not deign to look the jurors in the eyes.

"I wanted to know, frankly, who was on trial here," Lawler would recall. "I knew I was going to hold Texaco responsible, but

I wanted to know if I could give more because of the impact of other individuals.''

At 2:22 P.M., Judge Casseb called the lawyers together to read another note from the foreman: "To what extent is Texaco liable for the actions of Lipton, Winokur and Boisi?"

The correct answer was *not at all*. Texaco was not remotely liable for their actions in this trial; if Pennzoil ever resurrected its lawsuit against the Getty group in Delaware and happened to win a judgment specifically against the Getty entities, only then, under the indemnities, was Texaco liable.

Jamail immediately recognized that Pennzoil's strategy had paid off: The jurors, overdosed with the indemnity argument, were taking out their wrath on Lipton, Winokur and Boisi through Texaco. "I think you just ought to say, 'Follow the evidence and the charge,'" Jamail told the judge.

Dick Miller too grasped what was happening. But while he had not been closely involved in the preparation of the charge, he recalled its containing an instruction that a party is accountable only for its own agents.

"A request for anything additional, I guess, is denied?" Miller asked.

"At this time," Judge Casseb said. "No objections?"

"This is fine with us, Your Honor," Miller said.

Texaco had just committed the most costly oversight in courtroom history.

That day, after spending the month riveted to the $47 mark, Pennzoil's stock price jumped to nearly $50 a share.

"Well, this sure doesn't clarify it," Rick Lawler recalls thinking when the bailiff delivered the judge's ambiguous response. "We just went on thinking, 'We're going to hit Texaco with this,' because these three entities had done something we considered horrendous."

After another two hours of discussion, the jurors ended their first full day of deliberations at 3:58 P.M. with the issue of punitive damages on the table. They knew that the next day they would be reporting a verdict; the only question was whether it would be $7.53 billion, $15 billion or somewhere in between.

That evening Dick Miller missed a heartbeat.

The instruction that he *thought* was in the court's charge was not there after all. The sentence he recalled—that a party is responsible only for its own representatives—was contained only in some language that Texaco had proposed, but that Judge Casseb had refused to adopt. "We just didn't remember," Miller says. "There's no other way to put it."

First thing the following morning, he and his co-counsel implored Judge Casseb to re-instruct the jury with the following note:

> The court has considered your earlier question: To what extent is Texaco liable for the actions of Lipton, Winokur and Boisi?
>   . . . The court now instructs you that a party is only responsible for the actions of its own employees, agents or representatives acting within the scope of their employment.

"The request will be denied," Judge Casseb said.

Jim Shannon had conceived a plan.

He entered the jury room a few minutes before nine o'clock Tuesday morning and took a seat next to the one used by Susan Fleming, the accountant who was still holding out on Special Issue Number One, the issue of whether Pennzoil had a contract. Shannon and Fleming had exchanged harsh words already, but Shannon vowed that this morning he would break through the "linear thinking" he presumed she had come to depend on in her career as an accountant.

A year after the verdict, Fleming did not wish to discuss her exchange with Shannon or anything else involving the case. "I'm trying to forget," she would say.

But Shannon's recollection, which Lawler generally confirms, is distinct. He grabbed a sheet of paper and drew a line down the middle. On one side he listed the terms in Pennzoil's memorandum of agreement; on the other the terms outlined in the press release issued by Getty Oil. They matched.

He then showed his fellow juror the list of demands that Lipton had presented to Texaco in behalf of the museum—demands that included a guarantee that no matter what happened, Texaco would see to it that the museum got $112.50 for its shares, the same price Pennzoil had agreed to. He read aloud the "representation and warranties" section of the museum's contract with Texaco, in which Lipton had specifically refused to warrant that Pennzoil did not have a claim to the museum shares.

"No representation is made with respect to . . . the Pennzoil agreement," the museum's contract stated.

Susan Fleming changed her position. So too did the other woman holding out. "I was very gratified," Shannon says.

The verdict was now unanimous on Special Issue Number One and soon everything else. Only one obstacle remained: one woman juror, a fifty-six-year-old insurance worker for Harris County, continued holding out for the full $7.53 billion in punitive damages.

But Shannon had conceived a way of dealing with that as well.

He had become convinced that the jury needed a rationale—a formula of some kind—to pick the damages figure. These were punitive damages at stake. *What precisely are we punishing?* They were also referred to as exemplary damages. *What kind of example are we trying to set?*

Shannon had recalled the closing line in Irv Terrell's share of the summation for Pennzoil. "Just don't let them get away with these indemnities," Terrell had implored the jury. "Please don't."

That was it: The jury could punish Texaco for each of the indemnities it had awarded.

One billion for the indemnity to the museum. One billion for the indemnity to Gordon. And one billion for the indemnity to Getty Oil—even though Getty Oil's indemnity did not include the specific mention of the Pennzoil deal.

Foreman Lawler not only found the logic appealing, but recognized that the figure represented a nearly perfect compromise between the high and low figures that some of the jurors had been bargaining for the prior afternoon. And although the figure was less than half what Pennzoil was seeking, you could argue that linking it to specific evidence of "misconduct" would send a far more powerful message to corporate America than rubber-stamping Pennzoil's entire request.

The "twelve angry men," as Shannon would later call the jury, were now unanimous on every issue.

Hugh Liedtke was already standing vigil for the verdict with his friend and lawyer Joe Jamail when the lawyers were told that the jury was ready to come out. "Hang on," Jamail told him. "We're gonna take you on a rocket ride."

The jury filed stoically into the box. Lawler could feel his heart pounding as he handed the verdict to Carl Shaw, the bailiff, who handed it up to Judge Casseb, who began reading the questions and answers aloud.

Special Issue Number One: *Do you find that Pennzoil had a contract?* "The answer is, 'We do.'"

A murmur swept the courtroom.

"Is that the answer each one made?" All twelve jurors raised their right hand.

Special Issue Number Two: *Do you find that Texaco interfered?* "The answer is, 'We do.'" Again, all twelve raised their right hand, and the murmur built to a low roar. Knees were beginning to get weak throughout the courtroom.

Special Issue Number Three: *What sum do you award Pennzoil for actual damages?* "The answer is, 'Seven point five three billion dollars.'"

A gasp went up from the crowd.

"I want order in the court, please!" the judge said.

With the jurors continuing to raise their hands in unanimity, the judge read the remaining answers, including the $3 billion figure in punitive damages. Fortunately for Judge Casseb, Lawler had expressed the money figures with the word "billion." "If they had put zeroes," the judge would later recall, "I don't think I could have read it."

Joe Jamail leaned over the back of his chair to address Liedtke. "That suit you?" Jamail asked.

Liedtke was speechless. He threw his arms around Jamail.

Outside the courtroom, the jurors and lawyers began exiting through a gauntlet of television cameras, boom microphones and lights. "We won't tolerate this sort of thing in corporate America," Lawler proclaimed. Shannon chimed in, "The idea that 'anything goes' is dead."

The lawyers remained so floored they could barely muster replies.

"I'd say we got our asses whipped," Miller said. "At least they let us keep our briefcases."

And when asked to cite the major factor in the case, Joe Jamail replied, "I was."

After throwing back a few beers with his client and co-counsel, Joe Jamail returned to his office to find a banner in the reception area emblazoned, WELCOME HOME, $10.5 BILLION MAN. Jamail spent the remainder of the afternoon basking in the glory. Governor Mark White called. Coach Royal called. So did Willie Nelson.

*   *   *

By that evening, Hugh Liedtke had regained his speech and showed up at Jamail's house with his wife Betty Lyn, and Baine Kerr and his wife, Mildred. Jamail's cook, Joyce, turned out hamburgers while the men drank shots and beers and relived the satisfaction of winning justice. By about ten o'clock, the guests were getting ready to leave.

"Joe," Liedtke said, "I gotta tell you something." He paused, then smiled. "You left four and a half billion in punitives on the table!"

Jamail let loose a deep and hearty laugh. "Fuck you, Chairman!"

The expletives were also flying in the office of Texaco Chairman John McKinley in White Plains, but not from McKinley's mouth. He sat in stunned and silent disbelief, listening to his general counsel, William Weitzel, delivering the "message" sent by the jury one thousand seven hundred miles away. Weitzel then got on the company's public-address system and reported the same news. Nobody could believe it.

On November 19, 1985, when the verdict hit, Texaco's net worth totaled $13 billion. The value of all its shares in the stock market barely totaled $9 billion.

Was Texaco dead?

History's greatest saga of corporate brinksmanship—bigger than the Getty takeover or any other, bigger even than the record-setting trial itself—was about to begin.

· 25 ·

The psychologist Elisabeth Kübler-Ross, who specializes in people confronted with death, has identified five stages of reaction among her patients: denial, anger, bargaining, depression and acceptance. She might as well have been talking about Texaco. Like a patient convinced that his rare tropical disease is actually a misdiagnosed case of flu, Texaco refused to recognize the judgment was anything but a mistake, a fluke—something that would pass as quickly as a three-day virus, as soon as some judge, any judge, even Judge Casseb, made the most cursory review.

"This jury verdict will have *no impact* on the way we run the company," President Al DeCrane declared. Everywhere throughout the 150-nation Texaco organization, the same order went out: *Conduct business as usual.*

J. J. McGraw, an executive responsible for obtaining leases on drilling acreage, went ahead with negotiations on a $100 million transaction with a group of other companies. "The verdict won't interfere with the deal," he said.

K. J. McKee Jr. continued swapping $60 million worth of crude oil a day entirely on credit. "I just can't envision anything that will stop us from doing that," he said. "The whole verdict is just out of phase with the real world."

J. V. Woolly, a division accounting manager, instructed his

820 employees to report the slightest hints of concern or anxiety among royalty owners and other Texaco creditors; days passed and none was heard. "I just perceive it as a case of North versus South," he said. "I see no cause for concern."

The outside world was telling Texaco not to worry, either. From *The Washington Post:* "It's an absurd verdict—the kind of verdict that is bringing the whole American tort system and its wide-open verdicts into disrepute. It's also a killer verdict, intended to put the defendant out of business." From the *Sun Herald* of Gulfport, Mississippi: "The tort system reached the zenith of absurdity in Texas last week. . . . It is a disturbing commentary on a legal system that permits such foolishness." From the *Fort Worth Star-Telegram:* "\$10.53 *billion?* Come on, now!" From Michael Schwartz of Wachtell, Lipton, quoted in *The Wall Street Journal:* "It's like the inmates are running the asylum." The *National Law Journal* asked "whether the real lessons of *Pennzoil* had to do with a runaway jury."

Absurd . . . foolishness . . . inmates . . . runaway jury. . . . What could Texaco have to worry about?

Even among those who found merit in the issues of the case saw nothing but silliness in the damages award. *The Los Angeles Herald-Examiner* called it "an honorable decision," notwithstanding the "outrageous penalty." Professor Tom Dunfee of the University of Pennsylvania's Wharton School took up the case in his class on business responsibility. "This is business ethics on the front page," he explained. But while there may have been "bad ethics on the part of Texaco or Getty or probably both," he believed that the damages were "almost capital punishment."

More importantly, the worldwide business community refused to believe that the amount, or anything remotely close, would ever be paid. "The main perception is that the judgment will be reversed or whittled away," said an official of Caltex Petroleum Corporation, a worldwide partnership of Texaco and Chevron. "It is difficult to believe that it is the intention of any U.S. court to ruin a large company like Texaco," said an oil regulator in Britain. "It doesn't affect our confidence in entering a joint venture with them," added a French oilman in Bogota, Colombia.

Even the stock market—after driving Texaco's shares down 11% to \$34.75 within two days of the verdict—bid the price back up to \$35.12 on the third day. Standard & Poor's, whose business is assessing the financial health of corporate America, saw no reason whatsoever to reduce Texaco's bond rating; the appeals, after all, "could delay the final resolution of the lawsuit for several years."

As part of its business-as-usual strategy, Texaco bravely went ahead with the annual show-and-tell presentation for Wall Street stock analysts that it had unfortunately scheduled for November 20, the day following the verdict. John McKinley threw away his canned remarks and spoke from notes about the case, expressing every assurance that the verdict would quickly go away.

"We feel very confident that this is just an outrageous travesty and that there is no way it can withstand judicial review," added Bill Weitzel, the general counsel. Texaco, he said, would go all the way through the Texas appeals court system, on up to the U.S. Supreme Court, if necessary—which it doubtless wouldn't be. "We don't expect to have to follow that chain," Weitzel said. Why, Judge Casseb himself had the power to reverse the verdict, or knock it down to size, at a hearing scheduled to begin within three weeks. *This thing could be over any day now.*

As he was pinned to the wall by reporters and stock analysts seeking more information, Weitzel brushed aside questions about a certain procedural detail pending in the Texas court. Jamail had made a cryptic comment to the press—something about Texaco having to post an appeals bond in the full amount of the judgment. "We'll make sure we have the money up front," Jamail had said. "I'm sure they can afford it. And if not, Pennzoil is perfectly capable of running Texaco." Weitzel dismissed the bond issue out of hand; regardless of the court rules in Texas, he indicated, this case was *special.* "The award is so excessive and the case is so different from any other proceeding that we don't think it will be required," Weitzel said of the bond.

Once again, Texaco was denying the strength and cunning of its enemy—just as when it failed to answer Pennzoil's original complaint in Delaware, just as it had throughout the trial in Houston. Texaco had no idea what was coming.

In 1973 former Houston Astros infielder Bob Aspromonte was jump-starting a car when a small "pop" went off, sending a battery cap flying into his right eye, permanently and severely impairing his vision. His lawyer, Joe Jamail, won him a judgment of $675,000 from Sears, Roebuck & Company, claiming that the retailer had failed to include the proper safety warnings with the jumper-cables it had sold to the ballplayer.

Sears announced it would appeal, but quickly let down its

guard. Under the law of Texas, and many other states, the winner in a lawsuit is free to take possession of the loser's assets unless the loser has posted an appeals bond equal to the amount of the judgment. These bonds—called supersedeas bonds because they supersede the effect of the judgment—are routinely available for a fraction of the face value from insurance companies. Just as court bail preserves the freedom of an armed robber while his or her case is on appeal, supersedeas bonds permit corporations to maintain full control of their assets until the last gavel has been lowered in the appeals courts.

Sears, however, failed to post the bond, and Jamail seized the moment.

With a pack of photographers and two sheriff's deputies in tow, he burst through the doors of the Sears store on Main Street near downtown Houston and marched into the manager's office, waving a court order giving his client control of the store. *Maybe we'll have a big sale today,* Jamail announced. *Maybe we'll give everyone the day off with pay.* An hour later, Sears surrendered, telling Jamail it would abandon its appeal and write a check for the full amount of the judgment.

From the moment he heard the jury verdict, Jamail began wondering where Texaco was going to buy a bond to cover *its* judgment, if it were upheld and formally entered by his friend Judge Casseb. Including interest, which would pile up at the rate of $3 million *every day* through the two-to-three-year duration of the appeal—Texaco would have to post a bond of $12 billion or more. And before it had ever set foot in the appeals courts.

It could never be done. Irv Terrell of Baker & Botts made some inquiries and learned that the worldwide insurance industry's bonding capacity totaled just $700 million. Within about ninety days, if Judge Casseb entered the verdict, Pennzoil would have the right to have county sheriffs all over Texas putting red tags on Texaco oil wells, filling stations, refineries and pipelines. Those assets would then belong to Pennzoil, to return to their original owner only when, and if, the judgment were overturned.

Jamail began imagining himself in John McKinley's office at Texaco instead of in the manager's office at the Main Street Sears store. *Mmm*, he thought. *Maybe I'll rename it Jamaco.*

In addition to "business as usual," Texaco adopted another strategy in the days immediately following the judgment: "strength in numbers." After permitting Dick Miller's tiny law firm to stand

alone against Jamail and his huge back office at Baker & Botts'
firm during the trial, Texaco now swung to the other extreme.

First it fired up its longtime New York litigation firm—Kaye,
Scholer, Fierman, Hays & Handler, whose representative had
been relegated to sit in the bleachers at the trial. Kaye, Scholer
was instructed to gear up for a Herculean appellate effort. Within
days, the firm had twenty lawyers on the case.

Then Texaco engaged Houston's Fulbright & Jaworski, one of
the largest law firms in the West. Over the years Fulbright had
established a reputation as Houston's most battle-ready law
firm—a partnership where, in 1985, the clerks still wore uni-
forms, where crew cuts remained in fashion and where the trial
lawyers pulled out every stop to win. Texaco was unaware, however, that Fulbright's reputation for attacking judges as well as
adversaries had provoked resentment among the ranks of the
Texas judiciary. "Unfortunate," one Texaco lawyer said after
learning of Fulbright's image.

Texaco didn't stop at hiring Fulbright & Jaworski. For a flat
retainer rumored to total $2 million, it took on Rusty McMains of
Corpus Christi, Texas, a bearded, long-haired and heavyset thirty-
nine-year-old prodigy of Texas appellate procedure who had decided to become a lawyer when he was a junior high school
student watching episodes of *Perry Mason*. McMains had won a
large reputation for influence among the appellate justices of
Texas—a reputation exceeded only by Jamail and a few other
lawyers in the state. McMains denies he received a $2 million fee
from Texaco, but quickly demurs: "I won't say whether it's more
or less."

Which among Texaco's law firms—Miller, Keeton; Kaye,
Scholer; Fulbright & Jaworski; and McMains' firm—would take
the lead? None, as it turned out, for David Boies had also made
the scene. (*Boies*, not to be confused with Geoff *Boisi*, Getty Oil's
investment banker from Goldman, Sachs.) Boies was rapidly
gaining—indeed, cultivating—a reputation as the hottest litigator
in New York.

He was every bit as eccentric as the case deserved. Boies was a
top partner at the whitest of Wall Street's white-shoe firms—
Cravath, Swaine & Moore, a firm of such self-conscious deco-
rousness that it kept a candelabrum on the reception desk. On
some mornings, however, Boies' office was a veritable garbage
heap, littered with empty pretzel bags and Pepsi cans, loose
change, perhaps a bottle of Marriott Hotel shampoo, and other

wreckage of nights spent poring over drafts of legal papers. His office abounded with signs of his passion for gambling: a miniature craps table on the credenza, a pile of scratch-and-win coupons from a fast-food chain, a package of unopened poker chips amid the rubble on his desk. In fact, the desk itself was a poker table.

In his mid-forties, Boies could easily pass for a man ten years younger—even twenty years younger, taking his attire into account. When he wore a suit it was apt to be a $90 special from Sears. He bought his dress shirts from L.L. Bean, wore an old parka in the rain and took his work home in a burgundy backpack.

Boies was still aglow from successfully defending CBS against libel charges by General William Westmoreland; a map of Vietnam hung on the wall of his office. Boies had earlier extricated IBM from a $350 million judgment to Telex Corporation—to that time, the largest judgment ever obtained in a U.S. court.

"It's like in the Western movies," George Vradenburg III, the general counsel of CBS, would remark. "There always seems to be a younger, faster gunslinger coming into town. Right now, David's got the hot hand."

But hot enough for Joe Jamail and his Texas gang?

Three days after the jury verdict, Boies was attending to his terminally ill mother in California when Bill Weitzel, the Texaco general counsel, called to offer him the case. It was implicit, Boies would recall, that Cravath would be lead firm in the defense, even though Texaco's other New York firm had been working for the company more than thirty years. Boies' assertion, however, came as a surprise to the lawyers at Kaye, Scholer, who would later say they were "shocked" and "insulted" at being shoved aside for the hotshot Boies. "It doesn't surprise me with this bunch of sorority girls," said Jamail, whose command of the Pennzoil team remained undisputed—even if Baker & Botts was handling most of the work.

The hiring of so many new new lawyers immediately diminished the role of Miller, Keeton, Bristow & Brown, and Dick Miller himself. He and his colleagues, particularly Keeton, would be instrumental in educating the new Texaco lawyers on the case, but before long the role of Miller, Keeton would all but vanish. Dick Miller would never acknowledge the hurt.

"My father was a guy who could take a loss," he once said, speaking in another context. "If the screen door broke, he was liable to spring a gasket, but if he lost a tank car of molasses, he lived with it." A lot like Dick Miller himself—on both counts.

"A trial lawyer has to accept a willingness to accept defeat,"
he would explain after the verdict. "Sure, it's hard to get beat in
public. But you protect yourself—so you can try tough cases next
week. When a lawyer lacks confidence, you can smell it. You can
see his yellow streak. I know that if I had won, I'd have gotten
most of the credit.

"The only reason I do this is because it's exciting," Miller
would continue. "I'm sixty and I feel thirty-five. And I'm not
through knocking dicks in the dirt."

The week following the verdict, Texaco decided to "send a message" of its own, this one intended to reach the bench of Judge
Solomon Casseb. If the judge had the slightest inclination to uphold the verdict, Texaco would see to making him painfully aware
of the consequences.

Within a week of the verdict, after discovering that it might
eventually have to post a $12 billion bond after all, Texaco had
begun backing away from its business-as-usual bluster. Newspapers that had found it impossible to obtain interviews immediately
after the verdict suddenly found Texaco executives at their door-
step all but begging to be interviewed. "It was kind of curious,"
said John Talton, business editor of the *Beaumont Enterprise*, the
largest newspaper in the Texaco-dependent region of Southeast
Texas. "After the story broke, we tried and tried to get interviews, but they wouldn't arrange them. And then all of a sudden,
they decided to call us."

In Dallas, Houston, Beaumont and elsewhere, the message was
the same: Texaco, a cornerstone of the Texas economy, could
crumble under the weight of the history-making verdict unless
Judge Casseb lifted the threat. Some 14,000 Texas jobs and $500
million in annual Texas wages were in jeopardy. Royalty payments of more than $300 million to 70,000 Texas landowners
hung in the balance. The pensions of its 12,500 Texas retirees
were threatened. And the investments of the company's 319,000
U.S. stockholders—one-tenth of them in Texas—could be all but
wiped out. "Under this enormous economic threat that hangs over
Texaco's head, we can't go on with the kind of investment program we have planned, including two major investments for the
Port Arthur refinery," Vice Chairman Jim Kinnear warned in the
Houston Chronicle.

But the public-relations campaign boomeranged, striking fear
in the hearts of Texaco's employees, sapping morale and causing

anxiety over job loss. The threats angered Judge Casseb, who
months later would recall his resentment of the "Texaco media
blitz."

And most of all, a candid statement by President Al DeCrane in
one interview began to cause doubt in the business world, which
had initially rallied around Texaco as if it were a friend in need.
DeCrane had sat down for an interview at the *Dallas Morning
News* on Monday, November 25. "If a twelve-billion-dollar bond
is required—Texaco doesn't have twelve billion dollars, and in
my opinion probably can't get it—then we'd have to look for
some heroic measure," DeCrane said in the interview, "whether
it's Chapter Eleven or whatever."

Texaco had dropped its bombshell. Under Chapter 11 of the
U.S. Bankruptcy Code, Texaco could forestall Pennzoil from ever
collecting the judgment. But at the same time everyone with a
financial stake in Texaco—banks, stockholders, employees, oil
companies, drilling contractors sitting on Texaco holes, pipeline
partners, pensioners—would be relegated to the unhappy status of
unsecured creditor, hoping to receive a bankruptcy settlement of
the greatest possible cents-on-the-dollar. It would be far and away
the biggest bankruptcy case in history.

Bankers who had derided the Texas judicial system as a kan-
garoo court suddenly realized that the tiger had teeth. A group of
banks with a $125 million short-term loan to Texaco demanded
immediate repayment. Officials at Manufacturers Hanover Trust,
one of Texaco's lead financial institutions, immediately began
worrying that Texaco would begin bouncing checks. Southern
California Edison accelerated Texaco's payment schedule for
electricity from a monthly to a weekly basis. Money-market funds
and other institutions holding about $200 million of Texaco's
commercial paper—its corporate IOUs—demanded that the company take them back. Texaco had the right to refuse, however,
and did so.

The reaction in the stock market was far less forgiving. As the
tidal wave of sell orders gathered momentum, the throng of floor
traders crowding the Texaco trading "post" on the floor of the
exchange grew to forty. "Nobody here's eaten lunch. Nobody's
even gone to the bathroom," floor trader R. Peter Rose said.
"I've never seen anything like it." When the dust settled, Texaco's shares had closed at $31.50 each—down from $39.25 just
before the verdict hit. In barely a week, nearly $2 billion in losses
had been inflicted on the investing public.

But for every terror-struck seller of Texaco stock there was a steel-nerved buyer—someone who felt that Judge Casseb was ready to knock the verdict down to size. "We don't believe Texaco is guilty," explained a portfolio manager at First Pennsylvania Bank in Philadelphia. Smith Barney, Harris Upham & Company, the brokerage institution that boasted of earning money the old-fashioned way, refused to abandon its old-fashioned belief that justice will always be done. "We don't believe the award is commensurate with the damages Pennzoil suffered," one of the firm's analysts said publicly. Batterymarch Financial Management, one of Wall Street's shrewdest and most successful portfolio managers, held on to each of its two million Texaco shares—worth more than $60 million. "It's usually not a good idea to sell into the heat of such emotion," an official explained.

Trading in Pennzoil's shares was every bit as frenetic. The day after the jury verdict, when Pennzoil shares had risen more than $10 a share to $58, Salomon Brothers recommended dumping them, on a bet that the award would be disallowed or cut back on appeal. But even Kathryn Dera, a widow whose nephew was a Salomon Brothers analyst, refused to sell. She saw *Pennzoil v. Texaco* as "a David versus Goliath situation." A week later, with Pennzoil up to about $63 a share, Salomon realized that some quagmires are too deep to fathom. "The legal issues and options . . . are enormously complex and beyond our ability to analyze sensibly for investors," the firm told its clients.

"Everything depends on the psyche of just one judge in Texas," said Phil Dodge, an associate of Kurt Wulff at Donaldson, Lufkin, which had begun urging investors to unload Texaco shares. Even the Wall Street bond-rating agencies, which had initially declined to reduce their Texaco ratings, began to waffle. Standard & Poor's acknowledged that if Judge Casseb formally entered any judgment greater than $1.5 billion, it could spell trouble. Moody's cited "a very small but real possibility that Texaco's liabilities may be frozen for an indefinite time." Still, Moody's said, a judgment for the full amount remained practically unthinkable.

Texaco's expanding rabble of lawyers was still trying to act like a team. "There was a considerable amount of undirected activity," one of them would recall. "Some people were doing one thing, some people were doing another thing, and a lot of it was irrelevant."

The bond issue split the lawyers along regional lines. Maybe, just maybe, the New York lawyers thought, Texaco could escape the life-threatening requirement of a $12 billion bond by somehow getting the case into *federal* court. After all, wasn't there a federal Constitutional issue at stake here? Didn't the Constitution guarantee the right to due process? How could Texaco pursue its right to an appeal if it were already out of business? Boies and the other New York lawyers liked the idea; it was chancy, but suitably brash.

The Texas lawyers, however, unanimously opposed it. Seeking redress in the federal court system would be a slap in the face of the Texas courts, where the company would be undertaking its appeals of the judgment—if the judgment were entered. It would be one thing if the federal courts were a sure bet, the Texas lawyers said. But in fact a federal-court foray was a long shot. "Is there a federal statute that says if you just get fucked to the wall, you can get into federal court?" Dick Miller asked at the time. "I don't think so."

As near as Texaco could reckon at the moment, the bond crisis remained several weeks away; under Texas procedure, Pennzoil could not begin seizing physical possession of Texaco's assets for about ninety days, giving Texaco's lawyers plenty of time to invent a solution. For now, the lawyers had to concentrate on the next week's judgment hearing, and that presented a crisis all its own.

Suddenly, dozens of lawyers who were wholly unfamiliar with the twenty-four-thousand-page trial transcript were working alongside lawyers who had suffered through every word. The new lawyers were dismayed at the state of the record: Why no damages evidence from Texaco? Why were certain objections never made? Why was certain evidence never incorporated into the record—even if only outside the jury's presence, so that the appeals courts could see what the jury was never allowed to see? Why was Marty Lipton called as a witness? Why was the transcript never retyped into a computer so that the lawyers didn't have to contend with more than one hundred volumes of testimony? The new lawyers *had* to know, yet the questions heaped additional anxiety on an already anxious situation. "Questioning why they did certain things is bound to produce a little tension," Boies explained at the time. "They feel like they have a lot invested in the case and we've only been in it ten days."

Disagreements erupted over the length of the brief that Texaco would submit to Judge Casseb. *This judge is never gonna read this thing*, some of the Texas lawyers said. *How can you say that?* one of the New York lawyers asked. *If it's one hundred twenty-five pages long, that's just one page per hundred million dollars!*

There were other problems. David Boies, the master of the group, was shuttling between New York and California as his mother's illness worsened, requiring him to coordinate the effort from a distance. Hurt feelings persisted at Kaye, Scholer over the entry of Cravath's Boies into the case. Annoyance grew at Cravath that Kaye, Scholer did not go out of its way to convenience Cravath with copies of the documents in the case.

It was, in fact, miraculous that the lawyers could accomplish what they did. But Texaco desperately needed more time. Pennzoil's rush to judgment had to be forestalled, if only briefly.

On Monday, December 2, just three days before the judgment hearing was scheduled to begin, Jim Sales, a short-cropped and stocky former Marine from the Fulbright firm, showed up before Judge Casseb, begging for only a four-day extension so Texaco could have more time to prepare its written motions and oral arguments.

"I vehemently object to this!" Jamail shouted before the judge. Sales' request was instantly denied; the judgment hearing would begin as scheduled on Thursday.

In fact, Texaco's predicament was worse than even its lawyers had imagined. Far worse.

Although Pennzoil was not free to begin seizing Texaco's assets for about ninety days, it *was* free to begin placing liens on those assets—immediately after any judgment was entered. Texaco did not realize this. Despite having so many lawyers involved in the case—some would later say *because* of that—the provisions of the lien rule had remained unclear. But the fact was that the very second Judge Casseb signed the judgment—if in fact he did—Pennzoil could begin filing the liens on every Texaco-owned asset in Texas. Texaco would continue to own the assets, but Pennzoil would have first claim on all of them. Texaco couldn't borrow money against them. Its business partners would be cast under a pall, for every asset that Texaco jointly owned with another company might also become subject to a lien by Pennzoil.

Pennzoil, for its part, was well aware of its right to file the liens—and began plotting to do so at the first opportunity. Instructions were issued to begin taking inventory of Texaco's assets—

anything on which Pennzoil could put a lien, if the judgment were entered and if Hugh Liedtke gave the order. "We'll file liens against their properties wherever we can," a Pennzoil staff member told a reporter.

The stock traders began gathering at 6 A.M. Thursday in State Court No. 151, and by 7:30 A.M. every seat, and much of the standing room, had been taken. About a dozen traders were already milling about with cellular mobile telephones. Robert Mayes, who had placed bets in the options market on a victory by Pennzoil, flew in from Pensacola, Florida, just to be close to the action. "Houston is a lot closer to home than Las Vegas," he explained. One local lawyer, hired by a large New York investment firm to observe the court proceedings, said his client wanted him to report immediately any facial expression discernible on the usually impassive judge. "If this guy sneezed, they wanted to hear about it," the lawyer would say. The mystery made handicapping the game all the more interesting. Said Don C. Bustos of Duff & Phelps in Chicago: "All the wise men in the country can't get inside his mind, and that leaves lots of room for missed guesses."

At 8:57 A.M., three minutes ahead of schedule, Judge Casseb entered the courtroom for the first of two days of arguments over whether to enter history's largest court award. Nearly forty lawyers were crowded into the small bullpen at the foot of the judge's bench.

For two days, the four-and-one-half-month trial was replayed in telescopic form, with all the evidence once again trotted before the court: the argument over the meaning of *subject to*. . . . The fight over the meaning of *stop the train*. And so on. All the while, the arguments were framed by statutes and past cases. For every case that Texaco's Boies cited from memory, Pennzoil's Kronzer shot back at with citations from his equally superior memory. Boies and other Texaco lawyers went out of their way to bring up issues of federal concern—that Texaco's offer for Getty Oil, for instance, was precisely the form of competitive bidding that federal securities laws are designed to promote. Boies wanted to lay all the groundwork possible to establish some sort of federal jurisdiction in the case.

But the most passionate arguments arose around the numbers: $7.53 billion in actuals and $3 billion in punitives.

"What we have confronting the court today is the most devas-

tating specter of disaster in all of legal history,'' said Texaco's
Gibson Gayle Jr., the head litigator at Fulbright & Jaworski. ''We
have a situation here which threatens the existence of a major
American company which has its roots in the same state where its
destruction is now threatened.'' Said Jamail: ''The amount of
zeroes behind a judgment has never played a part in our system as
I know it—never played a part in whether something is right or
something is wrong.''

Five hundred million in actual damages—that was the absolute
most to which Pennzoil was entitled, Texaco's Dick Keeton argued. But the time to introduce evidence had passed, Jamail insisted; during the trial, ''Texaco chose to sit on its hands'' instead
of contesting Pennzoil's number. But that still didn't justify
the jury's finding, Keeton insisted. ''In some fashion they were
confused as to what it was they were finding. . . . What does
Pennzoil want? A billion barrels for free—indeed, with bonuses
laid on top?''

Keeton was missing the point, John Jeffers of Pennzoil countered. ''When you're talking about the loss of a billion barrels,
when you're talking about the loss of an opportunity to Pennzoil to
be transformed in one transaction into a major oil company''—
Hugh Liedtke's dream again—''. . . you are, make no mistake
about it, talking about a multibillion-dollar injury. *That's* what we
have suffered.''

As the courthouse throng grew, Keeton told Judge Casseb that
the largest award of punitive damages in New York history was $5
million—and that that amount was slashed on appeal. Pennzoil's
Terrell gave the judge the same old dose of the indemnities. ''If
ever there was a central feature of this whole lawsuit, it was
Texaco's willingness to put a cloak of protection around people
who ought not to be protected.''

Keeton replied: ''He's doing the same thing that happened
throughout this trial and, I believe, led to the result that if the jury
was punishing somebody, they were punishing what they perceived to be the actions of Mr. Lipton and Mr. Winokur and Mr.
Boisi. 'Okay, the indemnity transforms Texaco into *them* and we
don't like *them,* so that's three billion dollars for Texaco.' No
way, Your Honor. No way! . . . They're not commissioned to
render a verdict to 'send a message,'' Keeton said of the jurors.
Replied Pennzoil's Randall Hopkins: ''They gave us less than half
of what we asked for in punitive damages. But they *did* want to
send a message—and that's the function of exemplary damages.''

On Friday, Judge Casseb dismissed everyone for the weekend. He would render a decision on the verdict at 1:30 P.M. the following Tuesday.

Roger Gray: It is good to have you along with us on this Friday on *A.M. Houston,* December sixth. Let's check in with Sandy Rivera in the newsroom and see what's happening. Good morning, Sandy.

Rivera: Good morning to you, too, Roger. Texaco says if it is forced to pay a ten-billion-dollar judgment to Pennzoil, it will be destroyed. Visiting Judge Solomon Casseb listened to attorneys from Texaco and Pennzoil yesterday once again, but did not rule on the matter. It will be in court again this morning. Texaco was told to pay the record amount after a jury ruled last month that the company illegally merged with Getty Oil at the expense of a Pennzoil deal with Getty.

Roger: Big, big news in the business pages in the last few weeks. . . . Two of the gentlemen with us were involved in that. Rick Lawler, to my right, was the jury foreman. Also with us is one of the jurors, Jim Shannon, and they both sat in that jury box and made that ten-and-a-half-billion-dollar decision against Texaco. . . . Quick question before we go to our phone calls. Ten point five billion dollars. All Pennzoil was going to pay was three point five billion to buy only a percentage of the stock, not the whole company. Why did you award them so much?

Lawler: Well, you have to take into the damage considerations here as far as what was shown in the case and the evidence that was brought in. We are basing this on the barrels of oil that they had the opportunity to purchase and that is exactly what they showed. Pennzoil brought in experts for two, three weeks.

Gray: So even though they were gonna only spend three point five billion to buy into Getty, what they were going to get would be worth seven point five billion in oil reserves?

Shannon: Well, Roger, the law allows you to be put in as good a position as you would have been if the deal had gone through.

Gray: Okay.

Shannon: . . . Texaco never disputed the Pennzoil damages during the trial. So when we went into the jury room and we found that in fact there was a contract, what we had to go on was the damage testimony presented in court.

Gray: Texaco never questioned it, you said?

Lawler: Never.

*   *   *

On Saturday, the oil ministers of OPEC gathered in Geneva, and
the Saudis were hopping mad. A group of upstart nations with
relatively new oil fields—Britain, Norway, Mexico and the Soviet
Union among them—were opening the floodgates and grabbing
worldwide market share, loosening the stranglehold that OPEC
once held on world prices. Something had to be done. OPEC had
to reassert itself.

Yet another vise was about to close around Texaco.

By Saturday, Texaco's Texas lawyers had finally awakened to a
decidedly grim fact: Entry of the judgment *would* enable Pennzoil
to file liens on every single asset owned by Texaco. Immediately.
"It was like a death sentence," one of the Texaco lawyers says.

The lawyers began reviewing the boilerplate provisions in lend-
ing agreements and other financial documents; liens would trigger
defaults, defaults would trigger still more defaults. They discovered that the disaster could spread far beyond Texaco; liens
could be filed on oil wells, pipelines and anything else that the
company owned with partners, making them also subject to a
financial claim by Pennzoil. "We had no idea the liens would be
such a problem," one of the lawyers admits.

The consequences were almost too horrible to imagine, but
Richard Brinkman, one of Texaco's top finance men, did anyway.
After a career with one of the most creditworthy institutions in
the world—after helping to raise nearly $10 billion for the Getty
Oil deal in only a matter of hours—Brinkman heavy-heartedly
prepared a memorandum forecasting just what would happen if
Pennzoil laid claim to billions in Texaco assets. Such a move, he
wrote,

> . . . will lead Texaco on a downward financial spiral as sources
> of operating capital evaporate. . . . Secured financing, using
> property subject to the liens, will be impossible. No credit source
> will extend unsecured financing for fear of the repercussions of
> insolvency. The result will be that, in very short order, the
> company will no longer be able to finance its day-to-day opera-
> tions. It will be forced into bankruptcy.

Initially, Texaco's top managers refused to believe that
Pennzoil—that even Liedtke and Jamail—would have the au-
dacity to file liens on Texaco like some used-car dealer hiring a

repo man to chase down deadbeats. It flew in the face of common corporate decency, they thought. *Surely Pennzoil will be reasonable about this,* the executives said. But Texaco's lawyers convinced them otherwise. "It is too much leverage," Boies insisted. Casting the cloud of financial default over Texaco was too powerful a weapon for Pennzoil to ignore; by wielding that weapon, Pennzoil could extort a settlement on practically whatever terms it chose, Boies said.

After more than a year of doing battle with his old friend Jamail, Dick Miller also was convinced that Pennzoil would tighten the noose. "Given the total circumstances," Miller told an interviewer, "we can count on Pennzoil to be as ruthless as the law permits."

It was like a bad dream, like something out of a Hitchcock movie. The executives of Texaco felt like Cary Grant in *North by Northwest*—an innocent party mistaken for a notorious character, hanging from the face of Mount Rushmore by the fingertips with Joe Jamail and Judge Casseb stomping on their knuckles.

What should Texaco do? Offer Pennzoil money not to file liens? Just go ahead and file Chapter 11 immediately? Risk incurring the enmity of the Texas courts by asking a federal district court for an injunction against the liens? Offer to settle the entire case? Texaco was divided like the State Department and the Pentagon: Some lawyers and executives wanted to negotiate with Pennzoil, others wanted to fight it out, in the bankruptcy courts if necessary.

John McKinley spent the day listening, analyzing and probing. In the end, he decided, Texaco would set into motion *every* strategy available to it and then let Pennzoil make the next move.

After a secret meeting of the Texaco board, three plans were in motion:

1. Before Tuesday's hearing, now just thirty-six hours away, the lawyers would fly to Houston with top management, seeking to persuade Liedtke not to file liens. They would promise him that Texaco's assets would remain in place while the appeals dragged on, rendering liens unnecessary. If promises didn't work, they would offer him money.

2. Failing that, they would take the extraordinary step of offering to restore Liedtke's deal for three-sevenths of Getty Oil in exactly the terms he had bargained for: $112.50 a share, or a total of $2.7 billion. Texaco could ill afford to give Liedtke back his billion barrels of oil—but neither could it tolerate having *all* of its assets under the shadow of a claim by Pennzoil.

3. Finally, if necessary, Texaco would adopt the Doomsday
Plan. Since bankruptcy would be inevitable if Pennzoil filed liens,
Texaco at least could file bankruptcy before Pennzoil had the
chance to elevate itself to the status of a secured creditor. Harvey
Miller of New York's Weil, Gotshal & Manges, one of the nation's most prominent bankruptcy-law experts, was assigned to
have the bankruptcy papers typed up and ready to file at the first
sign that Pennzoil was going to slap on liens.

By Sunday evening, a contingent of Texaco lawyers and executives had reached Houston, ready to see whether an old card-
player like Hugh Liedtke was prepared to call the bluff on the
biggest bankruptcy in history.

· 26 ·

Like superpowers fearing imminent nuclear attack, Texaco and Pennzoil swung into readiness alert on Monday, December 9. Although both sides wished to negotiate a truce, neither wanted to be unprepared should the diplomatic efforts fail.

Pennzoil created the corporate equivalent of a distant-early warning system, engaging lawyers throughout Texas to stand watch over county courthouses in case Texaco filed any deeds pledging its assets to bankers, or otherwise putting them beyond Pennzoil's reach. The local lawyers were also standing by to receive certified copies of his judgment, which they could use to file liens on whatever Texaco assets Pennzoil had designated. To expedite delivery of the judgment to outlying courthouses, a fleet of Pennzoil automobiles was gassed up and ready to go. Pennzoil had identified about $5 billion worth of potential targets; only a signature from Judge Casseb and a final order from Hugh Liedtke stood in the way of an attack.

Texaco established its own NORAD-style defenses. Terrified that Judge Casseb might enter the judgment prior to Tuesday's hearing without informing anyone, Texaco stationed a representative in the clerk's office of the Harris County Courthouse to look at every piece of paper that came in. It prepared an elaborate communications network for deployment the second Judge Casseb entered the judgment, if he did.

A code was established: If the judgment were entered and Texaco had obtained no protection from liens, David Boies would whisper "White Plains" to an associate. The associate would race to the hallway, where phone lines would be held open to corporate headquarters and to the federal court in White Plains. The federal judge would be asked for an immediate injunction against the liens. But if it appeared that Texaco had no choice but bankruptcy, the code word would be "Foley Square," for the federal courthouse in Manhattan. There, the bankruptcy lawyers would be standing by an open phone line—with a copy of Texaco's Chapter 11 petition locked inside a briefcase.

And if the courthouse crowd proved too unwieldy for Boies to dispatch his associate to the telephones, then hand signals would be used to relay instructions outside of the courtroom.

As much as he relished his courtroom victory, Hugh Liedtke did not wish to see Texaco in Chapter 11. Any oilman who shared Liedtke's acute sense of oil patch history—and his great desire to leave his own imprint on it—would hardly wish to end his career as the man who threw Texaco into bankruptcy. Despite Texaco's lingering reputation for hardball dealings among the independents of the oil patch, Texaco was too big—too important to too many lives—in Houston. Liedtke would not want that on his head.

But Liedtke also had his self-interest at heart: He wanted money from Texaco, big money. And he knew that if Texaco slipped into bankruptcy court, Pennzoil's $10.53 billion claim—if it were upheld by Judge Casseb—would be relegated to the same status as every invoice for drilling mud that Texaco had purchased in the last thirty days. Liedtke wasn't about to let himself become the largest unsecured creditor in bankruptcy history.

As their strategists made their secret preparations for war, Liedtke, McKinley and several of their aides spent nearly eight hours in face-to-face negotiations in the offices of Baker & Botts. Besides avoiding the Doomsday scenario, each side had a secondary objective in the talks: obtaining the biggest club possible to wield against the other side in any attempt to settle the entire case. McKinley and his men believed that if Pennzoil could *never* file liens or seize control of Texaco's assets, then Pennzoil would be powerless to "extort" an exorbitant settlement. Liedtke knew that by putting a time limit on the truce, he would keep the pressure on Texaco to cough up "realistic" reparations for its $10 billion crime.

The Texaco group offered Liedtke $50 million in cash—no

strings attached, no questions asked—for a simple promise that he would never install liens as long as the case remained on appeal. But $50 million was pocket change in this case—less, in fact, than three weeks' interest on the judgment. Liedtke rejected the cash out of hand.

He was, however, willing to give Texaco a *little* breathing room—a week, perhaps, maybe even thirty days, during which the two companies could try to reach a peaceful solution to the whole case. Back and forth went the proposals until nearly 10 P.M., when Liedtke finally put his signature to a standstill agreement: Pennzoil would not attach liens for fifteen days and Texaco would not seek bankruptcy-court protection. McKinley and his men returned to Texaco's Houston office building—the same seventy-year-old building that Buckskin Joe Cullinen had intended as world headquarters for the Texas Company before it moved to New York.

The Texaco group looked at the document bearing Liedtke's signature, passed it around, bantered for a while and reached a decision. The proposal was no good. It put off the inevitable for only fifteen days. Texaco would not waive its right to seek shelter in bankruptcy court—or for that matter in any federal court.

Tomorrow, Tuesday, was judgment day, but the hearing was not scheduled to begin until 1:30 P.M., leaving plenty of time for Texaco to resort to yet another alternative that was approved by the board at its secret meeting over the weekend.

Reluctantly, Texaco would offer to give Liedtke his billion barrels back.

Monday was an eventful day in the oil patch for another reason. That morning *The Wall Street Journal* had reported that the world oil market was on the verge of a price collapse.

GENEVA — The Organization of Petroleum Exporting Countries, in a move that may prove a turning point for world oil markets, is abandoning a four-year effort to prop up oil prices by restricting oil production.

Instead, OPEC officials said, the group is determined to secure and expand its share of world oil markets, even at the risk of triggering a price war with other oil producers.

Oil prices fell $1.20 a barrel on the news. It was one of the biggest declines ever. Talk of an all-out price collapse, something once considered unimaginable, began taking on currency.

State Court No. 151 was long since full at 9 A.M., four and a half hours before Judge Casseb was scheduled to call the hearing to order. That morning, a mobile-phone store near downtown Houston had rented its last cellular telephone. The speculators finally had their chance to witness the main event firsthand and phone in their stock orders ahead of the public. But like spectators able to see only one half of the playing field, the crowd had no idea what was happening outside the courthouse.

Hugh Liedtke was fuming. He had left his bargaining session with John McKinley at ten o'clock the night before, after signing a fifteen-day standstill agreement; now his lawyers were informing him that Texaco had refused to sign. "We gave them everything they wanted!" Liedtke bellowed.

Liedtke received some other disturbing news. A company scout stationed at the Jefferson County Courthouse about one hundred miles from Houston had picked up two warranty deeds filed by a Houston-based vice president of Texaco. In one, Texaco had switched the ownership of its Port Arthur refinery—one of the world's largest, where John McKinley had started his career in the grease lab—to a Texaco subsidiary uninvolved in the Pennzoil case. The second deed had transferred ownership of the company's Port Arthur chemical plant. To Pennzoil, it appeared that a game of hide-the-refinery was under way.

Thus, Liedtke was in no mood to play Mr. Nice Guy when Jim Kinnear, the vice chairman of Texaco, showed up unannounced at the guard station in the Pennzoil North Tower at 11:15 A.M., requesting to see the Chairman. "I'm finished negotiating with him," Liedtke said, instructing Jamail and Moulton Goodrum of Baker & Botts to deal with the visitor.

But in essence, Kinnear had arrived with a white flag. He presented the Pennzoil lawyers with a letter from John McKinley:

> Texaco hereby proposes, without prejudice and for settlement-discussion purposes only, an overall settlement of the controversy between our companies. This proposal would put Pennzoil in essentially the same position it would have been in under the contested arrangements. In other words, Pennzoil would ultimately obtain, directly or indirectly, three-sevenths of the

Getty oil and gas assets, and a like share of its liabilities. In exchange for this interest, Pennzoil would pay an amount calculated based on the price of $112.50 per share which it proposed to pay for the three-sevenths of the Getty stock, subject to adjustments as generally outlined below.

Texaco would even *reduce* the price to take account of all the Getty oil-in-the-ground it had pumped in the nearly two years following the acquisition. And it would lower the price to account for the sale of ERC, the ESPN network and the other Getty Oil assets it had unloaded. "We sincerely hope," McKinley said in his letter, "that Pennzoil does not desire to impose needless punitive destruction upon Texaco, and that it only desires to be put in the position it contends it would have been in, but for Texaco's alleged interference."

Hugh Liedtke's once-in-a-lifetime opportunity had appeared for the second time.

Liedtke would later insist that the proposal was entirely too vague to so much as merit a response. "They had decimated Getty," he would explain. And indeed, although Texaco offered to make every adjustment necessary to replicate Liedtke's original deal, Texaco could never succeed in reassembling the people and pride that had made Getty Oil, in its day, one of the stars of the oil industry.

But it was difficult not to believe that Liedtke had something else on his mind. That very day, with OPEC in greater disarray than ever, oil prices were plummeting more than $2 a barrel—on top of the $1.20 they had fallen the prior day. On the other hand, that $10.53 billion judgment would be frozen—in fact, it would increase at the rate of $3 million *every day* at an annual interest rate of 10%.

Now, not even a billion barrels of oil-in-the-ground looked as good as $10.53 billion in the bush.

After reading McKinley's settlement offer, Liedtke left the office and walked with Jamail to the county courthouse, where it was nearly time for Judge Casseb to announce his decision.

Because two intruders had been caught in the courtroom the night before, necessitating a bomb search this morning, Judge Casseb was led through a different entrance into the courthouse.

"All rise," said Carl Shaw, the bailiff. "The One Hundred and

Fifty-first District Court is now in session, Judge Casseb presiding.''

Judge Casseb surveyed the sweaty throng that had been waiting hours for this moment. All six hundred eyes were cast forward to the bench. ''I felt like President Reagan,'' he would say later. Except for the companies' lawyers, no one in the courtroom had the slightest inkling of the Doomsday scenario ready to be played out.

''Gentlemen, anything further to present to the court?''

''Your Honor, we feel it's been fully submitted,'' Jamail said.

''Counsel?'' the judge asked, addressing Boies. Every heart in the courtroom was beating like mad. Unlike the jury verdict, this was the real thing.

''Your Honor, there *is* something I wish to advise the court of,'' Boies said. Then he issued the first public notice that bankruptcy was imminent, instead of merely a possibility after the ninety days when a bond would be required. A murmur swept the courtroom.

''Shhhh!''

''What'd he say?''

''Shut up!''

''Did he say *Chapter Eleven*?''

Boies continued, telling Judge Casseb that the companies had been negotiating to prevent any liens or bankruptcy filings. ''We've not been able to achieve such an agreement that was mutually acceptable,'' Boies said. ''We made, this morning, an attempt to have Pennzoil consider a resolution of the entire matter, which involved transferring to Pennzoil—''

Jamail was immediately on his feet.

''Your Honor, I would object to this!'' Jamail cried. ''It has nothing to do with the entry of this judgment. It's self-serving. We spent two days—''

Jamail cut himself off, refusing to be drawn further into any diversions.

''I don't want to get into that, Your Honor,'' he shouted. ''It's *totally improper* to discuss what went on by way of attempting to negotiate something with these people.''

''Well,'' the judge asked, ''is this a matter that counsel desires to talk to the court in chambers about?''

''No, Your Honor. *It is not*,'' Jamail insisted.

Casseb was undeterred. He wanted to get to the bottom of this. ''I'm going to see the lawyers in chambers.'' And with that, thirty-six attorneys, instead of the two or three that he expected, followed Judge Casseb out of the courtroom.

"What happened?" everyone in the courtroom asked.

They would have to wait and see. Nobody was leaving this show.

The scene was surreal: nearly forty lawyers in an overheated office, with loudmouthed reporters and stock traders practically clinging to the other side of the door. Jamail made a point of speaking first.

"To give you some background of this, Judge Casseb—"

"I didn't ask anything," the judge snapped.

If Pennzoil owned Judge Casseb, as several Texaco lawyers believed, he did not show any signs of it now.

"Let me make this statement on the record," the judge said. ". . . I believe that it may be presumptuous on some people's part to think that I'm going to make a ruling today and they know what the ruling is going to be, so I just want to tell y'all that ahead of time. So in that spirit—and this is why I just want to talk to the lawyers—I'm willing to listen to anything that may accomplish the objective that all of us want in our judicial system."

After recounting the fruitless efforts to reach a standstill, Jamail decided to do a little tattling.

"I asked them *specifically* yesterday if they transferred any properties. Here are copies of two deeds," he said, handing the judge the documents that switched ownership of the Port Arthur plants. *"Assets Pennzoil would be deprived of!"* Jamail said. "All this in the face of Mr. McKinley, Mr. Boies and Mr. Keeton saying last night that nothing had been transferred. . . . I have *never* been subjected to anything more subversive to the process. Why is Texaco entitled to special treatment?"

Texaco's local lawyers were embarrassed; they had been told nothing of the refinery switch.

"Let me say, Joe," Keeton said apologetically, "I don't know anything about it or what is involved. But I can tell you right now," he added, turning to the judge, "everybody has worked hard, Your Honor, and that's the status of it: still no agreement; people trying."

Boies chimed in, fumbling to describe the technicalities of the Texas lien procedure, with which he was unfamiliar. Keeton, his own co-counsel, immediately cut him off. "David," he said, "I'm not sure you're enough in this loop to be able to make an agreement.'' But Boies plunged ahead, telling the judge that Texaco only wanted protection from liens and from the $12 billion bond requirement for as long as the case remained on appeal.

"We're hung out to dry, Judge!" Jamail whined.

The judge had heard enough.

He ordered the two sides to reach a standstill for as long as he continued to have technical jurisdiction over the case—a period of about ninety days. When Texaco's lawyers questioned whether Pennzoil would abide by the order, Judge Casseb assured them he would throw Jamail in jail if necessary.

"Y'all get it typed up."

"They can't transfer assets?" Jamail asked.

"Getty assets," the judge answered.

"How about Texaco assets?"

"They cannot," the judge said, "except in the regular course of business. I can't stop them from selling gasoline. Then I'd have no gasoline for my car, because I use Texaco."

Judge Casseb stalked from the office and back into the courtroom.

"The court is going to be in recess for approximately one hour," he announced to mystified onlookers, and was gone.

For three and a half hours, the lawyers shuttled between a conference room on one side of the court and a jury room on the other. Fans were brought into the meeting rooms to cool the overheated lawyers and executives. From the courtroom the jacketless John McKinley could be seen in one of the conference rooms, his arms akimbo, a grim look on his face. Standing in suspenders amid all the gawkers, Al DeCrane, Texaco's president, talked into the phone on the bailiff's desk—the only phone not already staked out—trying to find out if any other assets had been switched. (A Texaco executive later swore in an affidavit that the refinery switch was a long-planned move intended as part of a plan to consolidate the assets of Getty Oil.)

Stock traders didn't let their ignorance of what was happening prevent them from speculating on what might be. Some purchased Pennzoil shares, reasoning that if a settlement was occurring, it was bound to benefit Pennzoil. But other traders watched with dismay as the minutes ticked by, until the time of the final closing bell on the floor of the New York Stock Exchange. Their legal "insider-trading" scheme had been foiled by the clock.

Finally, at 5:15 P.M., the standstill was negotiated, typed and incorporated as paragraph seven of the judgment. For as long as

Judge Casseb had jurisdiction over the case—a period that lasted until after he had ruled on any motion by Texaco for a new trial, a period of about ninety days—Texaco could not "pledge, encumber, convey or otherwise dispose of any of its assets other than in the routine and ordinary course of business." In effect, the nation's fifth-largest corporation was a ward of the court.

Pennzoil, for its part, could install no liens. If it did, Judge Casseb would vacate the judgment and order a new trial held.

The full jury verdict of $10.53 billion was formally entered with interest. The total came to $11,120,976,110.83.

"Before I sign it," Judge Casseb said in open court, "I want to say to all of the lawyers that I, in particular, certainly do appreciate your cooperation and your help in trying to work out the language—so that we can proceed to once again tell the American people that their confidence in the administration of justice under our system of laws, and the proceedings in our courts, still remain the way we all want them: as guaranteed under our Constitution.

"With that I'm signing it." He scratched out a signature at the bottom of the two-page document with his dime-store pen. "And thank you all, ladies and gentlemen.

"The court is in recess."

Most of the Texaco lawyers and executives vanished as if into thin air, leaving Boies to deal with the retinue of reporters who still had no idea what had just occurred. Although Texaco had managed to forestall bankruptcy, at least for a while, the day was otherwise a complete loss. It still faced the daunting prospect—indeed, the impossibility—of posting a $12 billion bond. And it would take many months, possibly years, to have the judgment reversed or reduced.

Liedtke and Jamail embraced, as they had three weeks earlier when the jury verdict came in.

"How about a drink?" Jamail asked.

"Love to," said Liedtke.

As they jubilantly left the courtroom side by side, Jamail threw up a clenched fist in victory for a wire-service photographer. The verdict was intact; no appeals court can overturn a jury's finding of fact, only a judge's application of law. Texaco could not squander its assets. And Pennzoil now had a powerful upper hand in any settlement negotiations.

As Liedtke jostled his way through the throng of TV cameras, he announced that he was always open to a peaceful settlement. But after two days of slippery behind-the-scenes dealings, he took one more swipe: "The problem is, dealing with Texaco is like trying to frisk a wet seal."

Liedtke and Jamail, alas, had no idea of the little surprise that Texaco had already arranged for them.

The decision to uphold the verdict finally broke the dam around Texaco.

Shell Oil demanded letters of credit—a payment guarantee by a bank—before selling crude oil or refined products to Texaco. So did the Venezuelan government. Charter Oil canceled credit altogether. Marathon Oil demanded payment in advance for a load of butane. Atlantic Richfield issued a memorandum urging employees to deal with companies other than Texaco wherever possible. Mobil and Exxon insisted on balancing their positions with Texaco in the sort of trading and swapping of crude and products that oil companies engage in every day.

Banks everywhere began looking the other way. Chemical Bank refused to write Texaco checks for lease bids. Bank of America discontinued letting the company engage in "daylight overdrafts." Chase Manhattan refused to issue a $5 million letter of credit needed for a drilling program in Angola. Foreign banks expressed reluctance to so much as quote exchange rates to Texaco. An Italian bank withdrew all credit, forcing Texaco S.p.A. to operate on cash. The U.S. Coast Guard checked up on Texaco's offshore insurance policies.

Within a week, Texaco's stock fell from $32 a share on the morning of the judgment hearing to $27.625. Since the day before the jury verdict, Texaco's total stock-market value had now fallen nearly $2.8 billion—practically what all of Pennzoil was worth. The company that had spent $10 billion to buy Getty Oil less than two years earlier was itself now worth a paltry $6.6 billion in the stock market.

The crisis metastasized beyond Texaco to many of its business partners. A refining unit in Wales, 35% owned by Chevron, had its rating on $700 million of debt downgraded to "speculative." The $700 million Louisiana Offshore Oil Port—built by Texaco and others to increase oil imports as their U.S. fields dried up—had its entire slew of revenue bonds downgraded. Five pipelines owned by Texaco with other oil companies were wounded in the crossfire. "The market just isn't buying paper from any company that's backed by Texaco," one of the company's partners explained.

And all the while, oil prices continued to plummet. Texaco's day-to-day cash flow began trickling. Barred from the short-term borrowing markets, with $1.5 billion of commercial paper coming due for repayment, Texaco by mid-December was literally days from running out of cash.

For all the denial, anger, bargaining and depression that the employees and executives would feel, they never reached the last stage defined by Kübler-Ross: acceptance.

One true act of heroism—saving the company from outright financial collapse—would become a source of pride in the difficult months to follow. The financial department at Texaco transformed itself into something like an emergency-room express team attending to a cardiac victim. They began monitoring the company's worldwide cash balance practically on an hourly basis, expediting funds owed by subsidiaries as needed to meet the parent company's obligations. The finance group also refused to enter into routine deals where trade partners insisted on letters of credit, dealing only with companies willing to trust the company's credit.

They did everything possible to work around the onerous condition of Judge Casseb's paragraph seven, which restricted the pledging of assets. They fought hard for Pennzoil's approval of a plan to sell billions of dollars in receivables—money that Texaco was due from customers. As a result, payments from gasoline credit-card holders, business partners and others were deposited directly into the accounts of a banking group that had agreed to provide Texaco with cash up front. It was one of the largest financing arrangements of its kind in history—similar to the moves undertaken by Chrysler and International Harvester to stave off bankruptcy—and it was a humbling episode for Texaco. "It was quite an experience to go from a first-class credit to something no better than a Seventh Avenue clothing merchant dependent on receivables financing," Dick Brinkman, the finance man, would recall. But it worked.

"We took on two of the biggest calamities the world had ever seen—an eleven-billion-dollar judgment and a collapse in oil prices—and we stayed in business."

The world heaped plenty of abuse on Pennzoil, Judge Casseb and the jury, too. Newspapers that had apparently not taken the initial verdict seriously enough to comment now chose to. "Texas Common Law Massacre," the editorial page of *The Wall Street Journal* trumpeted. Wrote *The New York Times*:

It started as soap opera. Did giant Texaco unfairly grab Getty Oil after Getty Oil had agreed to sell much of its reserves to little Pennzoil? Did Texaco, of White Plains, N.Y., make a fatal error in allowing the case to be tried by a good ol' jury in Houston? Would Burgess Meredith play Pennzoil's down-home lawyer in prime time?

But now the tale has become a sad farce. . . .

The drive-time rock-and-roll show on KKBQ in Houston made some Christmas mirth of Pennzoil's windfall:

> On the first day of Christmas
> Texaco gave to me
> Ten billion bucks,
> Holy cow we're rich
> Oh, what a switch!
> Isn't life a bitch?
> We're the caroling boys of Pennzoil.

In fact, it was hard to keep from gloating. Irv Terrell of Baker & Botts proudly placed a beat-up toy Texaco gasoline truck—a gift from the children of a family friend—on the bookshelves above his desk.

But at 6 A.M. on Saturday, December 14, Terrell had a rude awakening. Two hours earlier, a package had been delivered to the guard desk at Pennzoil's North Tower. It was a summons.

On the very day that Judge Casseb had entered the judgment, Texaco had gone ahead with a suit against Pennzoil in *federal* court—a federal court in its hometown of White Plains—seeking to have the entire case yanked out of Texas.

Texaco, at last, had learned to play the game by Pennzoil's rules.

· 27 ·

Weeks before Judge Casseb had entered Pennzoil's judgment, Robert Silver, a bookish young associate of Boies at Cravath, Swaine & Moore, was poring over hoary judicial opinions on a daunting mission. Silver was trying to find some case, *any* case, to justify the invasion of a federal court into the activities of a state court, *something* that Texaco could hold up as a shield against the life-threatening Texas bond rules.

From the earliest days of the Constitution, the courts had viewed state sovereignty as a cornerstone of federalism. Certainly, Congress and the courts had steadily chipped away at the notion of ''states' rights'' as it was applied to justify racial segregation in Dixie. But for the most part, the U.S. government had little more right to poke its nose in the ministerial affairs of a state court than into the content of a Sunday sermon.

Then, Silver came upon a case called *Henry v. First National Bank of Clarksdale*. Aptly enough, it was a civil rights case.

In 1966 the National Association for the Advancement of Colored People launched a boycott of white merchants in Claiborne County, Mississippi. Their businesses suffering, the merchants sued for damages in a state court and won. Then the state court ordered the NAACP to submit an appeals bond of $1.5 million—an amount that would have bankrupted the civil rights groups. The

NAACP, in turn, filed suit in federal court, claiming that the bond requirement violated its Constitutional rights of due process and free speech. The U.S. Justice Department joined the litigation on the blacks' side, arguing that important federal concerns were at stake in the litigation. In the end, the U.S. courts declared the Mississippi bond rule unconstitutional, as applied to the facts of the case.

In 1986, the years of resolve by the NAACP would help prevent one of the world's most powerful multinational corporations from financial ruination.

Over the objections of the Texas lawyers, in early December the Texaco board approved making an end run into a federal court, even though it could very well offend the same Texas appeals courts before which Texaco was trying to preserve its right to appeal. Texaco remained perfectly free to pursue a federal-court suit even after negotiating the "standstill" on the lien issue. Texaco had promised Pennzoil and Judge Casseb that it would not pledge its assets to anyone for at least ninety days—but Texaco had said nothing about trying to protect those assets through an end run into the federal courts.

But *which* federal court? It would be an extraordinary request to make of any court, for the *Henry* case appeared to be the exception rather than the rule. Besides some of the most creative lawyering the courts had ever seen, success would require a bold judge—a jurist unafraid of creating a controversial precedent in what already was history's highest-stakes court battle.

Paul Curran of Texaco's Kaye, Scholer law firm happened to be a friend of just such a judge. And most convenient of all, the judge happened to be sitting in Texaco's backyard in White Plains.

He was Judge Charles Brieant, a Nixon appointee who had held several political offices in Westchester County and had become acquainted with Curran, a former U.S. attorney, through the bar and through party activities. Sixty-two years old, with a dashing handlebar mustache and a penchant for polka-dot ties, Brieant was known as a judicial pragmatist who placed the spirit of the law as high as the letter of law. "He believes in looking at the facts of a case before you begin trying to apply the appropriate legal principle," Stanley Teitler, a criminal defense lawyer who was the judge's first law clerk, would later say. "You can apply legal cases to either side of the case; the facts are what's critically important to him." Added another of his former clerks: "He's an individualist. He does exactly what he wants to do and doesn't worry what other people think.''

A practical judge, a hometown judge.

On December 17, without having held a hearing, Judge Brieant publicly issued a temporary restraining order blocking the Texas bond rules. Instantly, the $12 billion club available to Pennzoil in any settlement negotiations was jerked from Liedtke's grasp.

It was only a short-lived decree, however. It remained to be seen whether Judge Brieant would issue a permanent injunction— one that would insulate Texaco throughout the duration of its appeals.

That evening, when he put in an appearance on public television's *MacNeil-Lehrer Newshour,* Hugh Liedtke wasn't about to let anyone think that the upper hand had shifted to his nemesis.

Robert MacNeil: Mr. Liedtke, should we score today's decision a round for Texaco?

Liedtke: Well, if it is scored that way, why, I think it's a Pyrrhic victory and a very short one. . . . Our attorneys advise us the whole proceeding is without merit.

MacNeil: There are rumors around that an out-of-court settlement is in the works. Is that right?

Liedtke: We have been prepared for over two years to work out a reasonable settlement of our differences with Texaco. Hours have been spent on it to no avail. . . .

MacNeil: Coming back to the ten-and-a-half-billion-dollar judgment. Texaco's lawyers have said that would lead to the ''total destruction and obliteration'' of their company.

Liedtke (chuckling): Well, they have, I think, mistakenly frightened the financial community, their shareholders and their employees. The facts of life are that the appraised net worth of Texaco . . . is something in the range of twenty-three to twenty-six billion. . . . It certainly seems to me that twenty-three to twenty-six billion ought to be able to cover twelve.

MacNeil: Let me ask you this finally, Mr. Liedtke. In saying that Pennzoil is prepared to negotiate with Texaco, do you mean that you would accept a settlement ''in kind,'' as it were—oil and gas—of *less* than ten and a half billion?

Liedtke: Well, I think that almost goes without saying. When you negotiate, why, you avoid years of very expensive litigation and you consider some discount. But as time goes on—you know, it's costing them three million a day just to sit. Under New York

law it costs them ten-percent interest. So even if you're trying to
lessen it somehow, why, unless something's done fairly soon,
why, it's going to be fairly hard for us to find something that they
might find acceptable, as well as ourselves.

MacNeil: So the "bargain sale" has a time limit?

Liedtke: As I said, the wise thing to do is to negotiate. But the
other side of it is, we are prepared to fight as long as we have to.

Among Texaco's incentives to settle the litigation, one was particularly compelling: the personal liability that the judgment imposed
on the company's board of directors. Within days of the judgment, Texaco stockholders began slapping summonses on the
directors and officers of their company, claiming that the Texaco
board had recklessly exposed the shareholders to grave risk.

C. J. and Dorothy Kirk were among the first to file suit. C. J.
had not only been a Texaco shareholder for thirty years but an
employee for forty-six years, until his retirement in 1974. Texaco's takeover of Getty Oil "was a brazen and arrogant attempt to
quash Pennzoil's legally and morally binding agreement with
Getty," the Kirks claimed. They insisted that their company made
a "tacit admission of guilt" by agreeing to indemnify Gordon and
the museum from suits by Pennzoil. "The Texaco directors were
fully aware of the liability of Texaco," the suit said. "Despite this
knowledge, they made no good-faith attempts to settle this dispute
and thereby avoid the enormous expense of litigation and the
catastrophic verdict which has been returned."

Shareholder lawsuits are commonplace after a corporate catastrophe, often generated by lawyers working on a contingency-fee
basis. But besides threatening directors with financial liability,
shareholders' suits are a slap in the face—and Texaco had a board
whose members were hardly accustomed to insult. They included
Frank Cary, the retired chairman of IBM; Willard Butcher, chairman of Chase Manhattan; Robert Beck, chairman of the Pruden-
tial Insurance Company of America; Thomas Murphy, chairman
of Capital Cities/ABC Incorporated; William Wrigley of chew-
ing-gum fame; and Lorene Rogers, the retired president of the
University of Texas. Before long they would be named in more
than a dozen suits by the very shareholders who had elected them.

Pennzoil was delighted at the outbreak of shareholders' suits
against the big-name Texaco board. "Mr. Wrigley is going to
have to sell a lot of Chiclets before this is over," Joe Jamail
crowed.

* * *

Any peaceful solution to history's biggest court judgment would doubtless involve history's biggest out-of-court settlement, and perhaps its most complex. Ambiguity, in fact, would be a worthwhile ingredient; both sides would have an interest in structuring the settlement around a "fudge factor," such as a transfer of properties or the payment of long-term securities whose value was unclear, through "optics" or some such device. Texaco would want to lowball the dollar value of the deal to help defend itself from the lawsuits filed by its shareholders, while Pennzoil would want to enlarge the value to avoid accusations that it sold out on the cheap.

Clearly, the biggest obstacle to a settlement was the number of zeroes on the judgment. McKinley, continuing to protest Texaco's innocence as strongly as ever, was said to shudder at the thought of anything over $1 billion. People close to Liedtke, on the other hand, said he was convinced that Texaco could easily absorb a settlement of $3 billion to $6 billion.

Moreover, Liedtke didn't want cash. Cash payments would be taxable as ordinary income, and over the years Liedtke had never demonstrated much enthusiasm for sharing more than he had to with Uncle Sam. On the other hand, settling the case with the transfer of oil-in-the-ground at discount prices could potentially avoid a huge tax penalty. That solution, however, suffered from the worsening complication of steadily plummeting oil prices.

Liedtke, in short, would have to demand more oil-in-the-ground for each dollar of liability. "The price of poker," he said, "keeps going up."

Texaco and Pennzoil certainly didn't lack for advice. Boone Pickens publicly remarked that he was getting sick of hearing about the case. "It should be cleared up and there's no reason why it shouldn't be," he said. The solution he proposed: "Texaco buys Pennzoil." Texaco could pay a hefty premium to take account of the judgment value, and by exchanging its shares for Pennzoil's, the whole deal would be tax-free to Pennzoil shareholders. Or, he said, the reverse could work: Pennzoil acquires Texaco. "Personal feelings are something that CEOs don't have the luxury of letting get in the way of a deal," he said. "If you don't like the way you're treated, go tell your wife or kick the dog."

Philip Corboy, the winner of numerous multimillion-dollar settlements in airline crashes and other cases, also called attention to

the need for ego-suppression. "If I were counseling Texaco, I'd
say, 'Put your egos in the closet or get me some alter egos.' If I
were Pennzoil, I'd say, 'Mr. Texaco, we've scabbed you enough.
We will be reasonable.'"

Even Ron Luciano, who settled more disputes as a major league
baseball umpire than any hundred lawyers ever could, went public
with his own solution. "I'd settle it the way I would settle any
play, say a play at home plate where I had no idea what happened.
I would mentally flip a coin—not physically, because the crowd
would see—and see who came up heads in my mind.

"Right now I'm doing it," Luciano continued, "and Texaco
came up the winner. That's it. Texaco is the winner, and that
should end it."

Settlement talks began in earnest on Friday, December 20, 1985,
at the Four Seasons Hotel in Tulsa—Hugh Liedtke's old stomping
ground, chosen as a convenient point between New York and
Houston. Liedtke wanted to buy some of the old, heavy-oil reserves in Getty Oil's Kern River field at a discount—a big discount. Reams of property and production data were exchanged,
and both sides withdrew to caucus. The day after Christmas,
Liedtke dispatched a Pennzoil jet to pick up several of Texaco's
advisors for a quick follow-up meeting that occurred on an airport
taxiway in Nashville.

On Monday, December 30, the scope of the discussions shifted.
Liedtke expressed a willingness to consider selling *Pennzoil* assets, at a premium price.

Was it possible?

Rumors of a friendly takeover of Pennzoil—the whole company, yellow can and all—mounted in the face of denials that
Liedtke would ever consider such a thing. "Why would Liedtke
sell out?" asked Charles Andrew of John S. Herold Incorporated,
a consulting service that regularly appraised the value of both
companies' assets. "He may want to make more money for his
shareholders, sure. But his whole aim has been to build a great oil
company."

On the other hand, Liedtke was barely a year from turning
sixty-five. Having failed to get Getty Oil, was he now ready to
cash in his chips?

On Sunday, January 5, 1986, Liedtke entered the Hyatt Hotel in
Nashville for his first face-to-face meeting with John McKinley
since the ill-fated negotiations on the lien issue. The day got off to

an inauspicious start when Liedtke and his associates knocked on the door of the wrong suite, which happened to be occupied by some newlyweds. Finally Liedtke reached the right room. McKinley's associates, meanwhile, were hanging out in a suite of their own, munching hamburgers, watching an NFL playoff game, awaiting any summons from their boss.

Liedtke was startled when his adversary began speaking from notes, as if he were delivering a speech. Nobody outside of their inner circles would ever reveal exactly what occurred in the conversation. But the result quickly became clear: a communication breakdown rivaling the worst of those that had occurred between Gordon Getty and Sid Petersen. Yet people close to Texaco would later swear to hearing that Liedtke indicated a desire to sell Pennzoil lock, stock and barrel at a premium over its market value of about $2.5 billion.

The next day, the Texaco board spent seven hours meeting well into the night. A settlement proposal involving the acquisition of Pennzoil was prepared.

Liedtke himself had scheduled a board meeting for the following day to receive the proposal, and he was not in a particularly conciliatory mood by the time it arrived. For one thing, the proposal was more than an hour late, and Liedtke did not delight in seeing his directors sitting at the table drumming their fingers. Moreover, Pennzoil's stock was suddenly experiencing wild gyrations—rising an astonishing $20, to $80—on rumors that the takeover was really about to happen; Liedtke was convinced that the rumors could have emanated only from the other side.

Finally, the proposal arrived with a messenger from Texaco. As described by several sources, it contemplated the friendly takeover of Pennzoil for well under $90 a share, a premium of only about $20 a share—less than $1 billion—over Pennzoil's value in the stock market.

Jamail, who was present for the meeting, was asked for his comment. "Sheeeeit," he replied.

Pennzoil issued a terse announcement that it had rejected a settlement offer from Texaco and that the form of the proposal was entirely unacceptable.

Texaco's representatives were said by some to be scratching their heads. *Isn't this what he wanted?* they wondered. Yet Liedtke would say, "It was *exactly* what we told them *repeatedly* we wouldn't accept."

The stock market, however, was undeterred.

The following day—Wednesday, January 8—rumors, denials and more rumors of Pennzoil's imminent takeover by Texaco caused the most frenetic trading that anyone could remember in any major security—crazier, in fact, than Texaco's shares had yet experienced on their worst day. After opening down $8, Pennzoil shares rocketed back $17 to $91 a share. Trading was halted for an hour; there were simply too few sellers to accommodate the number of traders willing to buy at any price. Yet when trading resumed, the stock fell back to $78, and spent the rest of the day declining to $74.50, practically where it had begun the day.

Even the most experienced investors were hard-pressed to keep up with the action. One arbitrager would later say he saw nothing unusual about going to lunch when the stock was at $75 and returning to find it at $76, except when he learned that it had reached $91 in the meantime.

That afternoon, Liedtke called in the media.

He released copies of a letter to the SEC demanding an investigation of the wild takeover rumors—rumors that he suggested had been planted by Texaco operatives to put pressure on Pennzoil to accept a takeover bid. Then he lit directly into McKinley and his men.

"The honeymoon is over," he declared. Settlement talks will not resume "until Texaco management will face reality, and we don't believe they ever will. . . . There is a certain mindfix that has existed there for years. . . . They've got their board in a mess, their shareholders in a mess, their banks in a mess, and they're intransigent. So be it. . . . You've got a management that's lost a major case and everybody's in trouble. . . . What you've got up there is the war party and the peace party; so far the war party seems to be in the ascendancy. . . .

"They ought to run that thing like a business instead of a fiefdom."

In the weeks and months ahead, settlement talks did resume in fits and starts, but always with the same result: Pennzoil dismissing the proposal in the most outrageous possible terms. "The same old picked-over bones," Liedtke called one proposal to settle for an exchange of Getty Oil reserves.

Joe Jamail assigned himself to the job of squelching scurrilous settlement rumors. When a $600 million settlement figure began working its way through the rumor mills, Jamail said, "McKinley ought to take his six hundred million in suppository form."

So much for a peaceful solution.

Attention was refocused on the courts, where the remarkable decisions had only begun to occur.

Texaco spared no effort in assuring that its hometown federal court remained on its side in the effort to smash the bond and lien rules of the Texas courts. Not only was the survival of Texaco at stake, the company claimed, but even the military preparedness of the U.S. and the free enterprise system as we know it.

Texaco loyalists bombarded Judge Brieant with friend-of-the-court briefs. Lester Thurow of the Massachusetts Institute of Technology, one of the nation's best-known economics scholars, said that the jury verdict had announced "a new and dangerous rule," which he called the "Pennzoil rule." The Pennzoil rule could all but wipe out "competitive capitalism," imposing an "*in terrorem* effect on the national economy." Julius Katz, a business consultant and former assistant U.S. secretary of state, warned of reduced foreign investment in the U.S. that would "further aggravate our balance of payment problems to the detriment of United States foreign commerce." Kenneth Dam, an IBM executive and former deputy secretary of state, predicted that foreign governments and "opposition political groups" might seize Texaco assets abroad. Any bankruptcy by Texaco, a major source of fuel for U.S. fleets, would impair "United States military forces in remote and strategically important areas of the world."

The briefs filed for Texaco had the ring of familiarity from one to the next, making it plain that Texaco itself had undertaken a major lobbying effort to recruit and even to help write them. But some were volunteered. When Benjamin Hooks, executive director of the NAACP, read about the size of the bond, he immediately thought of the group's hard-won victory in *Henry v. First National Bank of Clarksdale*. "Alarm bells went off," Hooks would explain. "I saw jeopardy to the case we had won." The NAACP came down squarely on Texaco's side.

Texaco's own legal filings with the court, among other things, sought to dramatize that Texaco "was deprived of its constitutional right to a fair trial" in Texas. "The entire Pennzoil action was so pervasively infused with bias, prejudicial error and other irregularities as to constitute a denial of Texaco's right to a fair trial." The company cited the $10,000 campaign contribution to Judge Farris; Judge Farris' opening-day comment that the case was the biggest ever filed, "signaling to the jury that a large verdict would be in order"; Judge Farris' "character slurs" against Bart Winokur; Judge Farris' willingness to let Pennzoil distort the significance of the indemnities; Judge Casseb's unfamiliarity with the months of testimony that preceded him; the "gross misstatement" of New York contract law in Judge Casseb's jury instructions.

All of these allegations were made in an affidavit filed before Judge Brieant by Judge Brieant's own acquaintance—Paul Curran of the Kaye, Scholer firm. That relationship, however, did not prevent Texaco from also filing an affidavit citing reports that "Judge Casseb is a close personal friend of plaintiff's lead lawyer," Joe Jamail.

A new War Between the States arose from the legal quagmire. The attorney general of New York, Robert Abrams, came down squarely on Texaco's side, writing a brief that suggested the Texas verdict had contorted New York contract law beyond recognition. Pennzoil, for its part, was joined in the case by Jim Mattox, the attorney general of Texas, who cited the vital issue of state sovereignty in the federal case.

Finally, when the flurry of brief-writing had passed, Judge Brieant's opinion came down from the mountain. "The concept of posting a bond of more than $12 billion is just so absurd, so impractical and so expensive that it hardly bears discussion," he said. "This corporation should not be squeezed through the bankruptcy court . . . at least until appellate finality has attached to the judgment. . . . There is no way in the world that Texaco could pay the judgment without reorganizing or liquidating. . . . The imminent disruption to the national economy and to the interests of the public supports a finding of irreparable harm. . . ."

Judge Brieant ruled that a bond of $1 billion was more than sufficient, a requirement that Texaco readily satisfied by posting $1 billion of its stock in its Texaco Canada Incorporated affiliate.

But Judge Brieant's opinion went far beyond saving Texaco from the $12 billion bond. If Texaco failed to reverse the actual judgment in the Texas appeals courts, U.S. District Court Judge Charles Brieant indicated he might do it himself, taking the entire case into his own hands. The judgment burdened interstate commerce. It had made a mockery of federal tender-offer rules, which are intended to encourage competing tender offers of the very kind Texaco had made for Getty Oil. Said Judge Brieant: "The compensatory damages awarded by the jury are too large by several orders of magnitude. . . . Even if the judgment ripened into a final judgment in this vast amount of money—and assuming all the decisions down the road go against Texaco—Texaco would *still* have the federal claims that are pleaded in this action, which it could litigate on the merits.''

A federal court had made itself the ultimate overseer of the Texas courts! Now, not only was Pennzoil's $12 billion settlement club gone—so perhaps was its chance *ever* to collect anything approaching the huge award.

"We need to bronze this motherfucker and put him in our lawn!" Pennzoil's Joe Jamail cried. Pennzoil immediately appealed to the Second Circuit U.S. Court of Appeals.

Even "Tough Tony" Farris—back on the state bench in Texas while under treatment for cancer—was heard from on the issue. "Apparently, the federal judge in New York—I won't even dignify him by calling his name—never read the *Federalist Papers* and is dismally ignorant of the fact that west of the Hudson there's a republic," Judge Farris said on local television in Houston. "I hope that the United States Court of Appeals for the Second Circuit *stomps* him. That's S-T-O-M-P-S. Legally, of course," he quickly added.

Lawyers unconnected with the case also expressed outrage. Laurence Tribe of Harvard University, one of the nation's most noted constitutional scholars, said the decision smacked of a "Chrysler bailout." Although noted for his support of civil-rights and other liberal causes—at the time he was assisting a legal fight against sodomy laws in Texas—Tribe could not deter himself from opposing the support that the NAACP had thrown behind Texaco. "This is a truly fundamental threat to the sovereignty of the state judiciary and almost a caricature of long-arm judiciary," Tribe said. The day after reading some of Tribe's comments in *The Wall Street Journal*, Pennzoil hired him to lead the federal-court appeal.

Once again, with the stakes higher than ever, a round of lobbying and brief-writing was under way. Texaco went from one state capital to another, from Alaska to Florida, urging politicians to file briefs with the appeals courts. Texaco operatives won support by posing a simple question; in Alaska, for instance, where the company had a quarter billion dollars in assets, "They asked us whether Alaska might have an interest in whether Texaco were forced into bankruptcy," says Deborah Vogt, assistant attorney general. Texaco representatives lobbied officials of the State, Justice and Defense departments, and other cabinet agencies, both to urge their support for Texaco's appeal and to defeat any influence that Pennzoil might have, through Vice President Bush. (In the end, the federal government remained neutral.)

On February 20, 1986, the Second Circuit upheld Judge Brieant's ruling on the bond issue—but threw out almost everything else that he had to say about the case. Brieant's attack on the Texas judgment and his suggestion that Texaco could eventually retry the entire case was an impermissible appellate review of issues that remained the province of the Texas courts, the Second Circuit ruled.

Texaco had preserved the right to appeal without filing bankruptcy—but it would have to appeal in Texas. And already, Texaco's end run into the federal judiciary had already begun to poison the well.

On the very day that the Second Circuit published its mixed opinion, Judge Casseb conducted his final act of jurisprudence in the trial case. But there would be nothing mixed about the outcome of that hearing.

Before the initial burst of postverdict publicity had passed, Jimmie Paul, an employee in the Texaco comptroller's office in Houston, watched Jury Foreman Richard Lawler delivering the jury's "message" in one of his several television interviews.

That's my son-in-law, Paul realized.

Paul had not seen Lawler for nine years. In fact, he had not even spoken to his daughter—Lawler's wife, Cheryl—nor seen their two children—his own grandchildren—in years. "My former wife, Cheryl's mother, has expressed great bitterness toward me since the time of our divorce in 1972," Paul would recall.

Concerned that his son-in-law might have carried the family feud into the jury room, Paul immediately went to his bosses at Texaco, who put him in touch with Dick Miller. Paul's affidavit about the family conflict would become Item #1 in Texaco's upcoming motion for a new trial; Lawler, Miller said, had been duty-bound to reveal the potential bias he might bring into a case against Texaco. Lawler told Miller, however, that until two months into the trial, he'd never even known where his father-in-law worked.

Several days after the Jimmie Paul episode, Texaco's lawyers picked up *The Wall Street Journal* to read a reconstruction of the trial and the events leading to it. A section of the headline immediately caught their attention. It read, THE CASE OF THE JUROR'S WIFE.

It was Jim Shannon's wife, Susie, who had lost her job at
Conoco Incorporated after the company's takeover by Du Pont.
Dick Miller recalled asking jury candidates whether any harbored
some bias that might prevent them from fairly hearing the case;
one woman had approached the bench saying her husband had
worked twenty-nine years for Conoco. "We had some very bad
feelings," she had said of the merger. As a result, the woman was
excused from consideration. "Shannon did not respond to this
question, even though he should have done so," Miller would
declare. That was Item #2 in Texaco's motion for mistrial.

The *New York Times* would deliver Item #3.

In a profile of Judge Casseb—"Dapper Texas Judge for a Historic Case"—the *Times* paraphrased a comment by Casseb's son
and law partner, Solomon Casseb III, who had said that after
serving eight years in the 1960s, his father had returned to the
bench because he needed a few more months of judicial service to
qualify for a state pension.

A golden opportunity had suddenly been delivered to Texaco. If
the comments attributed to Casseb's son were true, then Judge
Casseb had been unqualified to hear the case in the capacity in
which he had been assigned, as a "retired" judge. That, said
James Sales of Texaco's Fulbright & Jaworski, would mean only
one thing: "It's a null and void judgment." It was a hypertechnical complaint, to be sure, but one certainly worth pursuing
with $12 billion on the line. Texaco moved quickly to get all the
evidence necessary to have the judgment thrown out, or at least to
have Judge Casseb removed from ruling on Texaco's pending
motion for a new trial.

Texaco's attorneys hit a stone wall at every step.

The State Retirement System refused to provide Casseb's judicial service record, which it deemed confidential. An administrative judge in San Antonio refused to provide records of any
temporary judicial service by Casseb in the years after he had
retired from the bench. Because military service could be applied
toward a state pension, Texaco tried to obtain Casseb's Army
personnel file. The Pentagon, however, doesn't distribute such
information, either.

The only way to get to the bottom of the matter was a full-
blown hearing, Texaco's lawyers decided. They would have to
subpoena the custodian of the state's employment records—and
perhaps Judge Casseb himself.

The Texaco lawyers first ran to Tom Stovall, the administrative judge who had appointed Casseb to begin with, pleading for the right to subpoena state records and conduct a full-blown inquest into Judge Casseb's qualifications. "It was like they were talking in a mausoleum," Judge Stovall recalls. "They said they felt duty-bound as lawyers to raise the issue." Judge Stovall, however, did not feel duty-bound to settle the issue; a copy of Casseb's official appointment as a "retired judge" satisfied Stovall that his old friend Sol *was* qualified to hear the case. "If I see a fellow's driver's license, I'm not going to go behind *it*," Stovall would later explain.

The effort to disqualify Casseb took on the aura of a crusade. Texaco asked a state appeals court for an order compelling a hearing into Casseb's qualifications; the appeals court refused. Texaco then went to the Texas Supreme Court; it too refused to order a hearing.

Case closed. "There was a circling of the wagons because we had attacked the brethren," a Texaco lawyer would later insist.

On February 20, 1986, in the last, bizarre act by the trial court in *Pennzoil v. Texaco,* Judge Casseb entered the courtroom for a hearing on Texaco's motion for a new trial. Jury foreman Rick Lawler, a subpoena in his hand, was ready to testify that the employment of his estranged father-in-law with Texaco had absolutely no bearing on the jury deliberations. Juror Jim Shannon, summoned to explain his failure to divulge that his wife was a victim of an oil-industry megamerger, was planning to do one better. In Shannon's breast pocket was the one-dollar bill that the Harris County constable had attached to his subpoena, under an old tradition of supplying carfare to subpoenaed witnesses. During his appearance on the witness stand—the witness stand next to the same jury box where he had spent four and a half months— Shannon wondered if he might be asked whether he had received any money for his role in the case. When the question came, Shannon intended to withdraw the dollar bill and proclaim, *This is all I have received for serving in this case!*

And Texaco, for its part, was ready to beseech Judge Casseb himself to take the witness stand and settle once and for all whether he was qualified to have entered the $11 billion judgment.

After more than a month of watching Texaco try everything possible to dilute his jurisdiction, the seventy-year-old jurist entered the courtroom, greeted the lawyers and confronted them with one of the strangest questions they had ever heard in a courtroom.

"This court," he said, "would like a reply from each of the respective firms of record representing Texaco Incorporated if they still, at this time, question and challenge the right and authority of Solomon Casseb Jr. to preside over any hearings at this time in this case."

Texaco's lawyers exchanged glances. They were dumbfounded. He was threatening to *not even hold the hearing!* It appeared to them that Judge Casseb had pushed them into a multibillion-dollar Catch-22: There was only one way to get a hearing on their attack against the judge, and that was to *discontinue* their attack against the judge!

Finally Texaco's husky Rusty McMains ambled to his feet. "The answer to your question," said McMains, "is yes." Texaco *was* continuing to challenge the judge's right to preside.

"All right," Judge Casseb said. "Any other firms?"

One by one they rose.

"Your Honor, David Boies from Cravath, Swaine and Moore. We likewise continue our challenge."

"Jim Sales of Fulbright and Jaworski. We do continue the challenge."

"Your Honor, Michael Peterson for Miller, Keeton. We join in the motion of our co-counsel and continue the challenge."

Pennzoil's Joe Jamail rose. "*We* have no challenge to the qualifications of this judge," he said.

Judge Casseb then returned his attention to the Texaco lawyers.

"That still being your position, this court is then going to let the law take its course, and y'all are excused."

Judge Casseb stalked off the bench, leaving behind more than a dozen speechless lawyers.

In history's biggest court case, there would be no hearing on the loser's motion for a new trial. For lack of judicial action, Texaco's motion died.

Things came full circle in early May. The Getty Museum sued Texaco.

The whole thing turned out to be a humiliation for both parties—and one that played beautifully into Pennzoil's hands. Texaco, in the course of defending itself against a shareholder lawsuit in Delaware, made a court filing that contained an oblique but astonishing suggestion. It said that just maybe, Texaco would refuse to perform on the very indemnities that had got Texaco into so much trouble. The filing referred to potential "claims that

Texaco may have against others," and to the remote possibility of
"an action against the museum and the trust."

How could Texaco possibly consider filing suit against the very
people it had agreed to hold harmless? Marty Lipton, after all, had
testified that obtaining the indemnity was an absolute requirement
of the deal with Texaco. Lipton's associate, Pat Vlahakis, testified
that the indemnity was intended "to protect the museum against
every contingency, no matter how remote." Even Richard
Keeton, one of Texaco's own lawyers, had said during the trial
that "the indemnity *requires* Texaco to pay" any legal fees in-
curred by the Getty group.

But there was a catch, Texaco said.

The indemnities protected Gordon and the museum for any
liability arising from the "January 2, 1984" memorandum of
agreement with Pennzoil. But the jury in Houston didn't find
Texaco liable for any contract as of January 2, Texaco said. Under
Special Issue Number One, the jury found that a contract had
come into being "at the end of the Getty Oil board meeting of
January 3, 1984." The indemnities do not apply to *that* "so-
called agreement," Texaco said, "because no such 'agreement'
was disclosed to Texaco."

When the museum learned that Texaco was apparently thinking
of reneging, it immediately filed suit in Superior Court in Santa
Monica, claiming that Texaco's actions smacked of a "desperate
effort" to slough its Pennzoil liability onto the museum and onto
Gordon.

Texaco—caught between litigation by Pennzoil on the one
hand and by its shareholders on the other—was only trying to
make it clear that it wasn't waiving its rights to do *anything* in an
$11 billion case. But the museum's suit made it seem as though
Texaco no longer believed quite so passionately in its claim that
Gordon and the museum were "free to deal."

As publicity over the museum's suit grew, the museum consented to issuing a joint statement with Texaco. "Both the museum and Texaco believe that the judgment in the Texas Pennzoil
case is totally wrong," they said. For the time being, they added,
the museum would keep the lawsuit against Texaco on the back
burner, awaiting the outcome of the company's appeals in Texas.

After all of Texaco's undertakings in early 1986—the effort to
have Judge Casseb vacate the verdict, the end run into the federal
courts, the motion for a new trial, the attack on the jurors and

Judge Casseb himself—the $11 billion judgment was still intact, subject to reversal only in the appeals courts of Texas.

The first court to hear the appeals was the Court of Appeals for the First Supreme Judicial District of Texas, where James Warren, one of the three justices assigned to hear the case, was preparing for a re-election campaign in November. His running mate—Frank Evans, the chief justice of the court—was married to a woman who worked in the law offices of Joe Jamail. Evans, however, wasn't involved in hearing the appeal.

Texaco filed a brief enumerating ninety reversible errors in the trial, from the first words from Judge Farris' mouth ("This is the largest case ever filed in Harris County") to the last words from Judge Casseb's mouth (when he refused to rule on Texaco's motion for a new trial). It accused Pennzoil of conducting a "win-at-all-cost" legal strategy that had "infected the entire trial."

"Texaco," the brief said, "was denied the rudimentary elements of a fair trial."

Pennzoil countered that "Texaco's alleged errors are nothing more than a record of its failures" to present a worthwhile case during the trial. Indeed, Pennzoil enumerated dozens of occasions in which it claimed Texaco missed its chance to make proper objections, to introduce evidence or otherwise to "preserve its record" on appeal.

When the time came for oral arguments before the appeals court, *Pennzoil v. Texaco* had lost none of its carnival-like proportions. The arguments were held in a 750-seat auditorium at the South Texas College of Law in downtown Houston and transmitted by closed-circuit television into an adjacent room holding the overflow crowd of 250. Color-coded tickets were issued to the press, the public and the participants. (Joe Jamail left his at the office and had to get a court official to vouch for his identity.) Observers had their purses and briefcases inspected for tape recorders and weapons at three checkpoints before they could enter the hearing room. A shoving match erupted between a courtroom artist and a newspaper reporter fighting over a ringside seat. At one point the public-address system went haywire, piping in several measures of a Brandenburg Concerto. A dozen lawyers from each side sat before the justices, with dozens more seated in the audience.

The arguments were so starkly opposing that it was hard to believe that the lawyers were discussing the same case. Richard Keeton of Texaco asserted that "Texaco did all that a reasonable

man could have been expected to do" to avoid interfering with
Pennzoil's property rights. But Simon Rifkind, an eighty-five-
year-old former federal judge from Pennzoil's New York law firm
of Paul, Weiss, Rifkind, Wharton & Garrison, called Texaco's
action "as brazen an interference with another's contract as ever I
have seen," in sixty years of practicing law.

Each side depicted the consequences of the case in the most
extreme possible terms. "The imposition of this outrage on Texaco would in my opinion make a mockery of our entire legal
system," said Texaco's Jim Sales of Fulbright & Jaworski. Jamail
shot back, "Texaco is trying to bring us back to the law of the
jungle, where anything goes."

For months the world awaited the outcome of the hearing.
When Wall Street learned that the court usually published its
opinions on Thursdays, traders began bidding up the price of
Pennzoil shares every Wednesday and back down on Friday. Millions of dollars in winnings and losses changed hands in the stock
exchange on fresh speculation about the outcome, millions more
still on rumors that settlement talks had been renewed.

Fully a year after the jury verdict, *Pennzoil v. Texaco* remained
one of the hottest crapshoots around.

That's where it stood in early December 1986: Pennzoil still with
an $11 billion state-court judgment, Texaco with a federal court
preventing Pennzoil from collecting while the case remained on
appeal. Already, a total of twenty-two judges had heard various
aspects of the case, and far more were bound to—for no matter
what the Texas appeals court decided, the losing party was certain
to bring yet more appeals and fight for every inch of its case.

The next step for the trial court judgment would be the Texas
Supreme Court, and although the state High Court justices weren't
obligated to accept an appeal from either side, if they did, they
would not likely need any introduction to Joe Jamail.

By one tally, in only two years, the magnanimous Jamail contributed sums to the election campaign of five of the nine justices
on the Texas Supreme Court: $35,000 to C. L. Ray (the largest
that Ray had received); $20,000 to Chief Justice John Hill (his
second-largest contribution in that period); $10,000 to Justice
Robert Campbell (his largest); $10,000 to Franklin Spears (his
largest) and $7000 to James Wallace (his second-largest).

Among corporate contributors to Supreme Court candidates,
one of the least stingy by far was the political action committee of

Houston's Vinson & Elkins—a firm that used to work for Texaco, but which began working for Pennzoil in the federal-court litigation. There was no disputing that Vinson & Elkins brought outstanding lawyering to the Pennzoil team, but it also joined with the credential of having contributed at least $100,000 to Supreme Court candidates over the prior three years.

In mid-1986 a committee of the Texas Senate was scheduled to hold hearings to probe the inner workings of the Supreme Court, an investigation that touched off widespread speculation—and, apparently, concern among some members of the plaintiff's bar. Although the investigation had nothing to do with *Pennzoil v. Texaco*, Jamail had little interest in seeing the matter aired to public inspection. In his behalf, two prominent Texas lawyers lobbied successfully to have the investigation swept behind the closed doors of the Texas Judicial Conduct Commission, where the outcome of the investigation would never be made public.

Calvert Crary, a lawyer who studied major litigation for the brokerage firm of Bear, Stearns & Company, publicly called the cancellation of the public investigation an "important investment issue" for anyone trading Pennzoil and Texaco shares.

In addition, on January 12, 1987, the U.S. Supreme Court heard oral arguments on Pennzoil's motion to overturn the bond ruling. The mere fact that the High Court would accept Pennzoil's appeal put a presumption of advantage on Pennzoil's side, since the court is rarely anxious to stick its nose into headline-grabbing cases. Pennzoil was also heartened by President Reagan's nomination to the court of Justice Antonin Scalia, who was well known for adhering to strict views toward the separation of powers. "On the federalism issue, which is all that's involved in the bond case, the present and future composition of the court is favorable to our position," claimed John Jeffers of Pennzoil.

But when David Boies of Texaco faced off with Laurence Tribe of Pennzoil over the bond issue, it was Justice Thurgood Marshall who appeared most ready to take sides. Texaco had relied on the Civil Rights Act—Section 1983 of the U.S. Code—to justify federal invasion into the state courts of Texas, and Marshall, the court's only black justice, appeared offended by the whole thing.

"Do you really think 1983 was passed to protect multibillion-dollar corporations?" Marshall asked. "There's no statute that says the fat cat wins and the small guy loses."

"Your Honor," Boies answered, "I think 1983 was passed to protect the rights of everyone."

Marshall also demanded Texaco's explanation for its choice of the hometown federal court in White Plains for litigating the bond issue.

"Could this case have been filed in Anchorage?" Marshall asked.

"I don't think so, Your Honor," Boies said.

"What's the difference? You admit you're [forum] shopping. . . . They do have federal courts in Texas!" Marshall barked.

During the proceedings, Laurence Tribe accused Texaco of creating a "Fortune 500 exception to federalism" and of demanding "Dow Jones due process" by striking fear in the courts that the Texas bond rule would touch off history's largest bankruptcy filing. "If you're rich enough and big enough to project your fears into the stock exchange tape, then you get a better deal. . . . We find principles of federalism and due process bent and twisted to get a better deal for a larger corporation."

Although lofty constitutional principles were at stake, politics, too, made an appearance at the hearing. Sitting conspicuously with the Pennzoil lawyers was Howard Baker, the former majority leader of the U.S. Senate, who, as a presidential aspirant, was now working for the Washington office of Houston-based Vinson & Elkins, one of Pennzoil's law firms. It hardly hurt Pennzoil to have a former Senator—who had been involved in confirmation proceedings on some of these very justices—sitting on its side.

Finally, on February 12, 1987, the first tribunal to review the underlying verdict itself—the first appeals court to evaluate whether Judge Farris and Judge Casseb had properly applied the law, the first to evaluate the damages award, the first to judge whether Texaco had a fair trial—held forth.

Once again, Pennzoil emerged victorious.

The Texas appeals court affirmed $8.53 billion of the jury's original $10.53 billion award. The justices knocked off $2 billion of the $3 billion in punitive damages that the jury had awarded because of the indemnities; $1 billion, the justices demurred, was "punishment" enough. With interest tacked on, the liability facing Texaco now totaled $10.2 billion.

A theme ran through the understated, 162-page opinion: *The jury has spoken.* Only the jury could decide the facts, and judge the credibility of witnesses. The court assiduously avoided even the slightest second-guessing. And this was the last court with power to review the facts of the case. Neither the Texas Supreme

Court or the U.S. Supreme Court could re-try the facts. They could only sit in judgment over the application of law.

Among the litigation handicappers at Pennzoil, the settlement odds were now more favorable than ever. "If the seriousness of this doesn't sink in now, these people don't have the collective IQ of a geranium," Joe Jamail said. But Al DeCrane, the newly elected chairman of Texaco, was talking tough, insisting that Texaco would never let Pennzoil use the judgment as a settlement club. "For a court to allow that kind of terrorism to carry over into some sort of negotiation is really kind of polluting the well."

And there were even wider ramifications to the Pennzoil-Texaco case.

· 28 ·

Did corporate America hear the jury's message? Did the Pennzoil case put the honor back in a handshake?

If anything, it began to render the handshake obsolete.

Some time after the verdict, Michel Zaleski, a New York investment banker, was winding up a day and a half of tough negotiations. At that moment of achieving an agreement in principle, he would recall, "I looked the guy in the eye, stuck out my hand and said, 'Let's shake on it.'"

"We can't do that," the lawyer on the other side replied sternly.

"I never thought a handshake was anything more than a moral commitment," Zaleski says. "But now people are afraid to make even a moral commitment for fear someone will use it against them."

The Wall Street Journal would dub the phenomenon "the Texaco chill." It crept into a news conference at which Frank Borman, the chairman of Eastern Airlines, was trying to justify the company's sale to Texas Air Corporation at what many considered to be an unfairly low price. A rival takeover offer would probably never succeed, Borman said—"even if someone offered more."

Why was that? "You probably have heard of Texaco and Pennzoil," he said. There was talk that Borman's comment helped convince Chicago's Pritzker family, which controlled Braniff Incorporated, to back off from an interloping bid for Eastern—one that presumably would have more handsomely rewarded that company's shareholders.

Lawyer Klaus Eppler was advising several directors of Sperry Corporation when they agreed not to oppose a tender offer by Burroughs Corporation, a transaction that created a new computer company second in size only to IBM. When Sperry issued a public announcement about its assent to the takeover, "I warned my clients that by doing so they had probably created an enforceable contract," Eppler would recall.

The Texaco chill began ranging beyond the takeover game and into more prosaic corporate transactions. Pratt & Whitney reportedly had a handshake agreement to supply United Parcel Service with $400 million of engines for twenty UPS jets. Then Rolls-Royce Ltd. slipped in with a bid of $25 million less. The resourceful negotiators from Pratt & Whitney breathed a mention of the Texaco case, one source said, whereupon UPS told Rolls-Royce to take their less expensive engines and fly home.

Where deals *have* fallen through, the Texaco case has emboldened the losers to seek the same redress that Hugh Liedtke did—even in cases beyond the borders of the U.S. In June 1986, Allied-Lyons PLC of London sued Hiram Walker Resources Ltd. in Canada for $6.49 billion (U.S.), claiming that it had a deal to buy Hiram Walker's liquor business. Hiram Walker made the deal while resisting a takeover by Gulf Canada Ltd. But when Gulf Canada ultimately succeeded in gaining control of Hiram Walker, it canceled whatever arrangements Hiram Walker had made with Allied-Lyons.

"Our lawyers have seen similarities in the Texaco-Pennzoil decision and this situation," said Sir Derrick Holden-Brown, chairman of Allied. Allied's damages demand was believed the largest ever filed in a Canadian court.

The Texaco chill certainly didn't freeze out competition, nor did it make the New Good Ol' Boys of Wall Street begin acting like the Old Good Ol' Boys of the oil patch. Nevertheless, investment bankers everywhere began treading more lightly—if only to become more diligent in determining the extent of their targets' freedom to deal.

"No longer can we say, 'We stole a deal fair and square,'" said investment banker Alan Rothenberg. And that was reform enough.

*   *   *

Out on the West Coast, the effect was more pronounced. *Pennzoil
v. Texaco* only heaped more fuel on the suspicion burning in the
Getty family.

The suit by Tara Getty, which sought to have Gordon neutralized as sole trustee, lingered long after Getty Oil itself was
gone. Eventually Gordon and his relatives came close to settling
their differences through the obvious solution of splitting the
Sarah C. Getty Trust into four pieces, each to be managed separately for the four main bloodlines that descended from J. Paul
Getty. This solution, which hardly seemed in keeping with the
intent of the departed Sarah Getty, required an act of the California legislature to bring off.

However, then the $11 billion judgment hit, and suddenly some
of Gordon's relatives got cold feet about the new arrangements.
As much as they might have enjoyed having independent control
of their shares, they did *not* want to receive one-fourth of any
liability that Texaco might try to pass along to the trust, if ever
Texaco sued Gordon for misleading the company into thinking he
was free to deal.

"The Texaco-Pennzoil litigation is casting a very deep shadow
over the family's financial affairs," says Vanni Treves, the lawyer
for Paul Getty Jr. in London.

But Gordon had the happy experience of gaining credibility as a
composer. "All told," *Newsweek* said of his "Plump Jack," a
cantata about Shakespeare's Falstaff, "there is a genuine dramatic
sensibility at work here." Gordon seemed perfectly content to
concentrate on music over the oil business. "Composers," he
says, "are remembered when the businessmen are forgotten."

Still, Gordon sometimes wondered what it would have been
like to take control of Getty Oil with Hugh Liedtke as his partner.
"I'd have enjoyed the fun of seeing what could be done with the
Pennzoil situation," he would say.

But Gordon could take pride in something that few real oilmen
could. His decision to deliver Getty Oil into the hands of Texaco
had become one of the most auspiciously timed major transactions
in oil patch history. After years of indecision over whether to buy
or sell, Gordon finally had sold—virtually at the top of the market, near the peak of oil prices, just before the world oil markets
had come crashing down amid the disarray in OPEC. By late 1986
the $4 billion that the trust received for less than half of Getty Oil
would be enough money to buy *more than* half of Texaco.

"I picked up my marbles," Gordon would say, "and took them home."

Judge Farris continued to boil. "They have savaged me," he said of Texaco's lawyers.

One member of Texaco's Houston law firm—Thomas McDade of Fulbright & Jaworski—had the misfortune of having a suit over the disputed ownership of some North Sea oil leases assigned to Judge Casseb's court. In citing precedents, the judge told him and the other lawyers in the case, "do not cite any decisions by White Plains, New York, federal judges"—a slap at Judge Brieant, who had dared to invade the sovereignty of the Texas courts on the bond rule.

McDade was appalled. "I truly hope that you have the courage to put aside any claimed bias and prejudice against Texaco," McDade wrote the judge.

The judge, in turn, was furious. "To put it in the vernacular, sir, for the purpose of this letter, I have more balls than any partner or associate in your firm. . . . *Do not* lecture me on COURAGE, sir!"

"Try to leave Texaco the hell out of it," McDade replied in yet another letter.

On September 29, 1986, Judge Farris died of cancer. Joe Jamail served as one of his pallbearers, not because he had ever been such a close friend of the judge, a mutual acquaintance would say, but because the two had become soulmates of sorts in Texaco's unrelenting attack on the $10,000 campaign contribution.

Although nearing his seventy-first birthday, Judge Casseb remained as vigorous as ever, gaining prominence on the bar association lecture circuit. Casseb filled the house at engagements in San Antonio and Los Angeles—and in both cases the members of his audience included Texaco attorneys equipped with tape recorders.

The Los Angeles speech was the real bell-ringer.

"We do things big in Texas," he said by way of introduction.

He made note of an affidavit that Texaco filed prior to entry of the judgment, in which the retired chief justice of the New York Court of Appeals stated that Judge Casseb's charge was unrecognizable as a matter of New York law. "He can have his affidavit back, and he can do with it as he sees pleased to do with it," Judge Casseb said.

He told a Jewish joke—to dead silence.

And he made it plain that as one might expect of any judge coming in on the tail of a four-and-one-half-month case, he didn't quite have all the facts straight.

He referred to Marty Lipton, the museum lawyer, as the recipient of an $18 million fee; it was actually Geoff Boisi of Goldman, Sachs who had accorded himself that distinction. He had Pennzoil suing over a billion gallons rather than a billion barrels of oil. He indicated that Pennzoil called two expert damages witnesses when in fact it had called only one. And so on.

Then, during the question-and-answer period, Judge Casseb gave a long, rambling prediction about the outcome of the appeals, and noted: "I feel that there is a good chance that perhaps I may have read the cases wrong and not have applied it correctly. And as I said, you know, it was my first experience in trying to analyze New York law after I had been forty-six years with Texas."

It wasn't a confession; Judge Casseb went on to express confidence that his judgment would be upheld. Indeed, the media were present for the speech and made nothing whatsoever of the remark. But in the expressionless reality of the court reporter's transcript, based on the recording made by Texaco's lawyer, the comment looked downright awful.

Texaco would see to it that the world knew. One of its public-relations operatives tried to leak the transcript and tape to the *San Antonio Light,* the judge's hometown newspaper. "I don't exist," the PR man whispered, handing over a copy of the transcript replete with yellow highlightings. "I think this is a story," he said, noting the misgiving that judge had expressed about his application of New York law.

The San Antonio newspaper, however, dismissed the judge's comments as mere ramblings and declined to write a story.

Finally the *Houston Chronicle* published an account, causing such a furor that Pennzoil took out oversize ads in national newspapers with the headline, WHAT JUDGE CASSEB *REALLY* SAID. A debate over ethics in public relations ensued. Copies of the recording moved to the top of the hit parade of jurisprudence. Pennzoil shares fell on news of the judge's remarks. And yet, like so many other sideshows in the case, this one would make not a whit of difference to the outcome.

Jim Shannon, whom Texaco lawyers were calling the "one-man jury," decided to write a book, and even hung a jacket design in his apartment. The book would be called *The $10 Billion Jury*. "I want to place the reader in the jury box," he would explain.

As he launched into his research, delving for the first time into the written record of what had occurred outside the jury's presence, Shannon found a few things disturbing. One was the resignation of Pennzoil's general counsel, Perry Barber, after the Getty Oil deal fell through. Texaco was convinced that Pennzoil had made Barber the fall guy, but Judge Farris had refused to admit any evidence concerning the circumstances of Barber's departure. Shannon read in Barber's deposition that the resignation was purely voluntary. But still, Shannon would conclude, "There was something rotten in the whole deal."

Shannon learned of some of the past accusations made against Pennzoil. "If Miller could have gone over every peccadillo in Liedtke's career, he could have made some points," Shannon would say. And he read a *Wall Street Journal* article, written shortly after the Getty oil deal fell apart, in which a Baker & Botts lawyer conceded that Pennzoil had failed to get a signature on its "lockup" option to buy eight million Getty oil shares. "It intrigues me," Shannon would say. "It's an open point that nags at me."

But Shannon's doubts were the musings of a curious mind—of someone seeking the truth as a writer in control of his research, rather than as a juror with no control over the evidence. Even after examining a full copy of the record, he would remain passionately convinced of the correctness of the jury verdict.

According to Shannon, one Texaco lawyer flew in from New York, took him to dinner and suggested a surefire formula for making his book a bestseller: Tell the story of how the jury had screwed up in history's biggest court case. Shannon declined the suggestion.

Joe Jamail continued attacking Big Oil, going up against Exxon in a property dispute, but for the most part he returned to his personal-injury practice. Nevertheless, Jamail remained vitally involved in the Pennzoil case, maintaining his role as lead counsel over a retinue of lawyers that had grown to include seventeen law firms or individual practitioners. Jamail wanted to be certain that his friend Hugh Liedtke collected a boxcar-sized judgment. And in doing so, of course, Jamail would be helping Jamail collect a boxcar-sized fee.

Among lawyers, the matter of Jamail's fee understandably remained one of the alluring aspects of the case, and Jamail was uncharacteristically circumspect on the subject. He had once actually told the *Houston Post:* ''Hugh Liedtke is my friend and it was a labor of love. I have never billed them for a thing.'' That did not mean, of course, that he would not bill Pennzoil once a settlement or judgment was collected—which, after all, is how all contingency-fee lawyers work. The common rumor was that Jamail's fee would total 20%. On the full judgment, that would total well over $2 billion, enough to make him richer than any Getty family member could ever dream of being. Jamail, however, denied that rumor.

Jamail made his most expansive comments about the fee arrangement in a speech before a Canadian bar group. ''I will be rewarded handsomely,'' he said, but, ''the amount depends somewhat on the outcome.''

And in any case, he said, the fee would probably be of such size that ''I plan to give Pennzoil some of the money.''

In October 1986 Texaco announced that John McKinley would retire at the end of the year after having already served nearly two years past the normal retirement date. The announcement ended speculation that McKinley would remain in charge until the Pennzoil case was somehow gotten rid of. McKinley had virtually completed his effort to ''turn the ship,'' yet because of the judgment the ship would be listing when he gave up the bridge.

The chief executive's position fell to Annapolis graduate Jim Kinnear, the former vice chairman, who over the years had become a friend and occasional hunting companion of Hugh Liedtke's brother Bill. Kinnear's rise was cause for some rejoicing at Pennzoil, not for his relationship with Bill Liedtke but because he was viewed as more easygoing than McKinley—as *someone you could deal with*. Kinnear did not hold himself out as any pushover, however. ''There is no clear path for them to get anything'' out of the litigation, he would declare.

Just the same, Kinnear did not seem the kind of leader to put protestations of innocence ahead of corporate survival. In the Korean war, he had had the delicate job of navigating a vessel through the heavily armed Inchon Harbor. ''It was my job,'' he would make a point of saying, ''not to hit the mines.''

In November 1986, Marty Siegel was visiting Marty Lipton when a federal marshal invaded the premises. He handed a subpoena to Siegel, who burst into tears. The jig was up.

It had been Marty Siegel, the government would claim, who had tipped off takeover speculator Ivan Boesky to the tension building in Getty Oil. By some accounts, Siegel had been drawn to Boesky by the magnetism of the arbitrageur's wealth. And initially, it appeared, Siegel had entered into the allegedly illicit arrangement with innocent intentions, merely to provide Boesky with the expertise of an investment banker whose general knowledge might prove valuable. But along with his insights, the government charged, Siegel eventually began to pass inside information—on the Bendix–Martin Marietta takeover, on the merger of Natomas and Diamond Shamrock and finally on the goings-on at Getty Oil, sometimes in a downtown alley, sometimes in a midtown coffee shop. In exchange, the prosecutors said, Siegel received surreptitious cash payoffs in a briefcase.

In February 1987, Siegel weepily pleaded guilty to criminal insider-trading charges, consented to a civil penalty of $9 million and agreed never again to function in the securities business. He was to be sentenced in April.

Boesky had already been charged. So too had many other figures of high finance—a young man working at Wachtell, Lipton, for instance, and a top partner in Goldman, Sachs. Between the jury's repudiation of investment-banking tactics in *Pennzoil v. Texaco* and the government's stunning dragnet against the New Good Ol' Boys of Wall Street, the takeover game had been tamed, probably forever.

Among the Old Good Ol' Boys, Hugh Liedtke got his honor back.

It happened at a banquet of the All-American Wildcatters Association, the old-boy outfit whose motto is, "My word is my bond." The wildcatters bestowed their annual B.F.U. Award— publicly, it stood for Biggest Foul-Up—in recognition of what was ostensibly the oil patch's most ignominious achievement of the year. But a crowning achievement could also merit the B.F.U. award as long as the event in question provided an excuse to poke fun at the recipient.

What better excuse than Hugh Liedtke's $11 billion court judgment against Texaco? After its decades of heavy-handed dealings with the independents of the oil patch, Texaco remained one company without a single member in the wildcatters' club, whose members thrilled at the success of their buddy Hugh Liedtke in bringing Texaco to its knees.

This was the same speech in which Liedtke was roasted for his

obsessive pursuit of the "yellow can"—from the time of his boyhood fascination with the yellow commode in the home of his
nextdoor neighbors, the Skelly family. With his tongue in his
cheek, a geologist from Dallas recounted "Hugh Baby's" career,
all the way from J. Paul Getty to Gordon Getty, embellishing his
tale with all the hyperbole and scatology he could muster. Finally,
he concluded:

Before his yellow can could be filled with champagne to celebrate, a slinking Texaco slid inside him and fouled Hugh Baby's
line. Getty flipped off the hook and slithered into Texaco's waiting net. Hugh Baby cried foul and took complacent Texaco to
court. Hugh Baby won!
   To this point, Hugh Baby had handled himself admirably. But
let us examine what has happened since he won his case. . . .
   When Texaco pleaded, "If we are forced to put up the bond,
we might have to go into bankruptcy, which could seriously
hamper our operations," Hugh Baby noted, "Texaco is the only
company I know of where bankruptcy would probably speed up
their operations."
   When McKinley tried this emotional plea—"I regret that I
have but one life to give for my company!"—Hugh Baby's un-
called-for response: "That makes two of us!"
   I must ask you, Hugh Baby. Is this love? Understanding? Is
this America? Democracy? You ask me, "What is democracy?"
I'll tell you what democracy is. . . .
   Democracy is your friendly Texaco dealer filling your tank,
cleaning your windshield, checking your tires, getting your kid a
Nehi soda out of the cold-drink box. Yes, and for all of us who
have had such pleasant trades and warm relationships with them
throughout so many years, democracy is truly Texaco, which we
have all come to think of as the little company with the big heart.

At this point, the house was coming down with jeering laughter.
   J. Hugh Liedtke, you have received the 1986 B.F.U. award!
After the applause had died down, after all the backslapping,
handshaking and storytelling had come to an end, what did Hugh
Liedtke really have? By 1987, even after postponing his scheduled
retirement, his dream of building a major oil company remained
unrealized. He had won the respect of his peers, but the derision
of others. In his effort to obtain justice, his company had been
accused of unjust acts. His battle to restore values to "a world

gone slipshod'' had been reduced to a battle of wits in the appellate courts. Judges and jurors had been sullied—not by him, but by the adversary he had engaged.

Somehow, it seemed, the tables had turned. ''If it takes the last breath in me,'' declared James Kinnear, the new chief executive officer of Texaco, ''I intend to get vindication.''

May the best man win.

· EPILOGUE ·

On April 19, 1987, the unthinkable occurred. Texaco became the largest bankrupt in history.

And in the end—even after wielding Chapter 11, the most powerful weapon available to corporate management—Jim Kinnear, the new chief executive of Texaco, would fail in his quest for "vindication." For in this case of superlatives—the biggest damages demand, the biggest verdict, the biggest bankruptcy—Kinnear would also have history's largest legal settlement jammed down his throat. Hugh Liedtke, a card player to the last hand, would walk away victorious. For intrigue, brinkmanship and cunning, it is a story that nearly rivals everything that came before.

Barely three months earlier, on January 1, Kinnear became president and chief executive officer with a mission to restore Texaco to its long-lost grandeur. Texaco would join the rank of Atlantic Richfield, Unocal, Phillips Petroleum and others that had radically restructured their operations—some of them under pressure from possible corporate raiders, others just because doing so made good business sense. In the scheme taking shape in Kinnear's head, Texaco would jettison unprofitable assets. It would expand moneymaking operations. The company would rethink its second-rate exploration strategy. Kinnear began appointing a new crop of young, vibrant leaders—many in their 40s, plucked from middle-level positions where they had shined and where they had never known the inbred, penny-pinching ways of Gus Long. In the years following Long's retirement from Texaco, John McKinley had begun to "turn the ship." Kinnear, the Annapolis man, would devote himself to navigating it full astern.

But Kinnear's grand plan would have to wait until he had shaken the lodestone of the Pennzoil judgment. Kinnear was every bit as confident of defeating Pennzoil—of obtaining his "vindication"—as he was of his corporate turnaround strategy. Kinnear was involved in the events of January three years earlier when Texaco had snatched Getty Oil. He had sat through virtually every hour of testimony in State Court 151. To be sure, the judgment had been upheld—unanimously—in the Texas appeals court, leaving only the Texas Supreme Court, where, in Texaco's view, Joe Jamail's contributions had already sewn plenty of name recognition for Pennzoil. *But all we have to do is get this goddamn case in the U.S. Supreme Court*, Kinnear thought. *Either that, or we settle with Pennzoil—at a price I can live with.*

Indeed, within eight days of becoming Texaco's chief executive, Kinnear offered to settle by purchasing Pennzoil's oil-and-gas assets for about $1 billion. It was a low-ball offer at best, one that by Pennzoil's reckoning didn't even reflect the value of the assets that Texaco would be acquiring. Kinnear didn't care. The judgment wasn't worth a plug nickle in his view. Let Liedtke hold out—and watch him get nothing!

It was four days later that Kinnear's chief lawyer, David Boies, stood in the marbled halls of the U.S. Supreme Court, receiving his tongue-lashing from Justice Thurgood Marshall over the bond-and-lien issue. Suddenly, the federal-court injunction that had protected Texaco for a year showed signs of weakening. If the Supreme Court dissolved the injunction, Pennzoil would once again have the freedom to seize Texaco's assets. Texaco might even have to post a $12 billion appeals bond. Once again, Texaco was standing face-to-face with its extinction.

Like a company building a war chest for a long union strike, Texaco began harvesting cash from all possible sources—selling receivables, for instance, and even mortgaging stock in foreign subsidiaries to obtain new credit lines. Cash swelled to more than $2 billion from the usual level of less than $1 billion. It would probably be late April, the company's lawyers determined, before the Supreme Court ruled on the bond issue.

They were wrong. By a vote of 9–0, the Supreme Court on Monday, April 6, found that Texaco had gone running into the federal court system too soon—without asking the appeals courts of Texas for protection from the bond-and-lien rules. In retrospect, it appeared that Texaco had made yet another stunning blunder—just as it had failed to answer Pennzoil's original suit in Delaware, just as it had failed to contest Joe Jamail's stratospheric damages demand. Suddenly, Texaco was exposed to the full weight of the bond-and-lien rules of Texas—and to all the wiles that Hugh Liedtke cared to engage. Now, only a settlement—and certainly not one like the low-ball offer of three months earlier—could save Texaco.

The Texaco board was immediately summoned to White Plains. That evening and the next morning, two Texaco jets touched down in Houston, ready to importune Hugh Liedtke to settle.

Texaco's lawyers had begun to view Liedtke as something of a Buddha—someone before whom his disciples made offerings, to which he would merely nod or frown without further response. For the next week in Houston, in secret meetings at the Intercontinental Hotel and at a local condominium tower, Liedtke reinforced the impression. He refused a check for $2 billion. He refused to accept stock. He refused a check for $100 million just for a promise that he wouldn't install any liens.

As Kinnear and his cohorts raced around Houston in a silver limousine, Texaco's creditors grew alarmed over a possible Chapter 11 filing. It was a replay of the panic that immediately followed the verdict—only this time it was worse. The world was no longer dismissing the judgment as a fluke, certain to be reversed by the first appeals court to hear it.

Manufacturers Hanover Trust and other banks told Texaco that they would cancel a $1 billion receivables-financing agreement if Pennzoil filed liens against more than $500 million of Texaco assets; at the same time, Pennzoil was publicly demanding some form of security for the entire $10.3 billion judgment. Two days later, the banks told Texaco that they would no longer purchase receivables at all—no matter what. Executives began worrying about a "run" on Texaco debt, in which an action by Pennzoil would trigger defaults under one set of debt securities that would trigger defaults under still more debt securities. Chase Manhattan Bank asked for time deposits to cover routine "daylight overdrafts." Morgan Guaranty Trust also demanded time deposits as security against any daily transactions. Bank of America refused to wire certain funds without first receiving wire-reference numbers proving that Texaco had incoming funds to cover the transactions. Banks in Japan, Italy, West Germany, the Netherlands and elsewhere began refusing to engage in foreign-currency transactions.

The oil patch was no more understanding. Some trade creditors and business partners quit paying their bills to Texaco. Sonatrach, the Algerian national oil company, canceled future deliveries of crude oil and natural-gas liquids indefinitely. Challenger Petroleum refused to sell any more oil to Texaco than Texaco was selling to it. Citgo threatened to quit doing business with Texaco altogether. Occidental Petroleum demanded cash up front for a cargo of crude, and British Petroleum refused to accept an order for fuel oil. Texaco's crude-oil supply was growing so short that by late in the week, the company began making contingency plans to close a major refinery somewhere in the U.S.

Liedtke was never convinced that Kinnear would plunge Texaco into bankruptcy. Quite the contrary. It would be too cruel an act to the company's employees, to its shareholders, Liedtke thought—a suicide mission, a kamikaze dive. But as the settlement negotiations wore on, Liedtke, by the end of the week, finally was naming a price—a vague one, to be sure, somewhere in the range of four to five billion dollars. The figure was as far out of Texaco's ballpark as the $1 billion Kinnear had offered earlier in the year was out of Pennzoil's ballpark.

Kinnear flew back to New York knowing the end was near. Saturday, well into the night, he pondered the company's options with the board of directors. Early Sunday morning, a letter arrived from Houston with a formal settlement offer from Hugh Liedtke. It demanded $4 billion. There was no telling what Pennzoil might do if the offer were rejected. The company was festering with fear that Pennzoil would hamstring Texaco's assets. And fear quickly began to border on paranoia; some of the company's advisors began to worry that Pennzoil itself would throw Texaco into bankruptcy proceedings—on Pennzoil's home turf in Houston.

That's it, the board decided. Let's do it.

Harvey Miller, Texaco's lead bankruptcy counsel, departed from Texaco headquarters and drove to Larchmont, N.Y., to the home of Howard Schwartzberg, the only bankruptcy judge sitting in the corporate haven of Westchester County. At 11:22 A.M., he presented the judge with Texaco's petition for Chapter 11 protec-

tion. With the stroke of the judge's signature, Texaco was safe—
and bankrupt, legally if not financially.

Jim Kinnear drove into the city to meet Hugh Liedtke, who had
flown in from Houston. "I didn't think you'd do it," Liedtke said.

By midafternoon, Kinnear was standing before scores of reporters in a Manhattan conference room. "Out of necessity," he said,
"we've prepared for this decision for months. Still, when a decision had to be made, it didn't come easy. . . . This process has
been enormously painful," he said. "I love this company."

But Pennzoil, he told the press, "has turned the knob one too
many times. Texaco's ability to finance our business has virtually
ceased."

Back in Houston, Joe Jamail was asked for a comment. "It was
a totally irrational act," he said. And the worst was yet to come.

Robert Holmes a Court, the wealthiest man in Australia, knows a
bargain when he sees one—and in the months following the bankruptcy filing, Texaco's shares sold in the stock market at prices
befitting a bankruptcy sale. Holmes a Court made himself Texaco's largest shareholder, quickly snapping up 13 percent of the
company in the open market.

Had it been anyone else—Boone Pickens, say—a takeover, or
at least a proxy fight, would have been a foregone conclusion. But
to the surprise of the financial world—and to the disappointment
of Pennzoil—Holmes a Court turned out to be an ally, declaring
publicly that he considered Texaco an undervalued investment and
that he stood fully behind Jim Kinnear's efforts to get rid of the
Pennzoil litigation one way or another.

The Texas Supreme Court proved to no avail, however. Unanimously, on November 1, the court refused to hear the appeal with
a three-word declaration that the lower Texas courts had committed "no reversible error." Texaco lawyers were convinced that
if Joe Jamail's big campaign contributions did pave the way for
the decision, they certainly hadn't helped Texaco, either. Jamail,
of course, denied ever peddling influence, as indeed he may not
have. But the appearance of a conflict of interest on the court did
little to erase the lingering mistrust of the judgment among many
Texans and others.

Although Texaco was shocked at the dispatch with which the
court rejected its appeal, it never really thought much of its odds
there. But the U.S. Supreme Court, Texaco thought—now that
was a different story. Even while devastating the company earlier

in the year on the bond issue, the High Court seemed to invite an appeal from Texaco of the judgment itself. "If, and when, the Texas courts render a final decision . . ." the opinion said, "review may be sought in this Court in the customary manner." Jim Kinnear was convinced that simply by getting the case before those nine justices in Washington, he would practically be home free. And with so supportive a major shareholder as Holmes a Court, why couldn't he?

Then, on October 19 came Black Monday, the worst bloodbath, in dollar terms, in New York Stock Exchange history. Robert Holmes a Court was in desperate straights. He had to unload his Texaco stock—into a depressed market.

The month after the crash, Carl Icahn was traveling to Martha's Vineyard for the weekend when he read an account of Holmes a Court's travails. Like the Australian, Icahn had a keen eye for value. Unlike him, he had amassed a solid and much-feared record as a raider of American companies—a man who had sent the executives of U.S. Steel, Trans World Airlines, Phillips Petroleum and others diving for cover with his takeover threats. Icahn, as it turned out, also had picked up a small piece of Pennzoil—about 2 percent of the company's stock. And before long, he had plunked down upward of $1 billion in Texaco. It was a brilliant strategy: not only was the threat of a takeover—of either company—implied in his ownership, but a settlement would probably cause both stocks to increase.

As he bluntly put it to Kinnear, "I want to make money."

But Kinnear remained steadfast in his refusal to meet Liedtke's terms. "What are you going to do?" Icahn asked him at one point.

"Go fishing," Kinnear replied.

"You might not have a company when you come back."

"Well," Kinnear answered, "they can't take my fishing pole."

But Texaco, which had thrust itself into bankruptcy-court protection to maintain control of its destiny, quickly realized that these same legal proceedings provided a platform to a whole new series of players. Pennzoil, the creditors committee, a committee of shareholders, a group of lawyers suing the Texaco directors, and of course, Icahn—they all had a right to be heard in Texaco's bankruptcy case, and to maneuver behind the scenes on a bankruptcy-reorganization plan that would benefit them. Robert Norris, a cowboy whose face was known to millions as the Marlboro Man, also was a major Texaco shareholder, and took it upon himself to mediate. He demanded a final figure from Liedtke, who by this time said he was willing to take "more than" $3 billion. The following morning, the extremely literal Norris was on a plane to Houston. His offer: $3.001 billion. Quickly, other creditors joined in, casting their lot behind a "cram down," which is the law's term for a bankruptcy plan forced upon the debtor by an alliance of all other parties.

David Boies—the litigator depicted on the cover of *The New York Times Magazine* as the hottest lawyer on Wall Street—had the bad news thrust upon him when he was summoned to a meeting of more than two dozen creditor lawyers. The settlement of $3.001 billion was presented as a fait accompli. And there was more. All the company's takeover defenses would be dissolved—teeing the company up for a hostile acquisition that would most benefit Icahn and the remaining creditors. Boies wasn't even wearing one of his ninety-dollar Sears suits that day. But he tugged his pullover sweater over his head and stalked out of the meeting with Harvey Miller, Texaco's bankruptcy lawyer.

Kinnear had one card left to play: He could "bet the ranch," rolling the dice on the company's future by tying up the "cram down" plan with lawsuit after lawsuit until the company had a chance to get the case before the U.S. Supreme Court. Privately, the company's lawyers were advising him that Texaco had a one-in-five chance—maybe—of reversal, giving Texaco the benefit of absolutely every doubt.

Kinnear, in the end, succeeded in preserving the company's takeover defenses—and, in the most infinitesimal of victories, of knocking that last $1 million off the settlement figure, making the number a nice, round $3 billion. On Saturday December 19, just four years, almost to the day, after Hugh Liedtke launched his tender offer for Getty Oil, Kinnear signed the final surrender with the gold pen he had received on his twenty-fifth anniversary with the company.

Hugh Liedtke never did get his billion barrels of Getty oil back, yet he could spend every working hour counting out his settlement in one-hundred dollar bills and never reach the total.

"Good luck to you," Liedtke told him

"Good luck to you," Kinnear replied.

That same Saturday morning, I sat in a suite at the Plaza Hotel with Allanna Sullivan, my *Wall Street Journal* partner, awaiting an interview with Kinnear. Lawyer Boies had reserved the suite, though he hadn't slept in it, or anywhere else, for two days. His young daughter and son scampered from room to room until their mother, Mary, who is also a lawyer, gathered them up for some midtown sightseeing and Christmas shopping. Frank Barron, Boies's partner from Cravath, Swaine & Moore, arrived bemoaning his absence from a family outing to *Sesame Street on Ice*.

Then came Kinnear—wan, fighting off a head cold, sleepless after days of brutal negotiations.

"We're going to go out and run the best damn oil company in the world," he told us.

But would he get his chance? Icahn sat there still—Texaco's largest shareholder, plenty mad that the settlement failed to catapult Texaco's stock price as he had expected. A proxy fight for control of the company, and possibly a takeover, loomed.

Moreover, other shareholders were restive. By filing Chapter 11, Texaco had shaved $1 billion from the $4 billion settlement that Liedtke had demanded the morning the petition was filed the prior April. The shareholders, however, had lost close to that much in dividends. Sixty million dollars of their money had been paid to lawyers and advisors. (All told, on all sides of the case, some 350 lawyers had been involved from the beginning.) And the lost opportunities of the bankruptcy filing were incalculable.

The company would have to go deeply into hock to pay the settlement, reversing much of the progress made in reducing the $10 billion in debt Texaco had incurred to acquire Getty Oil in the first place.

Texaco could not go after Gordon Getty or the museum for the money—assuming that it could ever collect from either of them. In a delicious irony, as part of the bankruptcy reorganization plan, they received indemnities from any and all liability to anyone.

Texaco's reserves were still depleting rapidly.

The Internal Revenue Service came into bankruptcy court one day and announced that Texaco might owe $6 billion in back taxes because of Texaco's method of pricing much of the crude oil it had lifted over the years from Saudi Arabia.

The State of Louisiana was still claiming that Texaco had cheated on its royalties from the old Huey Long leases.

And Hugh Liedtke was laughing all the way to the bank.

"Do I think I could have gotten it for less?" Kinnear asked himself. "Yes, I do." But the bankruptcy process intended to protect the company had run out of control. "We gave the system a chance to work," he said, "and the system failed."

But could it have failed as utterly as Texaco thought—from Delaware to Houston to the court of appeals to the Texas Supreme Court to the U.S. Supreme Court, throughout the negotiating process, throughout a bankruptcy proceeding?

As my partner and I left the Plaza, I began to wonder. Texaco still owned Getty Oil. In effect, it had just heaped $3 billion on to the cost of the purchase.

J. Paul Getty, I thought, would be amused, if not delighted.

· SOURCES ·

Interviews

In addition to numerous sources interviewed for coverage in *The Wall Street Journal* (hereafter *WSJ*), the following individuals were interviewed by the author solely for this book. The list does not include many persons who have requested anonymity or dozens of individuals who cooperated in providing only a few comments or facts. Several of the sources listed below graciously consented to numerous lengthy interviews.

David Batchelder, Harold Berg, David Boies, Bill Bovaird, Ed Boyle, Richard Brinkman, Charles B. Cohler, Calvert Crary, Lee Daniels, Alfred DeCrane, Stuart Evey, Herbert Galant, Gordon Getty, Joseph D. Jamail, John Jeffers, Charles Kazlauskas, Stuart Katz, Gershon Kekst, Baine P. Kerr, James Kinnear, Jack Leone, Rick Lawler, J. Hugh Liedtke, William C. Liedtke Jr., Martin Lipton, Patrick Lynch, Richard B. Miller, John K. McKinley, Russell McMains, William Newsom, Sidney Petersen, Boone Pickens, George Pletcher, J. Hugh Roff Jr., John Selden, James Shannon, Martin Siegel, Michael Schwartz, Thomas J. Stovall, Harold Stuart, Vanni Treves, G. Irvin Terrell, Patricia Vlahakis, William Weitzel, Thomas Woodhouse, Kurt Wulff.

Cases

Ann Rork Getty v. J. Paul Getty, #D132475. California Superior Court, Los Angeles County.

Borden Incorporated v. Texaco Incorporated, #C-2-80-396. U.S. District Court for the Southern District of Ohio, Eastern Division.

C. J. and Dorothy Kirk v. John K. McKinley et al, #85-067412. 127th Judicial District, Harris County, Texas.

Gail Harris Getty v. Jean Paul Getty Jr., #715-305. California Superior Court, San Francisco County.

Getty Oil Co. v. Pennzoil Co., #7423. Delaware Chancery Court, New Castle County.

Gisela Getty v. Abbott Laboratories et al., #843986. California Superior Court, San Francisco County.

Gordon P. Getty et al v. Claire Getty et al, #836644. California Superior Court, San Francisco County.

Gordon Peter Getty v. J. Paul Getty, #570527. California Superior Court, San Francisco County. (Hereafter *Getty v. Getty*)

Howard Good and Moses Weiss v. Getty Oil Co., et al, #8333. Delaware Chancery Court, New Castle County.

J. Paul Getty v. Sarah C. Getty, #452,368. California Superior Court, San Francisco County.

The Matter of the Declaration of Trust of Sarah C. Getty Dated December 31, 1984, #P685566. California Superior Court, Los Angeles County. (Hereafter Sarah Getty Trust litigation)

Louisiana Power & Light Company v. United Gas Pipe Line Company and Pennzoil Company, opinion. 478 So.2d 1240 (La. App. 4 Cir. 1985).

Pennzoil Co. v. Getty Oil Co. et al, #7423. Delaware Chancery Court, New Castle County. (Hereafter Pennzoil's Delaware case)

Pennzoil Co. v. Texaco Inc., #84-05905. 151st Judicial District, Harris County, Texas. (Hereafter *Pennzoil v. Texaco*)

Pennzoil Co. v. Texaco Inc., #85-1978. United States Supreme Court.

Texaco Inc. v. Pennzoil Co., #01-86-216-CV, First Supreme Judicial District of Texas.

Texaco Inc. v. Pennzoil Co., #85 Civ. 9640. U.S. District Court, Southern District of New York.

Texaco Inc. v. Pennzoil Co., #86-7047. #86-7052. United States Court of Appeals for the Second Circuit.

Triad Chemical et al v. Texaco Incorporated, #J81-0178(R). U.S. District Court, Southern District of Mississippi.

Notes

This book is based almost entirely on interviews and court records. The dialogue has been culled virtually verbatim from the sworn testimony of dozens of individuals as well as from the author's own attempts to reconstruct certain conversations through interviews. Comments appearing in quotation marks are reported exactly as spoken, to the best of anyone's ability to recall them. Italicized comments, used to indicate paraphrased remarks or mental thoughts, are also based on testimony and interviews. They were not invented.

Some courtroom dialogue has been abbreviated for clarity—to remove stammers, repetition and non sequiturs—although this has been done sparingly. In most cases, italics within a quotation from a printed source are the author's.

The printed sources listed in the notes generally do not include interview subjects and standard reference and research sources, such as encyclopedias, dictionaries, directories, corporate public-relations materials and high school and college yearbooks. Nor would it be possible to list all of the several hundred newspaper and magazine articles reviewed in the research for this book. For the most part, only feature articles or pieces from which a direct quote or controversial fact was obtained are included in the notes.

CHAPTERS 1–2

(pp. 3–24)

Before *Pennzoil v. Texaco*, history's largest verdict was the $1.8 billion awarded in 1980 to MCI Communications Corporation in its antitrust suit against American Telephone & Telegraph Company. An appeals court, however, ordered a retrial of the case, which ultimately resulted in a verdict of $113.3 million, or about one-ninetieth the amount of Pennzoil's verdict. Before that, a record-breaking $259.5 million antitrust verdict, won by Telex Corporation from IBM in 1973, was reversed by an appeals court.

A fuller account of Gordon's acceptance of the offer by Texaco appears in Chapter 15. As noted in the text there, Gordon has a slightly different recollection of the conversation, although this version was confirmed by several other people present in the room.

The city-income figures were calculated from data in *The Book of American City Rankings*, by John Tepper Marlin et al. New York: Facts on File Publications, 1983.

Joe Jamail's service as a pallbearer for one of the judges is reported in a death notice by Earthman Funeral Directors in the *Houston Post*, September 30, 1986.

The love life, marriages, family life and early career of J. Paul Getty (hereafter JPG) are discussed in his autobiographies, the most detailed and revealing of which is *As I See It*. Further details appear in Lenzner's *The Great Getty* and Miller's *The House of Getty*.

The facts of Gordon's boyhood were obtained from Gordon and Bill Newsom and are told in "Oil Heir: Gordon P. Getty," by Robert J. Cole, *New York Times* (hereafter *NYT*), April 22, 1984; "Gordon P. Getty," *Current Biography Yearbook, 1985*, and an ABC News *20/20* interview by Barbara Walters aired August 30, 1984.

The details of Gordon's career at Getty Oil Company (hereafter GOC) are told in his deposition in *Getty v. Getty*, in his deposition in *Pennzoil v. Texaco*, and in the few publicly available segments of his deposition in the Sarah Getty Trust litigation. Many of Gordon's foibles in business were recounted in "How to Succeed in Business," by James McDonald, *Pacific Oil World*, November 1983. Former executives recall the circumstances of Gordon's arrest in the Neutral Zone somewhat differently, but as Gordon was present for the event and is a credible source, the author has relied on his version, as told in an interview and in his deposition in *Getty v. Getty*.

Gordon's letter from his father—"Your failure to be duly respectful . . ."—appears as an undated exhibit in the Sarah Getty Trust litigation. JPG's comments—"Your work at Tulsa . . ."—appear in the same letter. The comments attributed to Gordon's mother appear in a letter from JPG to Gordon dated September 4, 1963, which appears as an exhibit in Gordon's deposition in *Getty v. Getty*.

Gordon's experiences at Spartan Aircraft are told partly in "Getty Moves Along in Mobile Homes," by Jim Downing, *Tulsa Tribune*, January 27, 1960; "Gordon Getty to Learn About Mobiles Though He's Not New to Trailer Life," *Tulsa World*, January 19, 1960; and other local coverage.

JPG's comments to George Getty II—"He feels that he is entitled . . ."—were taken from a letter dated December 12, 1962. JPG Jr.'s comment to JPG—"You know how Gordon consults . . ."—appears in the September 4, 1963, letter cited above.

George Getty's comment to Gordon—"almost like Hitler or Napoleon . . ."—is recounted in a December 1, 1965, letter to JPG.

Gordon's comment—"I must be expected . . ."—appears in a February 1, 1966, letter to George Getty. George Getty's remark—"In view of your basic conclusions . . ."—is taken from a September 29, 1965, memorandum to Gordon. George Getty's comments—"I think it is a fine thing . . ."—appear in a June 17, 1965, letter to JPG. The remark—"bumming off friends . . ."—attributed to Gordon appears in a January 30, 1963, letter to JPG. George Getty's car memorandum to JPG is dated February 18, 1966.

CHAPTER 3

(pp. 25–41)

The career and family life of William Liedtke Sr. is based on numerous interviews with Tulsans who respected and knew him well, as well as on several articles, including "Tulsan Helped Shape State Constitution," by J. Bob Lucas, *Tulsa World*, April 29, 1964. Boomtown Midland is described in *Wildcatters*, by Oilen and Oilen. The account of the formation of Zapata is based partly on "Zapata Marks 25th Anniversary," in *Viva*, an employee newsletter published by Zapata Petroleum Corporation, 1979.

Some details of Pennzoil's history and the career of the Liedtkes appear in "Zapata Petroleum Corp.," *Fortune*, April 1958; "The Scrappy Mr. Pennzoil," by James R. Norman, Terri Thompson, Cynthia Green and Jo Ellen Davis, *Business Week* (hereafter *BW*), January 27, 1986; "J. Hugh Liedtke," by Charley Blaine, *USA Today*, April 22, 1986; "First Things First," *Forbes*, October 15, 1970; and King's *Bush: A Biography*. The account of the move by J. Hugh Liedtke into South Penn is based on interviews with former South Penn directors and on a number of *WSJ* articles appearing at the time.

The origins of the Texas Company are told in King's *Joseph Stephen Cullinan*, Sampson's *The Seven Sisters* and other works. Among the articles describing the problems faced by Texaco, or providing background on Gus Long or John McKinley, are "Digging In: Texaco Sheds Its Image As a Master Marketer, Stresses Exploration," by Steve Mufson, *WSJ*, October 20, 1981; "Texaco's Balancing Act," *Forbes*, April 1, 1961; "Texaco Restoring Luster to the Star," *BW*, December 22, 1980; "Texaco's Master Returns to the Helm," *BW*, September 19, 1970; "Texaco's Fire Chief," by Alvin A. Butkus, *Dun's Business Monthly*, January 1971; "Texaco's Lost Ground," *Forbes*, February 1, 1975; "Rebuilding the House That Long Built," by James Cook, *Forbes*, April 17, 1978; "Comeback at Port Fumble," *Fortune*, February 25, 1980; "Texaco's Single-Minded Boss," by Carol E. Curtis, *Forbes*, May 9, 1983; "Oil Exec in High Gear," by Max Heine, *Tuscaloosa News*, August 9, 1983; and a Gus Long profile appearing in *Fortune*, October 1970.

CHAPTER 4

(pp. 42–60)

George Getty's "Mount Olympus" comment appeared in "J. Paul Getty Dies at 83," by Alden Whitman, *NYT*, June 7, 1976. Gordon's "architectural integrity" quote appeared in the *Tulsa World*, January 19, 1960, cited above.

Gordon's discovery of and litigation involving the Sarah Getty Trust is detailed in the depositions, pleadings and judicial opinions in *Getty v. Getty*, from which most of the direct quotes from and about him on this subject were obtained. The account of the trust's origins is based on the record in the same court case as well as on JPG's autobiographies.

Gordon's demand for dividend payments by GOC appears in a January 9, 1963, letter to his father, included as an exhibit to Gordon's deposition in *Getty v. Getty*. JPG's comments about the Kennedys not suing each other and about his wish to have Gordon "work for a living" appear in a letter to Gordon dated October 10, 1963. JPG's "lone wolf" comments appear in letter dated August 9, 1963. George's comment—". . . you have lost a father"—is recounted in a September 16, 1963, letter from Gordon to JPG.

The vitriolic attack inflicted on Gordon by JPG's lawyers is based on numerous documents produced in *Getty v. Getty*. JPG's "keep killing my son" conversation is told in Lenzner's *The Great Getty*.

JPG Jr.'s former drug problems, which are confirmed by his lawyer Vanni Treves, are mentioned in, among other court documents, "Declaration of Gail Harris Getty Re Status of Seth M. Hufstedler as Guardian ad litem of Tara Getty," filed June 29, 1984, in the Sarah Getty Trust litigation, which also describes the circumstances of Talitha's death and of JPG Jr.'s subsequent self-imposed exile to England; and in "Trustee's Memorandum Analyzing Some of the Evidence to Date of Complicity between the Guardian and Getty Oil Company," filed December 28, 1983, in the same litigation.

The berating letters by JPG to George Getty were obtained by the author from Stuart Evey, George's former executive assistant, whose assistance the author gratefully wishes to acknowledge. The letters are variously dated December 31, 1968; January 2, 1969; May 25, 1972; and July 24, 1972. The account of George Getty's death is based partly on the Los Angeles Police Department Death Report of George F. Getty II, June 6, 1973; a press release by the Los Angeles County Department of the Chief Medical Examiner dated August 27, 1973; and on a report by Diane Davenport and Julie Diller, associate consultants to the Los Angeles County Department of the Chief Medical Examiner, dated November 9, 1973. Copies of these documents were also provided to the author by Stuart

Evey. That George Getty was stabbing the back of his hand with a letter opener was reported in Lenzner's *The Great Getty*.

The controversies over Pennzoil's natural-gas supplies and its spin-off of United Gas are detailed in, among other places, *Louisiana Power & Light Company v. United Gas Pipe Line Company and Pennzoil Company*, 478 So.2d 1240 (La.App. 4 Cir. 1985); "Plato Can Collect Sharply Higher Charge for Gas," by James Tanner, *WSJ*, June 7, 1978; "Pennzoil Co., Others Named by Bondholder in Class-Action Suit," *WSJ*, April 17, 1974; "FPC Studies Pennzoil's Spinoff of Unit for Possible Violation of Natural Gas Act," *WSJ*, May 15, 1974; "Unit of Southern Co. Files $134 Million Suit Over Gas Curtailment," *WSJ*, November 19, 1974; "Love Her and Leave Her," *Forbes*, September 15, 1974; "Why Pennzoil Failed at Diversification," *BW*, May 13, 1972; "Why Pennzoil Dumped United Gas," *BW*, March 23, 1974; "PSC Blames Firm For Causing Gas Shortage in La.," Baton Rouge, La., *State-Times*, December 13, 1972; and "Gas Firm Given Warning by Gov.," by Charles Layton, Associated Press, January 29, 1973.

Bill Liedtke's presidential fund-raising and other political activities are detailed in *Hearings Before the Select Committee on Presidential Campaign Activities*, 93rd Congress, especially the testimony of Hugh Sloan, June 6, 1973, and Maurice Stans, June 12, 1973; *Final Report* of the committee, June 1974; "Stans Scathes Report," by Bob Woodward and Carl Bernstein, *Washington Post*, September 14, 1972; and "Liedtke Linked to FPC Choice," United Press International, June 26, 1973, among others. The interview with investigators in which Bill Liedtke discussed recommending candidates for White House appointments occurred on September 6, 1973; a summary of the interview is available at the National Archives. Bill Liedtke says, and the record also suggests, that the Pennzoil aircraft used to deliver the Nixon campaign funds was making a regularly scheduled trip to Washington.

The insider-trading charges against the Liedtkes appear in *Litigation release No. 6414*, July 1, 1974, Securities and Exchange Commission; in an account of the investigation by Kenneth H. Bacon, *WSJ*, May 29, 1974, and in "Pennzoil Co., Investment Firm Charged by SEC," *WSJ*, July 2, 1974.

Texaco's decline is documented in numerous *WSJ* stories and in some of the other articles cited above. The life of the company's oil reserves, given in this chapter and elsewhere, is based on theoretical calculations by Arthur Andersen & Co.

CHAPTER 5
(pp. 61–75)

JPG's obsession with building oil reserves is described in "Getty Oil: The House that J. Paul Built," *Forbes*, March 1, 1974. Getty Oil's reserve rankings are based on 1979 figures in *Oil and Gas Reserve Disclosures: 1980-1983.* Houston: Arthur Andersen & Company, 1984.

George Getty's remarks about Sid Petersen were contained in a memorandum to JPG dated April 11, 1972, and provided to the author by Stuart Evey. Petersen's diversification activities are described in "Getty's Petersen: Seeking Gushers Outside of Oil," by James Flanigan, *Los Angeles Times,* July 6, 1980; "Changing of the Guard," by John Merwin, *Forbes,* October 29, 1979; "Getty Changes Draw Some Fire," by Robert Metz, *NYT,* July 25, 1980; and "Getty Bucks the Trend and Bids for More Oil," *BW,* October 29, 1979, as well as in several *WSJ* articles about individual diversification moves. The problems of Premier were described in the *WSJ* of April 24 and August 5, 1980. JPG's comment, ". . . minor importance," appeared in "Getty Bids for a Tight Empire," *BW,* January 11, 1958. "I think management . . ." appeared in *International Management,* May 1974. Petersen's comment, "I suspect . . ." appeared in *Los Angeles Times,* July 6, 1980.

Besides the articles cited above and other sources, information about Gordon's activities in adulthood appear in "San Francisco's Shy Tycoon," by Pat Steger, *San Francisco Chronicle,* September 30, 1983.

The Astor quote appeared in Epstein's *Ambition.* Gordon's trust income, based on 1983 figures, was calculated by multiplying the GOC shares held by the trust (31.8 million) by GOC's annual dividend rate of $2.60, divided by one third.

Ann Getty's remark, "I never come down . . ." appeared in *Vogue,* October 1977. Ann's interests and activities are also described in Steger, cited above; Miller's *The House of Getty;* "Courting Society: Friendship and Favors Win Wealthy Clients for a New York Broker," by Randall Smith, *WSJ,* July 16, 1985; and "The Boardroom Drama Now Playing at Getty Oil," *BW,* December 12, 1983.

Hays' boast about his purported "control" of Gordon is recorded in a memo to file by Robert H. Smith, executive vice president, Security Pacific National Bank, February 9, 1981, which appears as an exhibit in the Sarah Getty Trust litigation.

Takeover-related financing is discussed in "Huge Credit Lines May Presage Acquisitions," by Paul Blustein and Steve Mufson, *WSJ,* August 7, 1981. The "jackass" quote about Texaco appears in "Oil Gusher: Texaco's Bid for Getty, a Record $9.89 Billion, Belies Its Stodgy Image," *WSJ* news roundup, January 9, 1984.

Security Pacific's relationship with the trust is detailed in a report attached to a December 5, 1980, letter to Petersen from Frederick Larkin; in the Smith memo to file, cited above, and in other documents on file in the Sarah Getty Trust litigation.

CHAPTER 6
(pp. 76–96)

The background and detail of the Getty Oil war were derived almost entirely from interviews and court records. However, several outstanding articles have already revealed a number of details. Among them are "Scion's Struggle: Gordon Getty Missed Main Goal of Heading Oil Firm, Insiders Say," by Stephen J. Sansweet and Scot J. Paltrow, *WSJ*, January 13, 1984; "The Tangled Fight for Getty Oil," by Leslie Wayne, *NYT*, December 11, 1983; "Gordon Getty's Goal Realized," by Thomas C. Hayes, *NYT*, January 10, 1984; "The War Between the Gettys," by Carol J. Loomis, *Fortune*, January 21, 1985; "Getty's Billion-Dollar Surprise," by William Hall, *Financial Times*, January 5, 1984. Lenzner's *The Great Getty* and Miller's *The House of Getty* also reported some elements of the struggle.

The origins of JPG's art collection and the Getty Museum are told in *My Life and Fortune* and *As I See It; Handbook of the Collections*, and in, among many other articles and essays, "The Getty," from *The White Album* by Joan Didion (New York: Simon & Schuster, 1979). The rankings of the collections are those expressed by Harold Williams in his deposition in *Pennzoil v. Texaco*.

Gordon's use of GOC's stock price is described in publicly filed selections from his deposition in the Sarah Getty Trust litigation.

The terms of JPG's bequeathments are described in the document itself, in the *WSJ* of June 8 and June 10, 1976, and in the museum's *Program Review, 1981-1985*. The trust creating the museum's endowment was established by an indenture dated December 2, 1953.

GOC's reserve valuations and stock-market performance were taken from estimates of John S. Herold Incorporated and from various Wall Street research reports.

Texaco has always denied responsibility for the economic problems Louisiana experienced during the gas shortage. However, a far different picture emerged from dozens of interviews with chemical-industry representatives and local government officials conducted by the author for the *WSJ* in 1984. See also *Borden Incorporated v. Texaco Incorporated* and *Triad Chemical et al v. Texaco Incorporated*.

The comment, "It's just amazing how bad . . . ." appeared in *WSJ*, September 15, 1980. Texaco's violations of the Louisiana Code of Ethics are contained in a letter to Texaco from George Hamner, secretary of the Louisiana Commission on Governmental Ethics, dated September 27, 1979. The *Gris Gris* cover story on Texaco and Governor Edwards appeared August 15, 1985. For a fuller discussion of Texaco's political activities in Louisiana, see "Oil's Legacy: In Louisiana, Big Oil Is Cozy With Officials and Benefit Is Mutual," by Thomas Petzinger Jr. and George Getschow, *WSJ*, October 22, 1984.

Kurt Wulff's dealings with Gordon were recounted in an interview and in a deposition taken from Wulff on November 28, 1984, in the Sarah Getty Trust litigation.

Pickens' personal background is summarized in *Takeover Madness;* in "Boone Pickens: Company Hunter," by Peter Nulty, *Fortune*, December 26, 1983; and other sources. Willametta Keck Day made her comments about Pickens in interviews on March 8 and March 11, 1984. Her relationship with her brother and Superior Oil is described in "Drama in Texas: Keck Family's Feud, More Than Economics, Sealed Superior's Fate," by George Getschow and Thomas Petzinger Jr., *WSJ*, March 13, 1984.

Pickens recalls having two meetings with Gordon—in Los Angeles in the summer of 1982 and in San Francisco in the spring of 1983. But Gordon recalls only the Los Angeles meeting, and he believes that it occurred several months later than Pickens says. This accounts for the somewhat ambiguous timing of the Los Angeles meeting in the author's narrative.

The accounts of Gordon's meeting with Bass and Tavoulareas are based partly on Gordon's deposition in the Sarah Getty Trust deposition as summarized in Loomis' "The War Between the Gettys," cited above. His dealings with Robertson were discussed in interviews with Gordon, Sid Petersen and others; in a publicly available section of his deposition in the Sarah Getty Trust litigation; and in a letter to Gordon from Robertson dated October 21, 1982, included as an exhibit in the same litigation.

Papamarkou's relationship with Gordon and Ann is described in "Courting Society" by Smith, cited above.

The account of the inside-information discussion at the Bonaventure is based on interviews with Gordon, Tim Cohler, Petersen and two additional sources, and on portions of the depositions of Petersen and Gordon in the Sarah Getty Trust litigation. Gordon's giving of the company's internal study to an outsider was probably not illegal, despite what Petersen thought.

The background on Goldman was obtained from Geoff Boisi's deposition and his live testimony in *Pennzoil v. Texaco*, from Ferris' *The Master Bankers* and from "How Goldman Grew and Grew," by Irwin Ross, *Fortune*, July 9, 1984.

CHAPTER 7

(pp. 97–118)

Moses Lasky's accomplishments were described in UPI stories dated June 2 and August 17, 1983, an AP story dated December 3, 1979, and in *Facts on File*, December 21, 1979.

"Project Brutus" is an appellation entirely of the author's invention.

Harold Stuart's speculation about Gordon's desire to become chairman, as well as several other beliefs and actions attributed to Stuart, are contained in three undated, unpublished monographs written by Stuart and provided to the author by another source.

Harold Berg's "pie in the sky" remark appeared in Lenzner's *The Great Getty* and was affirmed by Berg in an interview.

The account of GOC's involvement in bringing about Tara Getty's suit is based on numerous interviews and court documents, including a fifty-two-page document, "Trustee's Memorandum Analyzing Some of the Evidence to Date of Complicity Between the Guardian and Getty Oil Company," filed December 28, 1983, in the Sarah Getty Trust litigation. O'Melveny & Myers declined to have any partners serve as guardian ad litem because the firm had a potential conflict of interest. A number of outside lawyers other than those mentioned in the book were hired by GOC to work on the co-trustee suit.

The account of GOC's July 8, 1983, board meeting is derived from the board minutes and from interviews. The personal data on Teets appears in "A Body Builder Lifts Greyhound," by Brian O'Reilly, *Fortune,* October 28, 1985. Gordon does not recall speaking with such a "twisted tongue" when making his "optimum way" remark.

Williams' experience at Norton Simon Incorporated was recounted by *BW,* February 28, 1970, and elsewhere.

Personal background on Geoff Boisi was obtained from his testimony, from interviews with other sources and from "Goldman's Merger Chief: Geoffrey T. Boisi," by John Crudele, *NYT,* November 3, 1985, and "The Superstars of Merger," *Time,* May 14, 1984. Goldman, Sachs refused to permit the author to interview Boisi except under unreasonable and uncustomary conditions, to which the author refused to agree. The author did, however, review six volumes of Boisi's testimony in *Pennzoil v. Texaco,* much of which was recorded on videotape.

The "Medberry affair" was recounted in numerous interviews, as well as in a confidential memorandum to file by Moses Lasky dated October 4, 1983. This and other memorandums and letters of Lasky were attached as exhibits to a document entitled "Memorandum of Points and Authorities in Opposition to the Trustee's Motion to Limit His Deposition," publicly filed January 25, 1985, by the daughters of George Getty in the Sarah Getty Trust litigation.

Gordon's confidential discussions with his lawyers in London are detailed in Lasky's October 4 memorandum and in a supplement dated October 13, 1983. Boesky's comment about Gordon appears in his book, *Merger Mania: Wall Street's Best-Kept Money Making Secret* (New York: Holt, Rinehart & Winston, 1985).

Gordon's payment of the medical expenses of his nephew J. Paul Getty III is discussed, among other places, in "Paralyzed and Blind from a Drug Overdose, Paul III Is the Star-Crossed Getty," by Suzanne Adelson and Maria Wilhelm, *People,* December 14, 1981.

CHAPTER 8

(pp. 119–136)

Gordon's reaction to his designation by *Forbes* as the richest American was described by Pat Steger, *San Francisco Chronicle,* September 30, 1983.

The correspondence among Claire Getty, Paul Getty Jr. and Gordon was attached as exhibits to certain pleadings in the Sarah Getty Trust litigation.

The accounts of the GOC November 11, 1983, board meeting and of the events involving JPG Jr., Tara Getty and Mark Getty were partly derived from the following documents: GOC meeting agenda; "Affidavit of J. Paul Getty Jr.," sworn before a U.S. consul in London in December 1983; "Declaration of Mark Harris Getty re Status of Seth M. Hufstedler as Guardian ad Litem of Tara Getty," filed June 29, 1984, in Los Angeles County Superior Court; and in the affidavit of Gail Harris Getty, cited above. The debunking of the "backdoor" description is based on interviews not only with directors but with employees who worked in the building at that time.

Tara's suit was filed as "Petition for Appointment of Corporate Co-Trustee and Instructions Re: Validity of Appointment of Successor Co-Trustee," by Seth M. Hufstedler as guardian ad litem, in the Sarah Getty Trust litigation. GOC filed its motion to intervene as "Petition in Intervention" in the same case.

In addition to his deposition in *Pennzoil v. Texaco,* Jim Glanville's background is described in "The Clash of Styles in Investment Banking," by Eleanor Johnson Tracy, *Fortune,* September 25, 1978, and in Auletta's *Greed and Glory on Wall Street.*

CHAPTERS 9–13

(pp. 137–207)

The account of the Getty-Sarofim-Liedtke conversations appears in Liedtke's deposition in *Pennzoil v. Texaco.* Siegel claims the January 1, 1984, meeting would have occurred without Sarofim's intervention.

Gordon's offer of an ambassadorship is reported in Lenzner's *The Great Getty.*

The wealth attributed to Taubman and Tisch was listed in *The Forbes 400,* October 28, 1985. The biographical details on both men were culled from a variety of *WSJ* stories, and additional material about Tisch was obtained from "The Sociable Bob Tisch Throws the Ultimate Power Breakfast," by Cotten Timberlake, AP, March 28, 1986; "The Tisch Touch," by Edwin Diamond, *New York*, May 26, 1986, and "Gambling on CBS," by Ken Auletta, *NYT Magazine*, June 8, 1986. Tisch declined to be interviewed. Ann Getty's role in recruiting the new board members is described in "Gordon and Ann Play Takeover," by Carol J. Loomis, *Fortune*, February 6, 1984.

The account of what occurred inside the boardroom during the January 2-3, 1984, Getty Oil directors' meeting was detailed through numerous interviews and passages of testimony, but is based primarily on the sixty-eight-page typed version of the handwritten notes taken by Dave Copley, the secretary of GOC.

The final fifteen-to-one vote by the GOC board was recorded in the meeting notes as a fourteen-to-one vote, failing to take account of the vote by David Mitchell, the director who had missed most of the meeting. To avoid confusion, the author has corrected all incorrect references to a fourteen-to-one vote that appeared in subsequent documents.

Bruce Wasserstein declined to be interviewed for this book, citing the risk that it might expose him to a further deposition if *Pennzoil v. Texaco* were retried. Wasserstein's deposition in Pennzoil's Delaware case remains one of the most detailed and candid statements anywhere in the litigation over Getty Oil.

Wasserstein's background is described in "Artful Adviser: How First Boston Corp. Turned Itself Around Amid a Merger Mania," by Tim Metz, *WSJ*, April 21, 1982. Wasserstein's Delaware deposition is the source of most of the conversations to which he was an observer or in which he was a participant; based on testimony of and interviews with other participants, his recollections appear to be uniformly accurate. Perella's background is described in Ferris' *The Master Bankers*.

Claire Getty's claims against Gordon are enumerated in "Ex Parte Application for Order Instructing Trustee to Immediately Refrain from Signing Any Legally Binding Agreements . . . ," filed January 5, 1984, in the Sarah Getty Trust litigation.

McKinley's recollections of the chemistry lab are contained in "Saddle Shoes and Army Boots," by John K. McKinley, published by the University of Alabama.

Morris Kramer's effort to rule out that Skadden, Arps had a conflict of interest is told in *BW*, August 13, 1984.

CHAPTER 14

(pp. 208–217)

The phone calls placed by McKinley just before and after the Texaco board meeting were reconstructed primarily from the testimony of Bruce Wasserstein, although testimony from other participants in the calls was also taken into account for corroboration or amplification. McKinley's own recollection of the calls was not nearly so specific as Wasserstein's; when shown one of the conversations as recalled by Wasserstein, McKinley would state: "I don't think I said those exact words, but those are reasonable words to say."

The account of the meeting at the Pennzoil suite, in which concern was expressed about rival bidders and a status report was given on Pennzoil's document-drafting effort, is based on the deposition testimony of Ward Woods of Lazard Frères, who was present. James Glanville of Lazard did not answer the author's numerous phone messages. Liedtke would never recall Texaco's being mentioned as a rival bidder.

CHAPTER 15

(pp. 218–230)

The account of Texaco's meetings with Gordon at the Pierre is based on copious testimony and numerous interviews, but for the most part the dialogue came from representatives of Texaco, including McKinley, the chairman, and Weitzel, the general counsel.

Among the issues that Texaco would recall hearing discussed was whether an additional provision of the Declaration of Trust of Sarah C. Getty enabled Gordon to participate in a "merger." This entire section of the document, however, written in Depression-era legalese, is virtually imponderable, and does not make clear that a trustee could engage in a transaction that would fundamentally alter the character of the trust. The provision in question states that the trustee shall have

. . . all the rights, powers and privileges of an owner, including, though without limiting the foregoing, holding securities in the trustee's own name or otherwise, voting, giving proxies, payment of calls, assessments and other sums deemed by the trustee expedient for the protection of the interests of the trust estate, exchanging securities, selling or exercising stock-subscription or conversion rights, participating in foreclosures, reorganizations, consolidations, mergers, liquidations, and voting trusts. . . .

Texaco's representatives did not recall Lipton's having a substantive discussion with them before riding the elevator to Gordon's suite.

Gordon does not recall blurting the words, "I accept! Oh! You're supposed to give the price first," although virtually everyone else who was in the room recalls that he did. Just the same, Gordon's rendition of his acceptance of Texaco's offer was also included in the reconstruction of the dialogue. Besides his comments in an interview with the author and in his deposition, Gordon's anxiety over Texaco's offer is detailed in a countersuit he filed March 8, 1985, *Gordon P. Getty v. Claire Getty et al*, #836644. California Superior Court, San Francisco County.

CHAPTER 16
(pp. 231–250)

The suit by William D. Warren, guardian ad litem for the minor children of J. Ronald Getty, was filed as part of the Sarah Getty Trust litigation. The fee request of $5 million by Irell & Manella, which represented Warren in the case, is part of the court record.

The comments by Donald Schmude against Liedtke were reported by the *Tulsa World*, January 21, 1984.

The account of Fairlawn Oil Services' suit against Texaco appeared in the *Providence Journal-Bulletin*, February 10, 1984. The amount of Pennzoil's stipend to Fairlawn is contained in a counterclaim filed by Texaco against Pennzoil in the 151st Judicial District, Harris County, Texas, on June 28, 1985.

Perry Barber, the former general counsel of Pennzoil, did not return phone calls in which the author was seeking, among other things, verification of his sharp remark to Michael Schwartz of Wachtell, Lipton.

CHAPTER 17
(pp. 251–267)

Liedtke recounted his phone conversation with Exxon's Kaufman in his deposition in *Pennzoil v. Texaco*. Kaufman declined through an Exxon spokesman to comment on the conversation.

In addition to interviews with numerous Texas lawyers and with Jamail himself, the following articles, among others, provided background information on Joe Jamail in this and other sections of the book: "Joe Jamail's Lesson in Tort Law Spells Trouble for Texaco," by Matt Moffett, *WSJ*, November 21, 1985; "Trouble Is My Business," by Mimi Swartz, *Houston City Magazine*, November 1980; "Texas' King of Torts Loves to Battle the Giants," by Kent Biffle, *Dallas Morning News*, February 11, 1979; "Joe Jamail: The $2.4 Billion Dollar Man," interview transcript in *Utmost*, University of Texas Law School, Spring 1986; "Houston Lawyer Joe Jamail Sued the $10.5 Billion Pants Off Texaco and Stands to Pocket a Record Fee," *People*, January 6, 1986; "The Aristocrats of Tort," *Newsweek*, December 24, 1973.

The opinions cited in the SCM case are *Reprosystem B.V. and N. Norman Miller v. SCM Corp.*, 522 F. Supp. 1257 (1981) and 727 F.2nd 257 (1984).

The negative comments of Gordon's nieces were contained in the record of *Gordon P. Getty v. Claire Getty et al,* #836644, California Superior Court, San Francisco County. Moses Lasky's comment about the fight over the trust's capital-gains liability appears in "Judge Socks Getty Heirs," by Mark Potts, *The Washington Post,* March 20, 1985. The scope of the museum's largess, including the comparison to the combined budgets of other major art institutions, is described in "The Golden Getty," by Paul Richard, *The Washington Post,* April 13, 1984.

The aftermath of the purchase of Getty Oil's publicly owned shares is detailed in "Wither a Windfall: Rather Than Splurge, Getty Holders Invest Most of Sale Proceeds," *WSJ* Roundup, April 25, 1984. The anxiety of Getty Oil employees following the merger is documented in "Uncertainty Racks Getty Workers," by Stephen J. Sansweet, *WSJ,* April 19, 1984; Sansweet provided the author with a copy of "The Lost Book of the Gettians" as well as other examples of employee dissent.

CHAPTERS 18–20

(pp. 271–322)

Background on judicial politics in Texas appears in *The Selection and Retention of Judges in Texas,* by Anthony Champagne, University of Texas at Dallas. National Symposium on Judicial Selection, February 21, 1986. The history of Houston's law firms is told in "Empires of Power," by Griffin Smith Jr., *Texas Monthly,* November 1973.

Dick Miller described his views of Pennzoil's case in interviews with the author and in "Pennzoil-Texaco Fight is Question of Honor," by Bryan Burrough, *WSJ,* June 13, 1985, and in a *Washington Post* piece by Thomas W. Lippman, July 7, 1985.

The account of the controversy over Jamail's $10,000 campaign contribution to Judge Farris is based on interviews with numerous lawyers and on the following documents: "Motion for Recusal or Disqualification," filed by Texaco in *Pennzoil v. Texaco,* October 1, 1984; "Affidavit of R. Michael Peterson," *Pennzoil v. Texaco;* "Texaco Inc.'s Opening Brief in Support of Its Motion to Set the Case for a Prompt Trial and to Enjoin Pennzoil Company from Further Prosecuting Its Action in Texas," in *Pennzoil v. Getty Oil et al,* #7425, Delaware Chancery Court, New Castle County; "Pennzoil's Brief in Opposition to Texaco's Motion to Set this Case for a Prompt Trial . . . ," in *Pennzoil v. Getty Oil et al, supra;* and "Texaco Inc.'s Reply Brief in Support of Its Motion to Set the Case for Prompt Trial . . . ," *Pennzoil v. Getty Oil et al, supra;* "Memorandum in Support of Application for Temporary Restraining Order," *Texaco v. Pennzoil,* #85 Civ. 9640, U.S. District Court, Southern District of New York.

Judge Farris' background is based on interviews with Houston lawyers and on "A. J. Farris, Oil Suit Judge, Dies at 65," by Patricia Manson, *Houston Post,* September 30, 1986; "Funeral Rites Scheduled for State District Judge," *Houston Chronicle,* September 30, 1986; and "God Help the Meter Maid Who Has to Go Before This Judge," by Paul Harasim, *Houston Post,* June 13, 1985. His rankings in the Houston Bar Association survey were calculated by comparing his score in the "poor" response to that of the other judges who were the subject of the same survey.

Old Hanging Oak is described in *Houston Home and Garden,* April 1986. The history of jury selection is discussed in *Judging the Jury* by Hans.

Some description of the boredom caused by Pennzoil's case is described in "Blood Feud," by Stephen J. Adler, *The American Lawyer,* October 1985. Other major articles written about the trial include "The $15 Billion Court Battle," by John Riley, *National Law Journal,* September 30, 1985, and "How to Lose the Bet-Your-Company Case," by Stephen J. Adler, *The American Lawyer,* January/February 1986.

The "four sinkers" controversy is partly reported in the trial transcript and was developed more fully through interviews with the trial lawyers.

CHAPTERS 21–24

(pp. 323–407)

The Louisiana judicial opinion about United Gas' contract problems is a dissent by Judge Jim Garrison, *LP&L v. United Gas Pipe Line and Pennzoil, supra.* Pennzoil strenuously disputes the allegations made in the judge's opinion.

Miller's comments about Lipton appear in "How Texaco Lost $10.53 Billion Case," *Corporate Control Alert,* December 1985.

Background on Judge Casseb was obtained from interviews; from Texaco's numerous court filings to disqualify him; from "Judge in Texaco-Pennzoil Trial Known As 'Jury Man' with No-Nonsense Style," by Matt Moffett, *WSJ,* November 27, 1985; and from "Dapper Texas Judge for a Historic Case," *NYT,* December 11, 1985. A number of the remarks attributed to him appear in a transcript prepared by Texaco of his April 2, 1986, appearance before the Los Angeles Bar Association.

The reconstruction of the deliberations is based almost entirely on the accounts of Jim Shannon and Rick Lawler. The author also relied on interviews conducted by a *Wall Street Journal* reporter with other jurors shortly after the verdict, on the reports of the trial lawyers who similarly interviewed the jurors at the time, and upon published comments by the jurors. Many months after the verdict, most of the remaining jurors declined to discuss the deliberations in any substantive way.

CHAPTERS 25–28

(pp. 408–467)

The aftermath of the verdict is based almost entirely on reporting by the author and *WSJ* colleagues. Much of this material appeared in "Joe Jamail's Lesson in Tort Law Spells Trouble for Texaco," cited above; "Texaco Shares Trade at Breakneck Pace," by Scott McMurray, Pamela Sebastian and Beatrice E. Garcia, November 29, 1985; "Who's In Command of Texaco's Team? Me! No, Me! No . . . ," by James B. Stewart, November 6, 1985; "In the Texaco Case, Judge's Countenance Is of the Essence," by George Getschow and Matt Moffett, December 9, 1985; "Texaco's Woes Start to Trouble Several Partners," by George Getschow and Dianna Solis, December 18, 1985; "Courting Disaster: How Texaco Turned Big Takeover Victory Into Bigger Legal Loss," by Matt Moffett, Thomas Petzinger Jr. and James B. Stewart, December 20, 1985; "Eight Wise Men Offer Eight Ideas to End the Texaco-Pennzoil Fight," by Steve Frazier and Thomas Petzinger Jr., January 17, 1986; "High-Stakes Poker: Texaco and Pennzoil, With Truce Expiring, Again Talk of Settling," by Thomas Petzinger Jr., Allanna Sullivan and James B. Stewart; and "Cautious Talks: Texaco-Pennzoil Case Makes Firms Careful About Merger Moves," by Peter Waldman, April 15, 1986.

Additional information appears in "Texaco Assails Jury Decision," by Richard W. Stevenson, *NYT*, November 21, 1985; "Emotions Run High Following Pennzoil Verdict," by Laurel Brubaker, *Houston Business Journal*, November 26, 1986; "Appeal Won't Slow Payment?" *Associated Press*, November 20, 1985; "$10.53 Billion Judgment against Texaco Largest in U.S. History," by Patricia Manson, *Houston Post*, November 20, 1985; "Texaco Case Teaches Ethics," by Pat Guy, *USA Today*, November 21, 1985; "A Tense Day of Waiting," by Wayne Slater, *Dallas Morning News*, December 11, 1985; and "The Elite Meet at Texaco-Pennzoil Trial," by Barbara Shook, *Houston Chronicle*, August 1, 1986.

David Boies' background is described in "The Litigator," by Cary Reich, *NYT Magazine*, June 1, 1986. Dick Miller's "fucked to the wall" comment appears in *Corporate Control Alert*, December 1985. KKBQ's jingle about the judgment appears in "The Getty Deal," Debra Whitefield, *Houstonian*, June 1986.

Henry v. First National Bank of Clarksdale is described in "The Felt Necessities," by Anthony Lewis, *NYT*, January 23, 1986.

Judge Farris' remarks about the Second Circuit opinion were made on KTRK-TV on January 13, 1986.

Judge Farris' correspondence with Thomas McDade appears as part of the record in *Cusack v. Texas Gas Exploration, et al*, #84-01481, 151st District Court, Harris County.

The account of Texaco's leak of the speech appears in "How Texaco Told of Casseb Speech," by Steven H. Lee, *San Antonio Light*, April 12, 1986.

The contributions by Pennzoil lawyers to Texas Supreme Court candidates, as well as the investigation into the affairs of Supreme Court jurists, has been described by the *San Antonio News-Express* and in "The Texas Bench: Anything Goes," by Stephen J. Adler, *The American Lawyer*, April 1986.

Gordon's comments about composing and business appear in "Fat Cat and Plump Jack," by Alan Rich, *Newsweek*, April 1, 1985.

Eppler's comments about the Sperry merger appear in "The Tripple [sic] Ripple Ice Cream Case," by Jill Andresky, *Forbes*, August 25, 1986.

Siegel's downfall is described in "Unhappy Ending: The Wall Street Career of Martin Siegel Was a Dream Gone Wrong," by James B. Stewart and Daniel Hertzberg, *WSJ*, February 17, 1987, and in "Wall St. Informer Admits His Guilt in Insider Trading," by William Glaberson, *NYT*, February 14, 1987. The government claimed that Siegel began passing inside information involving Getty Oil as early as September 1983; this does not seem possible, as Siegel was never engaged by Gordon and had had no contact with him until the following month. As this book went to press, this apparent contradiction could not be resolved, and Siegel himself was not speaking publicly about the matter.

Bibliography

Anderson, Robert O. *Fundamentals of the Petroleum Industry*. Norman: University of Oklahoma Press, 1984.

Auletta, Ken. *Greed and Glory on Wall Street*. New York: Random House, 1986.

Clark, James A., and Michel T. Halbouty. *The Last Boom*. New York: Random House, 1972.

———. *Spindletop: The True Story of the Oil Discovery That Changed the World*. Houston: Gulf Publishing Company, 1952.

Davidson, Kenneth M. *Megamergers: Corporate America's Billion-Dollar

Takeovers. Cambridge, Massachusetts: Ballinger Publishing Company, 1985.

Donahue, Jack. *The Finest in the Land: The Story of the Petroleum Club of Houston.* Houston: Gulf Publishing Company, 1984.

Epstein, Joseph. *Ambition: The Secret Passion.* New York: E. P. Dutton, 1980.

Ferris, Paul. *The Master Bankers.* New York: New American Library, 1984.

Getty, J. Paul. *As I See It.* Englewood Cliffs, New Jersey: Prentice-Hall Inc., 1976.

———. *How to Be Rich.* New York: Playboy Press, 1965.

———. *How to Be a Successful Executive.* New York: Playboy Press, 1971.

———. *My Life and Fortunes.* New York: Duell, Sloan & Pearce, 1963.

The J. Paul Getty Museum. *Handbook of the Collections.* Malibu, California, 1986.

Hans, Valerie P., and Neil Vidmar. *Judging the Jury.* New York: Plenum, 1986.

Johnson, Moira. *Takeover: The New Wall Street Warriors.* New York: Arbor House, 1986.

King, John O. *Joseph Stephen Cullinan.* Nashville: Vanderbilt University Press, 1970.

King, Nicholas. *George Bush: A Biography.* New York: Dodd, Mead & Company, 1980.

Knowles, Ruth Sheldon. *The Greatest Gamblers.* Norman: University of Oklahoma Press, 1959.

Lenzner, Robert. *The Great Getty: The Life and Loves of J. Paul Getty, Richest Man in the World.* New York: Crown, 1985.

Michel, Allen, and Israel Shaked. *Takeover Madness: Corporate America Fights Back.* New York: John Wiley & Sons, 1986.

Miller, Russell. *The House of Getty.* New York: Henry Holt & Company, 1985.

O'Connor, Richard. *The Oil Barons.* Boston: Little Brown & Company, 1971.

Oilen, Roger M. and Diana D. *Wildcatters: Texas Independent Oilman.* Texas Monthly Press, 1984.

Sampson, Anthony. *The Seven Sisters.* Toronto: Bantam Books, 1975.

Subcommittee on Securities of the Committee on Banking, Housing and Urban Affairs, U.S. Senate. *Impact on Corporate Takeovers.* Hearing record: April 3 and 4, June 6 and 12, 1985.

Williams, T. Harry. *Huey Long.* New York: Vintage Books, 1969.

· INDEX ·